ENCYCLOPAEDIA OF
Seed Production
of World Crops

ENCYCLOPAEDIA OF

Seed Production of World Crops

Edited by

A. Fenwick Kelly

and

Raymond A.T. George

JOHN WILEY & SONS

Chichester • New York • Weinheim • Brisbane • Singapore • Toronto

Other Wiley Editorial Offices

John Wiley & Sons, Inc., 605 Third Avenue,
New York, NY 10158-0012, USA

WILEY-VCH Verlag GmbH, Pappelallee 3,
D-69469 Weinheim, Germany

Jacaranda Wiley Ltd, 33 Park Road, Milton,
Queensland 4064, Australia

John Wiley & Sons (Asia) Pte Ltd, 2 Clementi Loop #02-01,
Jin Xing Distripark, Singapore 129809

John Wiley & Sons (Canada) Ltd, 22 Worcester Road,
Rexdale, Ontario M9W 1L1, Canada

British Library Cataloguing in Publication Data

A catalogue record for this book is available from the British Library

ISBN 0-471-98202-4

Typeset in 10/12pt Times by Mayhew Typesetting, Rhayader, Powys
Printed and bound in Great Britain by Bookcraft (Bath) Ltd

This book is printed on acid-free paper responsibly manufactured from sustainable
forestation, for which at least two trees are planted for each one used for paper production.

Contents

Contributors

SYED IRFAN AHMAD Director General, Federal Seed Certification Department, Islamabad, Pakistan

A. BOULD Technical Liaison Officer, Plant Varieties and Seed Division, Ministry of Agriculture, Fisheries and Food, UK

J.D.C. BOWRING Consultant to OECD; formerly Head, Seed Production Department, National Institute of Agricultural Botany, Cambridge, UK

W.T. BRADNOCK formerly Director, Plant Products Division, Agriculture Canada, Ottawa, Canada

KUDLIP R. CHOPRA Director, Research and Production, Mahendra Hybrid Seeds Co. Ltd, Jalna (MS), India

RAKESH CHOPRA General Manager, Production, Mahendra Hybrid Seeds Co. Ltd, Jalna (MS), India

R.J. COOKE Head, Applied Technology Department, National Institute of Agricultural Botany, Cambridge, UK

S.R. DRAPER Deputy Director, Marketing and Seeds Department, National Institute of Agricultural Botany, Cambridge, UK; Chief Editor, *Seed Science and Technology* (Journal of the International Seed Testing Association)

R.H. ELLIS Department of Agriculture, University of Reading, Reading, UK

B. GREENGRASS Vice-Secretary General, UPOV, Geneva, Switzerland

BERNARD LE BUANEC Secretary General, International Seed Trade Federation, Nyon, Switzerland

MOGENS LEMONIUS Regional Seed Programme Manager, FAO/DANIDA, Bangkok, Thailand

ROSA MESSINA CRUZ Jefe Departmento de Semillas, Ministerio de Agricultura, Santiago, Chile

TOURKMANI MOHAMED Chef du Service du Controle des Semences et de Plants, Rabat, Morocco

K. MTINDI Seed Service, Harare, Zimbabwe

A.A. PICKETT Cereals, Pulses and Oilseed Trials, National Institute of Agricultural Botany, Cambridge, UK

J.C. REEVES Director, Research and Technology, National Institute of Agricultural Botany, Cambridge, UK

NIGEL START Barley Group, Plant Breeding International, Cambridge, UK

JOSIAH WOBIL International Seed Technology Consultant, Accra, Ghana

Preface

The book is intended as a reference book and contains three sections. For the reader interested in general concepts and practical aspects Section I, entitled 'Principles', gives an account of the main points involved in growing a seed crop. Chapter 1 describes the structure of the seed industry and this is followed in Chapter 2 by contributions from the principal international organisations which are active in the seed sector and some examples of national seed legislation from around the world. Chapter 3 gives an account of the technical aspects of quality control during seed production, including contributions from acknowledged experts in particular aspects of this complex question. The next two chapters of Section I describe the practical methods used in seed production. Finally the methods used in maintaining a cultivar true to type are described by a specialist and the role of government in the provision of reserve stocks of seed against major disasters is discussed. Key words used in Section I are indexed at the back of the book.

Section II is devoted to descriptions of seed growing of a comprehensive list of agricultural and vegetable plant species. This list is arranged alphabetically by families, but within each family the order in which the species are entered is to some extent arbitrary. A first division is made into 'agricultural crops' and 'vegetables' or some species which are used for both. Within each of these divisions the species are listed with the more important ones first; subsequently the reader may be referred to one of these key species in the text on less important species. To assist in using Section II the species are listed at the back of the book in alphabetical order of both Scientific and English names.

Section III contains three appendices. First is a glossary of the more important technical terms used in the text. Appendix 2 is a list of useful addresses. The book does not describe any of those species which are propagated vegetatively but a list of the more important of these forms the final appendix.

The comprehensive nature of the book has only been made possible through the willing co-operation of contributors and we wish to extend our sincere appreciation to these individuals. A special word of thanks is due to those who do not have English as their first language but nevertheless contributed excellent texts. Seed production is a truly international activity where the emphasis is on co-operation. We are privileged to be part of it.

A. Fenwick Kelly and Raymond A.T. George

SECTION I

Principles

1

Structure of the Seed Industry

ROLE OF THE PRIVATE SECTOR

Seed marketing requires a rather sophisticated approach. First the farmers' or growers' needs have to be identified, and these are often dependent on the requirements of the consumer of the crops to be produced e.g. the food processor or supermarket. The genetic quality of a cultivar is thus of vital importance to successful marketing because this determines the potential yield and quality of produce which the grower of that cultivar may expect. Having determined which cultivar will meet the growers' needs for a specific market outlet, the next step is to assess the potential size of the market and the share of that market which can be captured by a particular cultivar. The potential size of the market (i.e. the total seed requirement for a particular crop for a particular country) was calculated by Rosell (1986) for 10 major food crops in 103 countries, based on the total area of the crops grown, an average seed rate and one-third, two-thirds or full replacement rates (i.e. the amount of seed which a farmer might require each year to renew the basis for home-grown seed). Although now somewhat out of date, these data provide useful background information and the concept remains valid.

There will also be a need for the provision of advice to the farmer on how to make the best use of the seed supplied, particularly if new cultural techniques are required. This has led to the establishment of a sophisticated marketing system for seed.

The present worldwide trend is for governments to divest themselves of responsibility for trading activities. There are now far fewer countries where the managed economy is maintained. However, this trend has also resulted in the development of various strategies for the protection of the consumer, which usually take the form of regulations to ensure either that goods offered for sale are of a certain minimum standard of quality or that the consumer is given information by the seller as to the quality of the goods offered for sale. For seeds, the consumer is the farmer or vegetable grower and this aspect is usually covered in seed regulations under national legislation, which is described more fully in Chapter 2 in the section on 'Examples of National Legislation'.

One aspect of quality of particular importance to the purchaser of seed is genetic quality. The seed is sold as being from a particular cultivar which the purchaser expects will give certain results under the conditions where the seed is to be

Encyclopaedia of Seed Production of World Crops.
Edited by A.F. Kelly and R.A.T. George. © 1998 John Wiley & Sons Ltd

planted. For this reason many countries have a system for approving cultivars before they can be placed on the market.

It is relatively easy to multiply seed of a cultivar, particularly of self-fertilised species, but to preserve its genetic purity requires that certain procedures are followed. These are often incorporated in a government-controlled seed certification scheme, but may equally be followed by the seed producer without government supervision. Seed production and marketing have to be covered by a system of quality control which includes all stages, from seed growing until the final delivery to the purchaser. Most private companies have an in-house quality control system which is more or less sophisticated depending on the size of the company and whether it is producing seed or is purchasing seed from other producers.

One consequence of the relatively easy way in which cultivars can be reproduced has been that plant breeders have had no protection when others multiply their new and improved cultivars. This has led to the introduction of 'plant breeders' rights' which give to the breeder the possibility of obtaining royalties from sales of seed of cultivars which they have produced, in a manner similar to the authors' copyright or the industrial patent. This aspect is described more fully in Chapter 2 in the section on 'The International Union for the Protection of New Varieties of Plants'.

Seed as a commodity presents certain characteristics in trade. It provides a means by which crop plants can be stored from one season to the next and also can be transported from one location to another. However, seed is a living entity and is therefore vulnerable if not handled correctly. It is also required to be delivered for use at a particular time, usually governed by the seasons of the year suitable for sowing a crop. Seed therefore poses particular problems both for the private trader and for the government regulator. Much of the work is seasonal, particularly when a producer is specialising in one or a restricted number of crop species. Producer and regulator both have the problem of providing worthwhile employment throughout the year. This is often achieved by deploying staff on seed

crop inspection during the growing season and on testing seed samples after harvest.

The genuine private trader will wish to work within any regulations in the country where the seed is produced and where it is to be sold, and those regulations should be designed to provide an environment in which the unscrupulous entrepreneur cannot prosper.

To work within regulations, however, will not usually be enough for a trader to prosper in the long run. The genuine trader will wish to monitor carefully the quality of the seed offered for sale because increased sales will depend on the results which customers obtain from previous purchases. Poor quality seed which leads to a disappointing crop will quickly erode the reputation of a seed trader. The profit motive, while a powerful influence in any private company, has to be tempered by a recognition that future trade is largely dependent on reputation and therefore on the quality of the seed offered for sale; strictly applied quality control is therefore an unavoidable cost in the production of seed.

It is against this background that the present structure of the seed industry has developed.

Historical Background

When farming, as distinct from food-gathering from the wild, became established, part of a crop was set aside as seed for the next season. This is still a possibility for farmers and growers today for a large number of crop species, but as crop production becomes more complex this option is not always available to farmers or growers. Crops which do not utilise the seed as the end product (e.g. cabbages) require specialist seed growing techniques which are not easily followed by all growers; some crops have been introduced to new regions where seed production is impracticable, e.g. maize is produced in northern latitudes as a fodder crop but it is not possible to ripen seed satisfactorily in those areas and it has to be imported from elsewhere; hybrid cultivars require particular production techniques. Farmers and growers also wished to improve their planting stock and this led to an exchange of seed between neighbours or the import of seed from particular areas which gained a reputation for production of good quality seed. This led in turn to the establishment of trade in seed, both locally and internationally.

These developments were accelerated by the expansion of colonialism by European powers in the 18th and 19th centuries as seed for crop production was transported to colonies. This often involved the production in a colony of crop species which had not previously been grown in that country (e.g. wheat in North America). Gradually this led to the establishment of an organised seed trade. The outline of this trade was generally that a seed producer would provide seed which was bought by a wholesaler for cleaning and packing; the wholesaler then traded the seed on to a retailer, usually a local company supplying other needs (such as tools or equipment, fertiliser and animal feed). Sometimes brokers might intervene in this supply chain or alternatively the seed producer might sell directly to the retailer or even to the customer, i.e. the farmer or grower.

Many of the seed producers began to select particularly good stocks of seed from within their crops and the advent of plant breeding at the beginning of this century provided an added spur to this activity. Specialist plant breeding enterprises were established either in government research institutes and universities or as private companies. In the private sector a plant breeder could be attached to the seed producer, the wholesaler or the retailer. Gradually, however, the breeder/seed producer became absorbed into the larger companies and the retailers expanded so that the boundaries within the seed trade became less obvious.

Today the seed grower will normally grow seed on contract to either a wholesaler or a retailer, although there is still some seed grown for own use or sale to neighbours without any contract, particularly in the developing countries. The wholesaler will normally have seed cleaning and packing facilities and market the seed either through an intermediary broker or direct to the retailer. A broker does not normally handle the

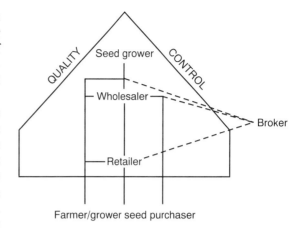

FIGURE 1.1 Seed trade channels

seed, but buys and sells on the basis of paper contracts only. The retailer will also have seed cleaning and packing facilities and may buy seed directly from the seed grower. Figure 1.1 shows a schematic representation of this concept.

The Large Plant Breeding and Seed Producing Company

Plant breeding is essentially a means of artificially creating variation within a crop species from which potential cultivars can be selected. The success of plant breeding depends in large measure on the number of potential cultivars which can be screened within a breeding programme. From the very large number of variants which are produced from the crossing programme each year (usually 1000 or more) only one or two will eventually become successful cultivars. Scale therefore becomes of crucial importance and seed trade has been concentrated into fewer, larger companies, many of which are multinational (i.e. they operate in several different countries). This is necessary to establish a wider market for their products so as to sustain the high cost of research and plant breeding. Thus a major part of the modern seed industry is 'research led'. This means that a company depends for success on plant breeding to produce cultivars which become popular with growers and farmers. Very often these programmes are also multinational, being conducted simultaneously in most of the countries where the company is operating. Obviously the extent to which there can be transference of material from one country to another will depend on the environment being suited to the particular crop species and also an appropriate market outlet.

In many instances, seed of agricultural crops is subservient to other interests in large companies. Some are concerned in the marketing of other commodities used by farmers and growers and others are selling agricultural or horticultural produce. The most influential among the former are the agrochemical companies, and of the latter, grain traders and those selling processed food. These companies operate internationally by setting up subsidiaries in countries other than that in which they originate, sometimes buying local smaller companies.

For vegetable seed many companies operate on the basis of seed alone, although some are controlled by interests in the processed food markets (e.g. canned soup). Vegetable seed is generally of higher value per unit weight and crop production fields or gardens are usually relatively smaller than those used for agricultural crops. Much vegetable seed, with the exception of peas and beans, is sold in small, usually sealed, vapour-proof packets; this enables companies to supply seed direct to local shops and so to cater for the amateur as well as the commercial trade.

Very few large companies whose sole interest is seed operate on a wide scale internationally. Competition is fierce and the company with other interests is better able to compete. There are, however, some companies which are able to achieve a standard of excellence in the cultivars of crops that they produce and this has enabled them to exist on the basis of trade in seed alone.

In developing countries it has been difficult for the private company to compete. Usually there is a government organisation responsible for marketing the products of government plant breeding institutions, which may be subsidised, and there is no provision for plant breeders' rights. As a consequence there has been little incentive for private companies. This situation is now being rectified in some countries by the creation of tax relief schemes and other incentives. However, the majority of companies which take advantage of these concentrate on high value crops dominated by hybrid cultivars; these usually require lower volumes of high priced seed for the smaller areas involved and the hybrid cultivar provides a protection for the company if the formula is kept secret. Consequently the large volume staple food crops receive less attention from the private sector.

Another more recent innovation has been the setting up of joint ventures between government and a private company. Two such ventures in Africa have recently been described. Wobil (1994) describes how the government of Swaziland set up a joint venture with Pioneer Hi-Bred International of the USA; the latter was selected from four applicants. The new company is known as Swazi-American (PHI) Seeds Ltd; the government owns 30%, represented by fixed and liquid assets, and Pioneer 70%, represented by new plant and plant breeding and management input. Agrawal and W/Marian (1995) describe a similar development in Ethiopia.

The Local Distribution Agent

Depending on the size of a country, it is not always possible to supply all seeds for use on farms or in gardens from a single location. The central supplier therefore has to set up a network of local agents. This can be through depots owned by the parent company, but often it is not possible to cover all areas of a country in this way and some system has to be devised to localise the distribution even more. To do this it is necessary to make use of agents who are trading with farmers and growers in other commodities. However, it is then necessary to impose some system of quality control to ensure that the seed reaches the customer in good condition. In some countries seed is supplied through local shops; this requires the supply of packages of a size that will meet a farmer's requirement without the need to open the sealed container from the main supplier, otherwise further storage of opened packages in local shops can cause rapid deterioration in the germination of the seed.

GOVERNMENT PLANT BREEDING AND SEED PRODUCTION ACTIVITY

The main influence of governments on plant breeding and seed production policies and activities emanates from each individual government's policy for agricultural development. This can be at either national or provincial level or even at regional level when policies common to several countries emerge as groups of countries from unions or partnerships resulting in the reduction or elimination of physical, fiscal or technical barriers.

The role of a government's agricultural policy relating to breeding and seed production is generally more pronounced in countries with a centrally planned economy than in those with a market economy. There is often a national seed policy within the framework of the overall national agricultural plan (Gregg and Wannapee 1985).

However, in most, if not all countries there is a clearly defined and stated government policy for agricultural development which takes into account the need for improved agricultural productivity as well as the wellbeing of the population. This policy will usually be influenced by factors such as the role and importance of agriculture in the national economy, identification of the wellbeing of the population (with special reference to the needs of the rural poor), and the individual needs for maintaining and improving specific crops or export products originating from crop sources. The latter may include, for example, high value horticultural crops such as tomatoes or industrial crops such as cotton. A further influence on the development of national policy is sometimes the desire to reduce imports of agricultural inputs such as seed and planting materials. In some countries it is the seed crop which is recognised as the source of an important export commodity.

In general, the ministry responsible for agricultural affairs will have a network of research stations or institutions which have specific terms of reference relating to the research and development of a crop, crop group or farming system. The national network is usually referred to as the 'national agricultural research system' (NARS). One of the remits for staff in this type of establishment will often be crop improvement, which usually embraces plant breeding.

Plant breeding policies in the public sector may range from the desire to produce an improved cultivar for a specific market outlet or environment (including pest or pathogen resistance) to the development of a breeding technique which does not necessarily result in the release of a new cultivar but will be made available to the seed industry at large.

In some countries it is considered more prudent to screen cultivars from elsewhere rather than concentrating on a breeding programme. However, screening of imported germplasm and existing cultivars from outside the country along with local material is usually a prelude to a breeding programme.

The levels of public sector involvement in plant breeding and seed production activities for individual countries are reported by FAO (1994).

Government Involvement in Seed Production

Despite the involvement of the private sector in seed production in a country, the government may be obliged to produce seed of a crop species in which the private sector does not have an economic interest. For example, in some African countries where hybrid maize seed requirements are adequately catered for by the private sector, there is public sector involvement to produce seed of the staple pulses. Governments are also frequently under pressure to maintain a seed supply to remote areas or to groups of farmers with low incomes and poor resources.

The public sector breeders normally retain responsibility for maintenance breeding of a cultivar which they have developed, especially

Pre-Basic Seed production and supervision of Basic Seed. In more advanced public sector plant breeding programmes the production of the final seed category for farmers to produce their next food or industrial crop is usually done by either contract seed producers or taken up entirely by the private sector. In some countries parastatal companies have been set up to produce and market seed. Joint ventures between government and a private company have been referred to in the previous section of this chapter.

Role of Universities in Plant Breeding and Research

Universities which are funded by the public sector often play a vital role in plant breeding and seed supply. A well run university department or faculty will have an active research programme. In applied biology or agricultural departments this will involve plant breeding or seed biology. In the latter case research into topics such as seed quality will provide important information for the furtherance of the seed industry. The emphasis placed on universities for agricultural development and implementation of national agricultural policy varies from one country to another. In some countries they are funded by government and are the main thrust for national policy, while in countries moving further to a free market economy the universities seek private sector funding and are therefore involved in seed programmes sponsored by the private sector. In practice, a university may take on some research sponsored by the public sector and also some funded by the private sector.

Extension and Advice

Traditionally the public sector has been responsible for providing farmers and growers with advice and information. Organisations which provide this service are usually referred to as advisory or extension services. The term 'extension service' originates from the time when universities or colleges provided an extension of their services

off the campus. Currently some so-called agricultural and horticultural 'extension services' are not operating from teaching establishments. For a history of agricultural extension, its methodology and management see Ban and Hawkins (1988).

The extension services funded by the public sector have remained as an important link between plant breeders and farmers. Information on new cultivars is usually distributed to farmers via extension systems using cultivar trials and demonstrations in local village communities in addition to all other forms of media. Conversely, the information feedback coupled with a close liaison over cultivar- and seed-related problems and requirements in the villages can act as a catalyst for plant breeding and seed-related projects.

Influence of Government Policies on Seed Industry Development

The most pronounced effect of government policy is when there is a change of emphasis from seed production by the public sector to production by the private sector. While on the one hand there is a clear policy statement by government, the various public breeding and seed production units may not be totally geared to winding down their responsibilities and transferring their activities to the private sector.

Government policy can encourage its agencies to create favourable conditions for the emerging private sector by tax incentives, credit arrangements and appropriate pricing systems. It can also influence the inclusion of plant breeding and seed production topics in the curricula at various levels of the education system.

The roles and responsibilities of governments in legislation and seed quality control are discussed elsewhere in this volume, with national examples in Chapter 2.

Government and International Co-operation

A national government, through its departments related to agriculture, can be the focal point for interactions and co-operation with other countries and the international organisations.

In some areas of the world, either because of relatively small seed input requirements or similar climatic conditions, countries in a region have co-operative seed programmes. For example, the Asian Vegetable Research and Development Center (AVRDC) has an African Regional Programme (ARP) which includes a 'Tomato Improvement Programme and the African Highlands' (Anon 1996).

The network of agricultural research centres supported by the Consultative Group for International Agricultural Research (CGIAR) provides co-operation at both regional and national levels, usually through recipient government agencies. Seed-related projects funded by bilateral or international aid organisations are frequently executed with a relevant national government's department as an active partner. The International Service for National Agricultural Research (ISNAR) was established by CGIAR to assist the National Agricultural Research System (NARS) in developing countries.

There has been an increasing interest in the setting up of regional networks in recent years; for example, in the Southern African Development Community (SADC) there is a collaborative network for vegetable research and development (Anon 1996).

CONTRACT SEED PRODUCTION AND THE ROLE OF SEED GROWERS

History

The development and release of high-yielding wheat cultivars in 1964 led to the concept of 'seed villages' in India. Farmers were not interested in wheat seed production, as the price difference between wheat seed and grain was so small that it did not even cover the normal additional costs involved in technical supervision, roguing, plant protection, processing and packaging. Hence a way had to be found to keep these costs to a bare minimum. Progressive farmers in Jaunti, a village near Delhi, were persuaded by the Extension Division of the Indian Agriculture Research Institute (IARI) to undertake wheat seed production. Technical expertise was given free of charge. Inputs were purchased in bulk and given on credit. Constant supervision and monitoring resulted in high yields per hectare and compensated for additional costs in roguing, processing and packaging. Because of the intensive promotional drive, their produce was sold at a fairly remunerative price. The following year the experiment was extended to a number of surrounding villages and the concept of 'seed villages' grew.

Through sustained efforts of various agencies, such as the Rockefeller Foundation, USAID, the National Seeds Corporation and state departments of agriculture in the early 1970s, and public and private sector seed companies thereafter, the concept – with changes to suit local conditions – was extended to many districts of major seed growing states.

A large cadre of progressive farmers residing in these villages are trained in the scientific techniques needed for quality seed production of most crops. Given proper motivation, encouragement and technical guidance, these seed growers produce quality seeds of improved cultivars and/or hybrids. Production of F1 hybrid cotton and vegetable seeds on thousands of hectares in the states of Gujrat, Maharashtra, Madhya Pradesh, Karnataka, Andhra Pradesh and Tamil Nadu is a classic example wherein the companies train seed growers and later train illiterate daily wage labourers in the art of hand emasculation and pollination of each bud/floret at field level. Through this system, an elaborate infrastructure to multiply large quantities of high quality seeds of superior hybrids and improved cultivars has been created in most hybrid seed producing states of India.

Identifying Areas for Seed Production

The consumer farmers throughout the world have become conscious of the physical attributes of seed.

They are now aware that diseased or shrivelled seeds often lose germination and vigour faster and that a plump and mature seed generally ensures good germination and establishes a healthy seedling. Hence the production companies carefully examine whether the growing conditions of the identified area are suitable for the production of high quality seeds of the chosen cultivar. Their objective is to ensure that seed produced should consistently have high germination, be vigorous and have good physical appearance. For economic reasons the chosen area should be reliable and should give high seed yield.

Important considerations while identifying such an area are:

- the right soil, optimum temperature and day length for ideal plant growth;
- assured water and power supply;
- no endemic weeds or seed-borne diseases;
- freedom from frequent strong winds or hail-storms at and after flowering;
- ready access by road and a good communication system;
- freedom from objectionable wild species of the crop, pollen from which can contaminate the seed crop at the time of flowering; and
- rain or even continuous high humidity during seed maturation period can cause mould development on the panicle, discolouration and substantial reduction of germination and storage potential and hence, wherever possible, the contract seed production is located and timed so that few rains occur during the seed development stage.

Compact Area Approach Leading to Seed Villages

The land holdings of individual farmers in developing countries are small. To facilitate production of commercial seed in adjacent fields in a compact belt, viable seed units are created by encouraging small but progressive seed farmers, with adjoining lands in the chosen production belt or village, to undertake multiplication under contract with the company. These individual seed farmers become contract growers or contract seed multipliers and the chosen area a 'seed village'.

The advantages of the compact area approach are:

- ease of finding isolation, as neighbours who do not wish to undertake seed production may be persuaded to grow the same cultivar (or male pollination line in the case of hybrids) or alternatively crops which will not cause contamination;
- ease of technical guidance and support, hence superior quality of produce;
- cost competitiveness because of collective plant protection measures, and effective, efficient and economical roguing;
- easy and cheap inspection, guidance, supervision and training;
- control of quality at threshing floor and sampling;
- to maximise the multiplication ratio, ease of organising credit facilities for inputs like fertiliser, pesticide and herbicide;
- economy on transportation to seed processing unit; and
- assured seed supply.

Options for Seed Procurement Contract

The seed company negotiates the procurement price to be paid, and terms of payment for seed being produced, with the representatives of the growers. Many procurement options are possible, but the following are those usually used.

(A) *Fixed price.* The procurement price of processed seed is negotiated and fixed on a kilogram basis. The seed is procured only if it meets the prescribed field and seed standards. Partial payment is made at the time the seed is delivered at the processing plant, and the balance 30–60 days later, on receipt of a satisfactory report from a recognised seed testing laboratory (STL), confirming that the seed lot meets the minimum prescribed seed standards. For certain crops, such as hybrid cotton, hybrid sunflower and hybrid vegetables, the contract also prescribes minimum genetic purity limits. Final payment in such cases is made only on receipt of a satisfactory grow-out test and STL report. The fixed price option is usually preferred for costly seeds, such as hybrids and vegetable seeds.

(B) *Premium over grain price.* For high volume, low price cereal seed crops, such as rice, wheat, maize, sorghum, oilseeds such as sunflower, and legumes such as cowpea, a procurement contract establishes a premium over the average wholesale price in a recognised grain market for the previous 7 or 15 days. The premium varies with the terms of payment, for example:

- 6–8% for cereals and large-seeded legumes, and 8–10% for oilseeds; full payment against delivery after rough screening to remove dirt and large contaminants; or
- 8–12% for cereal and large-seeded legumes, 10–15% for oilseeds; 50% payment on delivery at seed cleaning plant and balance on actual processed seed quantity after 30–45 days on receipt of a satisfactory report from the STL that the seed lot meets prescribed standards.

(C) *Profit sharing through pool.* The growers take an advance equivalent to 50% of the grain price and keep their seed in the company pool. The company deducts the costs involved in processing, storage and marketing, plus a previously agreed profit margin, usually 10% for the quantity sold. The balance is paid to the growers after deducting advance payment in proportion to their quantity in the pool. The unsold stocks are either carried over, sold as grain, or returned to the growers. This option is well suited to small farmer-owned seed enterprises.

(D) *Premium over negotiated price.* To encourage contract growers to produce seed of higher quality than the prescribed minimum, companies often pay a higher price for better quality seed. If germination is a problem, growers are paid on a graded scale basis on the percentage of germination above a specified minimum for each seed lot. For costly seeds, such as hybrid cotton or hybrid vegetables, a graded scale is based on the genetic purity and other quality factors. Seed samples

from each grower's lot are drawn to evaluate genetic purity, which indirectly also determines the quality of roguing and the care taken in post-harvest operations. This is often part of the contract, and the very fact that such 'grow-out tests' are conducted encourages the contract grower to do a better job.

The written contract is usually fair to both the contract grower and the company. The negotiated terms and conditions are clearly stated, specifying the preferred option, standards to be accomplished, the responsibilities of both parties, and terms of payment. The company is usually considerate to contract grower's needs and the latter makes all possible effort to produce high yields of quality seeds, which establishes the company's brand image and thus ensures that the symbiotic relationship between the two, based on mutual trust, integrity and monetary benefits, continues. If, due to any unforeseen natural disaster or calamity, seed yields are substantially reduced, the procurement rates are renegotiated to compensate the grower adequately.

The company provides Basic (foundation) Seed against cash or credit adjustable at payment time, gives technical guidance and frequently monitors production fields. Input costs and field operations are the responsibility of the contract grower. The company arranges to produce or purchase adequate quantities of high quality Basic Seed to plant the targeted area, at least one growing season in advance, and conducts a genetic purity or grow-out test. Basic Seed with high genetic purity reduces the time and effort spent on roguing and also ensures high genetic quality of the resultant seed. A germination test is also made to provide a recommendation on seed rate to ensure optimum plant density.

Seed Production Technology and Training of Contract Growers

Before introducing a cultivar and/or hybrid in the seed production chain, a systematic study on the ease of production, seed yield potential and infestation from major disease or pest, if any, is conducted under field conditions of the chosen area. In the case of hybrids, information is also generated on:

- number of days to flowering of each parent to ensure proper synchronisation;
- photosensitivity or thermosensitivity of either parent, which may influence the time taken for flowering;
- number of days the female parent stays receptive and information on its floral characters; and
- whether the restorer/male parent is a good or shy pollen producer.

Contract growers are supplied with appropriate and concise information in local language on how to grow and harvest high yields of good quality seed. Higher yields to the grower mean low procurement rate. This information is specific to each kind and cultivar and forms a technology package. It prescribes the minimum isolation requirement and step by step enumerates: appropriate time of planting; row-to-row and plant-to-plant distance; whether direct seeding or transplanting would give high yield; fertiliser dosage and stage of application; plant protection schedule; when and what to rogue; care during harvest; threshing and drying; and, for hybrids, the need, if any, for staggered plantings. For effective roguing and to ensure high genetic purity of resultant seed, the morphological characteristics of the cultivar (or the parents in the case of a hybrid) are also described.

Farmers in a village vary in background, skills, education, adaptability and integrity. Effort is made to choose progressive farmers who possess above-average ability, personal reliability, necessary resources and sincere interest in seed production. The company usually appoints a qualified and trained supervisor – who is provided with a motorbike – to train, motivate and guide its contract growers in efficient and scientific cultivation practices and roguing. The supervisor visits each field at seedling, grand growth, flowering and seed maturation stages of crop growth and monitors production of about 200–300 ha in a village or a cluster of villages. The supervisor is an important link in technology transfer from the company's laboratory to the seed field. As the contract grower gains experience, seed quality and production generally increase.

Quality Control

Commercial seed is produced from Basic Seed. This is the seed sold to farmers to sow their crop for seed production. Commercial seed must meet prescribed field and seed standards of trueness-to-type, germination percentage, and purity. If the company desires third party quality assurance, then the fees are paid and production fields are offered to the state seed certification agency, whose technical staff then monitor production. Certification is usually a voluntary process and may not be used by experienced companies having careful control of their production. In such a case farmers purchasing seed pay more attention to brand name than to the certification tag.

The quality control personnel of a seed company and/or certification agency who perform spot checks are empowered to reject any plot if they are satisfied that the isolation distance and quality or roguing do not meet prescribed standards. Grow-out tests (usually in the off-season) are made by drawing a representative sample from each seed lot, to determine genetic purity. Any lot which fails to meet the standards is rejected. The whole exercise of roguing is to avoid the chances of contamination and to ensure the production of pure seed. The term 'chance contamination' is not restricted to the field boundaries alone, but includes isolation distance also.

The first and final inspection reports of the company supervisor provide details about the contract grower's name, address, crop or cultivar, seed class, details of sowing, crop condition and likely date of harvest. Other reports concentrate on the quality aspects of seed production. The production office summarises information from the field. This information helps management to estimate likely yields, prepare a processing schedule, arrange for the necessary consumables and initiate marketing efforts, and allows the accountant to arrange for timely payments.

Seed Harvesting

Harvesting at the right stage of seed maturity is desirable to achieve optimum yield, minimise field deterioration and obtain high quality seed. The company supervisor helps the contract growers to determine the appropriate harvest time. Samples are taken from random panicles and the moisture content determined. The ideal level of moisture at the time of harvest depends on the drying facilities available to the grower and the company in that production area. If the panicles can be artificially dried and threshed mechanically, harvesting begins when the seed has attained physiological maturity; they are then dried to 12% moisture and stored in bulk until taken for further processing. If panicles are to be sun-dried, the harvest commences at 16–18% moisture.

Before harvest, the contract grower and the company supervisor inspect each field for plants of other cultivars, diseased panicles and, in the case of hybrid seed production, broken or lodged restorer parent plants, and remove them. Seed crop harvest in most developing countries is by hand. In the case of hybrids, the restorer parent rows are harvested first and removed from the field before beginning female parent harvesting.

Seed Drying and Threshing

Harvested panicles or cobs are spread on a concrete drying floor, preferably to a depth of 10–14 cm. They are continuously stirred with a wooden rake to encourage uniform drying. Meticulous care is taken during threshing operations. The identity of each seed lot is maintained by proper labelling. The threshing yards are kept clean at all times. The threshed or shelled seed lot is further dried by spreading on a concrete slab threshing floor and stirring frequently to reduce moisture below 12% before packing.

The production supervisor is responsible for ensuring that chance contamination during drying and threshing does not take place. Contamination is avoided by arranging separate threshing floors for each cultivar. The supervisor seals the bags and fills out a seed threshing report to a standard format, giving details about the contract grower, cultivar name and approximate quantity, and assuring that the seed has met prescribed field standards. The supervisor also ensures that the seed is not adulterated or poached by other companies in case of shortage. The grower brings the produced seed with the threshing certificate to the processing unit designated by the company.

Processing and Storage

It is desirable that the company's seed processing unit is central to the cluster of villages where seed production is organised and that it is connected to the main road and approachable in all weathers. In most developing countries, prescribed physical purity norms are not very rigid. Simple seed cleaners with a pre-cleaner attachment, a gravity table and a slurry treater are adequate to achieve prescribed seed standards. The processing machinery should be simple, so that a minor breakdown can be repaired locally. To keep the planned quantity of produced seed, a suitable seed storage warehouse is constructed, preferably attached to or in the vicinity of the processing facility. The storage is rat-proof and conforms to a recommended design, developed through seed technology research for longer seed life.

Finance

Cultivation of a seed crop, especially a hybrid seed crop, requires high investment for inputs as well as labour. Therefore, availability of requisite finance for contract growers must be assured. Most growers use their own finance or arrange it through their friends and relatives. However, for others the company helps to obtain a loan, to meet financial needs, in any of the following ways.

(i)   The contract growers collectively approach the rural bank in their neighbourhood for a seed crop loan. Bankers are usually shy in extending loans to farmers. The company therefore stands warranty, which means that out of the sale proceeds of the grower, the company pays the principal and interest to the bank and the balance to the grower concerned.

(ii) When a company has surplus funds, it may advance a loan collectively to contract growers on cross-guarantee, and charge bank lending interest. The cross-guarantee helps the company to recover the loan with interest at payment time, even if a particular grower's seed lot fails or is rejected.

(iii) The local financier and/or an input supplier in the village may advance a loan to the contract growers, on warranty from the company. The contract envisages that the company pays the financier first and the balance to the grower.

PRODUCING SEED FOR OWN USE

The development of agriculture and horticulture has been largely assisted by subsistence farmers for at least 10 000 years. During this period farmers and growers have maintained and improved their sowing and planting material by observation linked with relatively simple techniques; these have included retaining useful mutations and carefully selecting plant material by vegetative propagation or by seed. The effects of domestication on plants have been discussed by Schwanitz (1966). The advances which have been achieved by modern plant breeders in relatively recent times are described or discussed elsewhere in this work. However, there are still a large number of farmers in the world, especially in developing countries, who are either obliged, or even prefer, to maintain and produce their own seed supply. The production of seed for farmers' own use is frequently referred to as 'on-farm seed production'.

Sikora (1995) has estimated that only approximately 10% of farmers' seed requirements in developing countries are derived from formal seed supply systems, the remaining 90% or so being produced by the farmers themselves. Although 'on-farm seed production' is favoured by farmers in the Third World, it also occurs in some of the more developed countries. It is estimated that up to 50% of cereal seed sown by farmers in western Europe and North America is from own saved material. Kelly and Bowring (1990) put the figure for England and Wales at 30%.

Cereal farmers either retain part of their crop or grow a smaller separate field or plot specifically for sowing in the following season. A separate field or plot is used when producing seed of a new or replacement cultivar.

Reasons for On-farm Seed Production

The reasons why farmers rely on the production of their own seed are varied. They may live in areas isolated from normal commercial seed supplies. Other economic reasons include the lack of profit incentives for seed suppliers to bear the high distribution costs in remote areas or simply the problems of distribution over difficult terrain. Seed companies may tend to specialise in a small number of crop species such as hybrid maize, and therefore not be involved in production and marketing of other staple crops, especially those species with low multiplication rates. Many subsistence farmers prefer to wait until the sowing season is imminent and, if possible, to use feed grains left over from the previous harvest before seeking to procure the new season's seed requirements. It was reported that in the Southern African Development Community (FAO 1995), seed of supplementary staple crops such as cowpea, millet and indigenous vegetables is produced by farmers for their own use, while seed of cash crops, e.g. cotton and tobacco, is generally available from outside sources, as also is seed of hybrid maize. In some cases the improved cultivars do not offer farmers all the attributes or characters of traditional material for construction use or animal husbandry. The wider genetic base of local types is often preferred to the relatively narrow base of modern cultivars.

Poey (1982) reported that in the highlands of Guatemala it is common to find families who consider it prestigious to save their own seed; these farmers save and maintain their own sowing material which has the characters they desire.

Seed Exchange and Marketing

The seed produced by subsistence farmers as on-farm material will have a relatively limited distribution. While the primary reason is to be self-sufficient, some of the seed crop produced may be surplus to their immediate requirements and will usually be sold to neighbours or exchanged locally for other goods or services.

Some seed producing areas or 'pockets' have evolved from individual small farmers who either formed co-operatives or accepted contracts from outside organisations. Thus a more developed seed production enterprise may evolve from the relatively simple concept of self-sufficiency. This is often the pathway for seed industry development.

In some relatively isolated places in the world, small-scale farmers have been encouraged to develop seed production units which have become a vital seed source for other small-scale farmers in their immediate areas.

Future Assistance to On-farm Seed Producers

The fact that many small farmers in developing countries produce their own seeds despite numerous seed programmes has attracted a lot of attention (FAO 1995).

Those non-government organisations (NGOs) whose remits include agronomy have been playing a role by offering material assistance and advice to farmers who either seek or need to improve the quality of sowing and planting materials (Wilson 1995). Assistance from NGOs is frequently targeted at individual communities, especially the poor, or at discrete geographical areas (Musopole 1995). There has been a tendency to assist farmers to improve the quality of seed they produce and

to increase crop diversity; this in turn has improved crop rotation and also the range and seasonal availability of staple foods. The increased diversity of food crops has directly assisted nutrition and food security of the target populations, which is often the main objective of many NGOs. In addition, some NGOs see the protection of biodiversity as part of their role.

Rana and Bal (1982) have described the strategy developed for hill farmers in Nepal in which a seed multiplication system has been applied to encourage farmers to produce seed for local distribution. In Africa the development of 'seed growing villages' has been observed by de Lannoy (1988). Preliminary studies have been made in Colombia to use small farmers in a cottage industry approach, so that they establish small industries to multiply seed and distribute it to other small farmers (Anon 1982).

Disadvantages of On-farm Saved Seed

Farmers and growers attempting to produce crops from seed saved from hybrid cultivars will observe segregation when the seed is grown on, and in most cases this will be accompanied by a significant decline in yield or quality of the harvested crop. Although most farmers may appreciate the disadvantages of saving seed from hybrids, a clear warning should be given for those who may not understand this. It is generally accepted that a farmer who uses a hybrid cultivar or cultivars for production of the main staple or staples must receive sufficient income from the crop to enable the purchase of hybrid seed for the following season.

The decline in genetic quality of open-pollinated crop cultivars is more obvious in those which are cross-pollinated. In practice the deterioration will be mainly due to insufficient isolation and lack of satisfactory roguing. Cross-pollination between market crops and those destined for seed may occur when there is insufficient isolation between cultivars with a common period of anthesis. Other likely aspects of adverse seed quality include an increase in seed-borne pests and pathogens due to lack of specific controls and/or roguing out of infected mother plants. In some cases the harvested seed lot is likely to contain seeds of other crop species, weed seeds (including noxious and parasitic species) and other impurities. These impurities are especially likely in rice and crops with small seeds. Fresco (1985) points out that in the Sahel, in seasons with poor harvests the best sorghum and millet plants may have to be harvested as grain for food rather than seed, therefore seed will be collected from the more deleterious material. In the case of sorghum there may be a tendency to select seeds from the earlier maturing plants in the population in order to avoid insect damage which will occur later.

Although subsistence farmers have a long tradition of storing their seed successfully under locally prevailing conditions, there is always the possibility of rapid deterioration due to adverse conditions. An additional risk to the on-farm seed supply of subsistence farmers is that when prices are high, or when the individual farmer is in financial difficulties, there will be a temptation to sell for cash.

REFERENCES

Agrawal, P.K. and W/Mariam, W. (1995). Seed supply system in Ethiopia. *Plant Varieties and Seeds*, **8** 1–8.

Anon (1982). Improving seed availability for small farmers. *CIAT International*, **1**(2).

Anon (1996). CONVERDS: Collaborative Network for Vegetable Research and Development in the SADC Region. *TVIS Newsletter*, **1**(1), 32 (Tropical Vegetable Information Service of AVRDC, Taiwan).

Ban, A.W. Van den, and Hawkins, H.S. (1988). *Agricultural Extension*. Longman, Harlow.

de Lannoy, G. (1988). Seed production. In *Vegetable Production under Arid and Semi-arid Conditions in Tropical Africa*. FAO, Rome, 346–360.

FAO (1994). *FAO Seed Review* 1989–90. FAO, Rome.

FAO (1995). *Proceedings of Regional Workshop on Improved On-farm Seed Production for SADC Countries*. FAO, Rome.

Fresco, L.O. (1985). Characteristics of low external input cropping systems (with special emphasis on seed related aspects). In *Proceedings of the Seminar on Seed Production, Yaounde, Cameroon*. CTA and IAC Wageningen, The Netherlands.

Gregg, B. and Wannapee, P. (1985). Organization for effective management of a national seed programme for high- and low-input cropping systems. In *Proceedings of the Seminar on Seed Production, Yaounde, Cameroon*, CTA and IAC, Wageningen, The Netherlands, 1–21.

Kelly, A.F. and Bowring, J.D.C. (1990). The development of seed certification in England and Wales. *Plant Varieties and Seeds*, **3**, 139–150.

Musopole, E. (1995). Actionaid seed involvement. In *Proceedings of Workshop on Improved On-farm Seed Production for SADC Countries*. FAO, Rome, 47–49.

Poey, F. (1982). Quality and variety characteristics of seed saved by the small farmer. In *Improved Seed for the Small Farmer, Conference Proceedings*. CIAT, Cali, Colombia, 18–20.

Rana, P.N. and Bal, S.S. (1982). Experience with small farmers in the use of improved seed in the hills of Nepal. In *Improved Seed for the Small Farmer, Conference Proceedings*. CIAT, Cali, Colombia, 78–81.

Rosell, C.H. (1986). *Seed requirements of major food crops*. AGP/SIDP/86/9, Food and Agriculture Organization of the United Nations, Rome.

Schwanitz, F. (1966). *The Origins of Cultivated Plants*. Harvard University Press, Cambridge, MA, USA.

Sikora, I. (1995). Opening Ceremony Statement. In *Proceedings of Workshop on Improved On-farm Seed Production for SADC Countries*. FAO, Rome.

Wilson, A. (1995). An NGOs experience in small scale seed production. In *Proceedings of Workshop on Improved On-farm Seed Production for SADC Countries*. FAO, Rome, 42–46.

Wobil, J. (1994). Seed programme development in Swaziland. *Plant Varieties and Seeds*, **7**, 7–16.

Further Reading

Anon (1996). *Service Through Partnership*. ISNAR, The Hague, The Netherlands.

Chopra, K.R. (1982). *Technical Guideline for Sorghum and Millet Seed Production*, FAO, Rome.

Chopra, K.R. (1986). *Lead Paper on Problems in Seed Production, Certification and Quality Control*, First National Seed Seminar of Seed Association of India, New Delhi.

Chopra, K.R. (1992). *Organization of a Seed Network for Public and Private Seed Sectors in the Asia and Pacific Region*. Regional Consultation, Bangkok.

Chopra, K.R. and Reusche, G.A. (1993). *Seed Enterprise Development and Management, A Resource Manual*. FAO, Bangkok.

Chopra, K.R., Kharche, S.K. and Rakesh Chopra

(1992). *Pearl Millet Seed Production Technology.* Oxford and IBH Publishing, New Delhi.

Copeland, L.O. and McDonald, M.B. (1995). *Principles of Seed Science and Technology.* Chapman and Hall, New York, Chapter 15.

Cromwell, E. (1996). *Governments, Farmers and Seeds in a Changing Africa,* CAB International, Wallingford, UK.

Douglas, J.E. (1980). *Successful Seed Programmes: a planning and management guide.* Westview, Colorado.

George, R.A.T. (1985). *Vegetable Seed Production.* Longman, Harlow, Chapter 5.

Kelly, A.F. (1988). *Seed Production of Agricultural Crops.* Longman, Harlow, Chapter 4.

Kelly, A.F. (1994). *Seed Planning and Policy for Agricultural Production.* Wiley, Chichester.

Rosell, C.H. and Kelly, A.F. (eds) (1983). *Seed Campaigns: guidelines for promoting the use of quality seeds in developing countries.* FAO, Rome.

Rotzal, H. (1982). *Seeds.* FAO Plant Production and Protection Paper 39, Food and Agriculture Organization of the United Nations, Rome, Chapter vi.

Tripp, R. (1995). *Seed regulatory frameworks and resource-poor farmers: A literature review.* Overseas Development Institute, London.

Wright, M., Donaldson, T., Cromwell, E. and New, J. (1994). *The Retention and Care of Seeds by Small-scale Farmers,* NRI, Chatham, UK.

2

International Agreement and National Legislation

Part One: International

INTERNATIONAL UNION FOR THE PROTECTION OF NEW VARIETIES OF PLANTS (UPOV)

The expression 'variety' is used in the UPOV Convention and in the plant variety protection laws of most UPOV member states. It is accordingly adopted in this chapter, while the term 'cultivar' is adopted in other chapters. Article 1 (vi) of the 1991 Act of the UPOV Convention contains a definition of 'variety'.

The International Union for the Protection of New Varieties of Plants (UPOV) is an intergovernmental organisation with headquarters in Geneva. The acronym UPOV is derived from the French name of the organisation, 'Union internationale pour la protection des obtentions végétales'. UPOV was established by the International Convention for the Protection of New Varieties of Plants (the UPOV Convention), which was signed in Paris in 1961. The Convention entered into force in 1968. It was revised in Geneva in 1972, 1978 and 1991. The 1978 Act entered into force in 1981. The 1991 Act had not yet entered into force on 7 February 1998, but was expected to do so during the course of 1998.

The Purpose of the UPOV Convention

The purpose of the UPOV Convention is to ensure that the member states of the Union acknowledge the achievements of breeders of new plant varieties, by making available to them an exclusive property right in their varieties, on the basis of a set of uniform and clearly defined principles. To be eligible for protection, varieties have to be (i) distinct from existing, commonly known varieties, (ii) sufficiently homogeneous, (iii) stable, and (iv) new in the sense that they must not have been commercialised prior to certain dates established by reference to the date of the application for protection. Plant variety protection should be contrasted with the protection granted under the industrial patent system by many countries for inventions related to plants. Such inventions must be useful in industry (this includes agriculture), must represent a non-obvious advance over the existing technology for persons skilled in that

Encyclopaedia of Seed Production of World Crops.
Edited by A.F. Kelly and R.A.T. George. © 1998 John Wiley & Sons Ltd

technology, and must be new in the sense that they have not been communicated to the public prior to filing the application for protection.

The Effect of Plant Breeders' Rights

Both the 1978 and the 1991 Acts set out a minimum scope of protection. Member states may offer a greater scope of protection but may not offer less.

Under the 1978 Act, the minimum scope of the plant breeder's right requires that the holder's prior authorisation is necessary for the production for purposes of commercial marketing, the offering for sale and the marketing of propagating material of the protected variety. The 1991 Act contains more detailed provisions defining the acts concerning propagating material in relation to which the holder's authorisation is required. Exceptionally, but only where the holder has had no reasonable opportunity to exercise his right in relation to the propagating material, his authorisation may be required in relation to any of the defined acts carried out with harvested material of the variety.

The breeder's right, under the 1978 Act, extends not only to the protected variety itself but also to varieties whose production requires the repeated use of the variety, for example F1 hybrids. Under the 1991 Act, the breeder's right also extends to varieties which are 'not clearly distinguishable from the protected variety' and to varieties which are 'essentially derived' from the protected variety. The definition of 'essential derivation' in the 1991 Act is such that a derived variety will only fall within the scope of protection of the protected variety if it retains virtually the whole genetic structure of the initial variety. An essentially derived variety might result, for example, from the addition by genetic engineering of patented genetic information (a so-called patented gene) to a protected variety. The resulting modified variety cannot be sold without the permission of the owner of the protected variety.

The variety protection right extends only to the variety of varieties covered by the right. It does not confer rights to any single trait or characteristic of a variety taken in isolation. It should be contrasted in this respect with the right of the owner of a patent related to a plant. The rights of the owner of a patent are defined by the valid claims of the patent. A patent for an invention relating to a plant might claim a plant of a specified species with a particular characteristic, for example resistance to a particular herbicide. The patent owner's rights might cover all plants of that species with that characteristic.

Like all intellectual property rights, plant breeders' rights are granted for a limited period of time, at the end of which varieties protected by them pass into the public domain. The rights are also subject to controls, in the public interest, against any possible abuse.

It is also important to note that the authorisation of the holder of a plant breeder's right is not required for the use of a variety for research purposes, including its use in the breeding of further new varieties. The rights of farmers to save part of their harvest in order to sow the following season's crop also receives special attention under the UPOV plant variety protection system. Generally speaking, member states are free under all Acts of the Convention to define any such rights of farmers in their national laws. The great majority of UPOV member states permit farmers to save seed of the main sexually reproduced species under their national laws.

The agricultural, horticultural and forestry industries and the final consumer all ultimately gain from the additional stimulus that plant breeders' rights give to the creation of new varieties that are better suited to satisfy man's needs.

Reasons for the Protection of New Varieties of Plants

Protection is afforded to new varieties of plants both as an incentive to the development of agriculture, horticulture and forestry, and to safeguard the interests of plant breeders.

Improved varieties are a necessary and very cost-effective element in the quantitative and qualitative improvement of the production of food, renewable energy and raw material.

Breeding new varieties of plants requires a substantial investment in terms of skill, labour, material resources, money and time. The opportunity to obtain certain exclusive rights in respect of a new variety provides the successful plant breeder with a better chance of recovering the costs and accumulating the funds necessary for investment in further breeding. In the absence of plant breeders' rights, investment in plant breeding is problematic since there is nothing to prevent others from multiplying the breeder's seed or other propagating material and selling the variety on a commercial scale, without recognising in any way the work of the breeder.

Members of UPOV

By becoming a member of UPOV, a state signals its intention to protect plant breeders on the basis of principles that have gained worldwide recognition and support. It offers its own plant breeders the possibility of obtaining protection in the other member states and provides an incentive to foreign breeders to invest in plant breeding and seed production on its own territory.

It has the opportunity, through membership of UPOV, to share in and benefit from the combined experience of the member states and to contribute to the worldwide promotion of plant breeding. A constant effort of intergovernmental co-operation is necessary and this requires the support of a specialised secretariat.

UPOV Activities

The main activities of UPOV are concerned with promoting international harmonisation and co-operation, mainly between its member states, and with assisting countries in the introduction of plant variety protection legislation. A smoothly operating international trade in seeds and plants requires uniform, or at least mutually compatible, rules.

The fact that the UPOV Convention defines the basic concepts of plant variety protection that must be included in the domestic laws of member states leads, in itself, to a great degree of harmony in those laws and in the practical operation of the protection systems of the member states. Such harmony is enhanced, firstly, through specific activities undertaken within UPOV leading to recommendations and model agreements and forms and, secondly, through the fact that UPOV serves as a forum to exchange views and share experiences.

UPOV has established a detailed set of general principles for the conduct of the technical examination of plant varieties for distinctness, uniformity and stability, and more specific guidelines for the application of these general principles to some 150 genera and species. These documents are updated periodically and progressively extended to further genera and species. Their use is not limited to plant variety protection. The UPOV guidelines are also used worldwide as the basis for establishing varietal descriptions for the purposes of national listing and seed certification.

UPOV member states co-operate intensively in relation to the examination of plant varieties. Their co-operation is based on arrangements whereby a member state either conducts tests on behalf of others or accepts the test results produced by others as the basis for its decision on the granting of a breeder's right. Through such arrangements member states are able to minimise the cost of operating their protection systems and breeders are able to obtain protection in several countries at relatively low cost.

The UPOV member states and the UPOV Secretariat maintain contacts with, and provide legal, administrative and technical assistance to the governments of a growing number of states expressing interest in the work of UPOV and in the idea of plant variety protection. Regular contacts are also maintained with many intergovernmental and international non-governmental organisations.

Information on the development of plant variety protection legislation throughout the world is published in *Plant Variety Protection* (the UPOV Gazette and Newsletter published irregularly but usually at least four times a year).

Administration and Management of UPOV

The Council of UPOV consists of the representatives of the members of the Union. Each member that is a state has one vote in the Council. Under the 1991 Act, the possibility also exists for certain intergovernmental organisations, for example the European Union, to become members of the Union. The Council is responsible for safeguarding the interests and encouraging the development of the Union and for adopting its programme and budget. The Council meets once each year in ordinary session. If necessary, it is convened to meet in extraordinary session. The Council has established a number of committees, which meet once or twice a year.

The Secretariat of UPOV (called the Office of the Union) is directed by a Secretary-General. Under a co-operation agreement with the World Intellectual Property Organisation (WIPO), an organisation within the United Nations system, the Director General of that organisation is the Secretary-General of UPOV. He is assisted by a Vice Secretary-General. The Office has a small international staff.

Member States of UPOV

As of 8 February 1997, the member states of UPOV were: Argentina, Australia, Austria, Belgium, Canada, Chile, Colombia, Czech Republic, Denmark, Finland, France, Germany, Hungary, Ireland, Israel, Italy, Japan, Netherlands, New Zealand, Norway, Paraguay, Poland, Portugal, Slovakia, South Africa, Spain, Sweden, Switzerland, Ukraine, United Kingdom, United States of America, Uruguay. Many non-member states currently have proposals for laws to protect varieties before their legislatures. Belarus, Bolivia, Brazil, Bulgaria, Ecuador, Kenya, Panama, Russian Federation and Trinidad and Tobago have initiated with the Council of UPOV the procedure for becoming members of the Union.

Mexico has taken steps with a view to ratifying the 1978 Act.

Future Developments

The 1978 Act of the UPOV Convention will be closed to further accessions when the 1991 Act comes into force. The 1991 Act will come into force when five states, of which three must be party to an earlier Act of the Convention, have adhered to it. As at 8 February 1997, three states (Denmark, Israel, Netherlands), party to earlier Acts of the Convention, had ratified the 1991 Act while a further ten states (Australia, Belarus, Bolivia, Bulgaria, Ecuador, Poland, Slovenia, South Africa, Russian Federation, United States of America) had laws conforming with the 1991 Act and were in a position to ratify or accede to it. Furthermore, the European Union had enacted the Council Regulation on Community Plant Variety Rights which enables the grant of plant variety protection with effect in the 15 member states of the European Union pursuant to a single application filed with the Community Plant Variety Office. The Regulation conforms with the 1991 Act. Accordingly, by the time that the 1991 Act comes into force in 1998, it will already be established for the future as the international standard for plant variety protection systems.

Parallel to its implementation of the 1991 Act, the international community concluded, on 15 April 1994, the Marrakesh Agreement Establishing the World Trade Organisation (WTO). The Marrakesh Agreement entered into force on 1 January 1995. Annex 1C of the Marrakesh Agreement contains the Agreement on Trade-Related Aspects of Intellectual Property Rights (TRIPS Agreement). Article 27.3(b) of the TRIPS Agreement requires member states of the WTO to provide protection for plant varieties either by patent or by a *sui generis* (that is, a specialist) system for the protection of plant varieties. It can be anticipated as a result of the TRIPS Agreement that the great majority of countries of the world will introduce *sui generis* plant variety protection systems in the years immediately ahead. The

UPOV system of protection is the only widely accepted *sui generis* system for the protection of plant varieties. Accordingly, most countries seem likely to adopt the principles of the UPOV Convention for their purposes. This development will coincide with the introduction to the market-place of an ever-increasing number of varieties which incorporate patented traits or have been developed using patented processes. Intellectual property will be a subject of ever-increasing importance for plant breeders and seedsmen in the future. Varieties will frequently be protected by plant variety protection but will also fall within the scope of patents relating to one or more genes.

ORGANISATION FOR ECONOMIC CO-OPERATION AND DEVELOPMENT (OECD) SEED SCHEMES

Objectives

The OECD is an intergovernmental organisation comprising 27 member countries, all of which operate the OECD Seed Schemes.

The member countries of OECD include the United States, Canada, Australia, New Zealand, Japan and 21 European countries including 15 member states of the European Union (EU) and Turkey. Additionally, a further 19 countries from other continents have been admitted to the Seed Schemes.

The principal objective of the OECD Seed Schemes, which are administered by the Directorate for Food, Agriculture and Fisheries, is to allow seed produced under the schemes' rules and directions to be moved in international trade with an assurance that it has met specified minimum standards and has been controlled and inspected according to agreed conditions.

A further objective of the OECD Seed Scheme is to extend the principles, procedures and standards of seed certification to other countries of the world which could use the OECD model for their domestic certification schemes.

Historical Background

Following World War II there was an urgent need in Europe to improve the quality and yield of grassland production for animal feed. It was realised that this could be achieved only if there was a continuous supply of high quality herbage seed available.

In 1954, the European Productivity Agency (EPA) of the then Organisation for European Economic Co-operation (OEEC) initiated a programme of international co-operation and a detailed survey of national systems of seed certification. This identified considerable problems due to a wide range of practices in the production and control of seed in different countries and a general confusion in terminology.

The outcome was a standardisation of procedures and standards which led to the introduction in 1958 of the Herbage Seed Scheme, which was extended in 1973 to include oilseeds and became the OECD Scheme for the Varietal Certification of Herbage and Oil Seed Moving in International Trade (OECD 1996a).

A scheme for cereals was introduced in 1966 (OECD 1996b). Further schemes have since been added to include: sugar beet and fodder beet in 1968 (OECD 1996c); vegetables in 1971 (OECD 1996d); subterranean clover and similar species in 1974 (OECD 1996e); and maize in 1977, later extended to include sorghum (OECD 1996f). The schemes take account of seed production of both conventional and hybrid cultivars where appropriate.

Legal Framework

The Seed Schemes are voluntary seed certification schemes in the sense that there is no obligation for an individual country to participate, but if participating, the National Designated Authority (NDA) acting on behalf of its government is obliged to ensure that there is full compliance with the rules for all seed which carries its country's OECD label.

Each of the six schemes has Rules and Directions and a number of Technical Appendices which deal with definitions of terms used, the minimum requirements which must be met for the production of Basic and Certified Seed, label and certificate specifications and the procedures for the extension of the schemes to non-members of OECD. These are under continual review and are modified to take account of changes in certification procedures, advances in plant breeding and biotechnology and international developments or changes in government policy. Recommendations for changes are made at the annual meeting of representatives of all NDAs. These are then passed to the Committee for Agriculture for agreement before ratification by the OECD Council. The official languages of the OECD are English and French.

Countries which are not members of OECD are allowed to participate in the Seed Schemes and to contribute to the recommendations made at the annual meetings of the NDAs, but they do not have seats on the Committee for Agriculture or the Council. Annual meetings alternate between the headquarters of OECD in Paris and the participating countries, including the non-members of OECD.

As countries become more developed in their seed industry so the need for effective seed law becomes essential. A country seeking admission to one or more OECD Seed Schemes normally makes an application to the Secretary General of OECD through its appropriate government department. Before such a country is accepted it must demonstrate that seed can be produced which fully complies with the requirements of the scheme.

The production of OECD labelled seed in a country does not prohibit the production of seed under other national schemes. Many countries, however, find that running two or more separate schemes is very complicated, especially if the rules and standards are different, so that in practice many national certification schemes have been modified to make them fully compatible with the OECD schemes. In this way seed which is produced under a national scheme can qualify for an OECD label if it is required for export.

A particularly close link exists between the OECD and the EU regarding seed certification.

The object of this co-operation is to ensure, as far as possible, the equivalence of the OECD Seed Schemes and the EU Directives on seed. This allows member states of the EU to apply the OECD Seed Schemes rules and for seed to be traded between the EU and Third Countries. Similar links occur elsewhere in the world such as between OECD and ALADI which represents 11 countries in Latin America.

The OECD co-operates with many other international organisations working in the seed sector which have representatives at meetings as observers. Prominent among these are ISTA (International Seed Testing Association), FAO (Food and Agriculture Organisation of the United Nations), FIS (International Seed Trade Federation), UPOV (International Union for the Protection of New Varieties of Plants), AOSCA (Association of Official Seed Certifying Agencies) and ASSINSEL (Association Internationale des Sélectionneurs).

Achievements in the Technical Field

Although the various schemes for agricultural seeds differ in detail, all are based on the following principles.

(a) Cultivars can be admitted into a scheme only when official tests for Registration have established that the cultivar is distinct, uniform and stable (DUS), an accurate description is available and the cultivar has an acceptable value in at least one country.

(b) Each country has an official national catalogue of cultivars which have been accepted into a scheme following the tests referred to above and only seed of these cultivars is eligible for the OECD Seed Schemes.

(c) A list of species covered by the various schemes, together with a catalogue of eligible cultivars and their maintainers is published annually by OECD (1995a).

(d) The maintainer of a cultivar is responsible for ensuring that it is maintained true to type and that seed is available for multiplication.

(e) The government of each country participating in one or more schemes through its NDA is responsible for the supervision and implementation of these within its territory and for checking by means of control plot tests and by field inspections that the maintenance of the cultivar has been adequate and that seed has been produced to conform with published standards.

(f) Basic Seed is produced under the responsibility of the maintainer and is intended for the production of Certified Seed. It must conform to the appropriate conditions of the scheme which must be confirmed by official examination. Seed of the generation or generations before Basic Seed is known as Pre-Basic Seed and is at any generation between parental material and Basic Seed.

(g) Certified Seed is intended for the production of either subsequent generations of Certified Seed or of crops for purposes other than seed production.

(h) Seed crops are field inspected by trained inspectors of the NDA to ensure that there are no circumstances which might be prejudicial to the quality of the seed to be harvested. The crop must meet specific standards for isolation against mechanical admixture or out-pollination and the site for previous cropping. The cultivar must have the correct identity and satisfy minimum standards for cultivar purity. At least one field inspection is made after the emergence of the inflorescence but in some species inspections are conducted at other growth stages.

(i) The schemes require that control plots are available for examination at the time of field inspection. These plots are grown from samples of the seed used to sow each crop entered for certification. The control plot serves both as a pre-control of the growing seed crop and as a post-control of the seed produced at the last harvest. Their purpose is to check the level of cultivar purity of individual seed lots and that the cultivar has remained unchanged during the multiplication cycle. A Standard Sample is used as a reference and as a living description of the cultivar as it was at the time of registration.

(j) Laboratory tests can also be conducted on seeds and seedlings using samples of the same seed as that used in the control plot.

(k) Details of the methods used in field inspection and control plots and lists of morphological characters for 160 agricultural and vegetable species are included in guidelines (OECD 1995b). The characters may be used in conjunction with the official description to confirm that variants are correctly identified as not being true to the cultivar.

(l) All practicable steps are taken to ensure that the identity and cultivar purity of the seed are preserved between harvest and the sealing, labelling and sampling after processing.

(m) A fully representative sample is drawn from each seed lot by a person authorised by the NDA. One part of each sample is used for conducting the control plot test. The second part is used for the analysis of germination and analytical purity conducted where appropriate according to ISTA rules. These tests may be conducted either in the country of production or by the importing country.

(n) The success of the OECD Seed Schemes is measured at each annual meeting of NDAs by monitoring each country's certified seed production which is intended either for domestic use or for international trade. The species most widely certified within the schemes include *Zea mays, Glycine max, Helianthus annuus, Medicago sativa, Lolium multiflorum* and *Sinapis alba*.

Future Developments

With the development and introduction of genetically modified varieties (GMVs) the OECD is considering their likely impact on the Seed Schemes. The status of transgenic cultivars submitted for inclusion on National Catalogues may have to be declared to satisfy both environmental and food safety evaluations before listing and release. Some traits, such as herbicide tolerance or insect resistance, which are not expressed phenotypically could become part of the official description of such a cultivar and therefore clear labelling will be essential. Special tests for some transgenic cultivars may be needed to ensure that cultivar purity is maintained and to confirm continued expression of the transgene.

Following adoption by UPOV of the gel electrophoresis technique as an additional DUS character for maize, the use of such a character is under consideration by OECD for use in postcontrol tests.

The OECD Seed Schemes have recently been modified to permit the export of seed after field approval but before final certification. This seed must be sampled, the containers fastened and the contents identified with a special label stating that the seed is 'not fully certified'. This allows an importer to process the seed to local requirements under the supervision of the importing country's NDA.

Since the inception of the Seed Schemes, field inspections have been conducted by trained inspectors of the NDA, but there are now moves to allow this function to be conducted by authorised persons or organisations which would become accredited by the NDA. Temporary investigations have been set up in some countries to establish whether inspections carried out by an authorised inspector meet the high standards required by the OECD Seed Schemes.

Developments in seed science and technology, biotechnology, environmental and food safety measures, regulatory reform and the dissemination of intellectual property will undoubtedly add significantly to the OECD seed certification procedures in the future.

FOOD AND AGRICULTURE ORGANISATION OF THE UNITED NATIONS (FAO) QUALITY DECLARED SEED SYSTEM

The Food and Agriculture Organisation of The United Nations (FAO) has introduced the Quality Declared Seed system (FAO 1993) for use in countries where there are insufficient resources for a fully developed seed quality control scheme such as seed certification. It aims to ensure that the resources available in a country are used to the best advantage to enable farmers and growers to purchase seed which is of a satisfactory quality within a country and also in international trade.

The system is also designed to encourage and assist the development of technical expertise in a seed industry. It recognises the potential role of an extension service which is capable of both assisting in the development of Quality Declared Seed and also of demonstrating the advantages of the product to farmers.

Historical Background

The Quality Declared Seed system was developed as a result of an initiative put forward in 1981 at the FAO/SIDA Technical Conference (FAO 1981). The full text of the action proposals and recommendations was published as a separate document FAO (1982) and subsequently developed by successive consultations organised by FAO in 1985 and 1986. The Expert Consultation in 1986 resulted in the publication of Guidelines (FAO 1987). Further work was done by experts familiar with seed production of individual crops to produce Part 2, which is composed of standards applicable to individual crop species. The scheme was published as FAO Plant production and Protection Paper No. 117 (FAO 1993).

Legal Framework

The scheme has four basic points.

(i)   A participating country should establish a list of cultivars eligible for inclusion in the scheme.

(ii) Participating seed producers must register with the relevant national authority.

(iii) The national authority in each participating country will check 10% of seed crops entered in the scheme.

(iv) The national authority will check 10% of 'Quality Declared Seed' offered for sale in the country.

Variety description forms and passports were published earlier (FAO 1986). The term 'variety' is used throughout the scheme, but is used in the context of being synonymous with 'cultivar' as defined in Article 10 of *The International Code of Nomenclature for Cultivated Plants* (Anon 1980). The Quality Declared Seed Scheme (FAO 1993) outline the conditions which must be met and complied with in order to operate within the scheme.

In addition to an introductory section there are a further nine sections; these cover definitions, eligibility of varieties, register of seed producers, seed production, labelling, supervision by government, penalties, organisational framework and the format of the quality declared declaration.

In principle, a government desirous of authorising production of Quality Declared Seed on its territory has to have an established seed consultative committee. The duties of this committee include establishing lists of eligible cultivars. Participating governments must also have a Seed Quality Control Organisation whose duties include monitoring the relevant activities, maintaining a register of authorised producers and checking the stipulated proportions of seed crops and seed in the scheme which are offered for sale. An example of a suitable form of declaration to be completed by the seed producer and made available on request to the seed purchaser, or the seed quality control organisation, is outlined. This form includes details of the producer, crop species and cultivar; it also includes the seed testing history and results of the seed lot. Other information on the declaration includes details of any fumigation treatment, results of cultivar purity tests, number of generations of multiplications, numbers of specified weed seeds per unit weight, tests for seed-borne diseases and moisture content. The declaration contains a statement signed by an authorised signatory to the effect that the seed lot bearing the reference number has been produced according to Quality Declared Seed requirements and has met the specified standards at inspections and tests.

The FAO Scheme for Quality Declared Seed provides crop-specific sections for some 82 crops. These include eight cereals, nine food legumes, three oil crops, 13 forage grasses, eight forage legumes, two industrial crops and 31 vegetables. There is provision for additional crop species. The individual seed crop guidelines are each specific to either open-pollinated or hybrid cultivars.

The requirements and obligations for each individual crop include the following.

- *Facilities and equipment* with recommended storage, seed extraction, cleaning, bagging and weighing equipment plus other more crop-specific items of equipment and/or facilities to be specified according to local needs.
- *Land requirements* which include freedom from volunteer plants and in some seed crops, those which are related species.
- *Field standards* including isolation requirements, percentage varietal purity in the seed crop, standards for seed-borne and other diseases, general and specific weed tolerances for some crops and additional specifications especially relating to hybrid seed production.
- *Field inspections* – number and timings, with additional crop-specific instructions and inspection technique. These include inspection requirements before entering the field, in the field and instructions for completion of the inspection report and acceptance or rejection of the seed crop. There is also provision for recommendation of further remedial action before the final decision as to whether or not the seed crop is of the required standard.
- *Seed quality standards* – the minimum percentages for germination and varietal purity using national seed testing rules are stipulated. Additional requirements are specified by each country depending on local requirements; these include, for example, moisture content and seed-borne diseases. Other elements of seed

quality, e.g. analytical purity, may be included according to an individual country's local requirements and the testing is based on national rules.

Achievements in the Technical Field

The Quality Declared Seed scheme has only recently (FAO 1993) been published and it is as yet too early to assess whether or not it will be adopted by many countries. It has, however, provided a technically sound basis for the development of a system which takes account of the expertise currently available in many private seed companies which are becoming established in developing countries. In this way it has the potential to encourage the distribution of good quality seed with a minimum of dependency on scarce government resources.

Possible Future Developments

The Quality Declared Seed system was formulated so that individual countries could use it as a framework for both initiating a degree of seed quality control and for building a more comprehensive national system such as a seed certification scheme.

Further crop species not included in the current FAO Guidelines can be added according to the demands and needs of countries using the scheme. This would simply require further specifications for individual crop species to be drafted and added to Part 2 of the document.

The crop-specific standards in Part 2 can be modified according to regional requirements and the specifications for individual components of seed quality of a crop can be upgraded and added according to a participating country's needs and rate of seed industry development.

Governments of participating countries could introduce legislation to make it compulsory for seed producers to enter a Quality Declared Seed scheme.

ASSOCIATION OF OFFICIAL SEED CERTIFYING AGENCIES (AOSCA)

Objectives

To establish standards for genetic purity and identity and to standardise seed certification.

Historical Background

Starting towards the end of the 19th century there was considerable interest, in both Canada and the United States (US), in the development of improved cultivars. A concern of plant breeders and others, however, was that new cultivars often lost their identity. In some cases, successive growers would rename a cultivar, so that the same cultivar might have more than 20 names. In other cases, because there were no standards for seed production, cultivars became so contaminated that their identity and value were lost.

Organisations were established in Canada and in several US states to facilitate the production of pure seed of improved cultivars. One means to this end was the introduction of field inspections of seed crops. These inspections were initiated in Canada in 1904 and subsequently in several US states.

Those involved in the development of seed certification were mainly interested in the release and distribution of improved cultivars from universities and government research stations. Thus the major thrust at first was in developing methods of release and multiplication of cultivars and in encouraging their use by farmers.

Rules and terminology for inspected seed varied among different agencies because they worked in isolation from one another. Seed, however, was crossing state lines and the Canada/US border. Eventually a meeting was convened in 1919 to bring together representatives of seed or crop improvement organisations from Canada and the US. The objectives of the meeting included 'attaining higher and more uniform standards of seed requirements' (Hackleman 1961).

At a subsequent meeting in Chicago, in 1919, it was decided to set up an international organisation to standardise inspection requirements and raise standards so that seed would correspond in quality and purity wherever it came from. The International Crop Improvement Association (ICIA) was established in 1919 for 'any national, state or provincial organisation carrying out inspection and certification of seeds' (Hackleman 1961). Initially it included about 15 state agencies and the Canadian Seed Growers' Association (CSGA).

ICIA developed certification rules, initially without consistency among unrelated species. A committee was established in 1920 'to standardize the nomenclature and rules for the inspection and certification of pedigreed and improved varieties of seed and grain' (Hackleman 1961). This committee divided into two, one for grasses and clovers and the other for cereals.

The cereal committee established some fundamental concepts, many of which still apply:

- registration and certification based on lineage
- recognition of growers
- field inspections by qualified inspectors
- trials to establish identification and usefulness of cultivars
- records of pedigree of stocks in certification
- purity and germination standards
- sealing of seed.

The terminology for different generations, or classes, of seed differed between crop kinds. For the small-seeded legumes and grasses there were two classes: Registered and Certified. For other species, including cereals, soyabeans, maize and cotton, there were three classes: Elite, Registered and Certified. It was not until 1929 that uniformity was achieved with the adoption of the classes Foundation, Registered and Certified. It took even longer to get agreement on the colour of tags to be used on bags of certified seed. The use of the blue tag for first quality certified seed was approved in 1937.

Inevitably, since the organisation included so many representatives involved in the production and use of improved cultivars, ICIA became involved in the promotion of certified seed. This was achieved by co-operation with other organisations with similar interest in improved seed and its use.

One of the limitations on the use of improved cultivars of small-seeded legumes was the low level of seed production. Seed was multiplied in the areas where the cultivars were bred, which were not always the most suitable for seed production. ICIA led a study, which included the crop improvement agencies, the United States Department of Agriculture (USDA) and the seed trade, to develop a programme for the production and certification of small-seeded legumes outside of their area of adaptation and consumption. This programme, implemented in 1934, allowed seed production of these species to move to the north-west states where yields were higher and more consistent. It commenced the important principle of interstate certification in which seed produced and certified in one area is recognised as certified in another area.

As experience increased with the multiplication of cultivars of small-seeded legumes outside of their area of origin, it was realised that there was the possibility of genetic change. Tests were conducted co-operatively by USDA and several states to study this issue. It was concluded that the number of generations of multiplication should be limited, and with increased experience there were further limitations.

The first rules for certification were written for single species or groups of similar species but were not standardised between unrelated species. It was realised that there were a number of sections of the standards that should be the same for all species. A uniform standard for crop certification was published in 1946 covering all species in one book. This became a model that was used in other parts of the world.

The value of seed certification as a means of ensuring that cultivars retained their identity and purity was recognised by plant breeders and others involved in crop improvement in more and more states. At the same time, the importance of interstate trade in certified seed was increasing.

The membership in ICIA expanded until, by the 1950s, most states had joined.

Most of the US certification agencies were associated with universities or were associations of seed growers. Only in a few states were the state governments involved in seed certification. At the federal level, the US agencies had had no legal recognition, unlike the situation in Canada where CSGA was the national certifying agency, with inspection provided by government inspectors.

The US Federal Seed Act of 1939 included definitions of Registered and Certified Seed. Only officially recognised state agencies were permitted to certify seed moving in interstate commerce. State laws were changed in several states to grant formal recognition to the state certification agencies. Some agencies also decided that they should be legally incorporated, and this led to the decision to incorporate ICIA in 1951 under the General Not-for-Profit Corporation Act of Illinois.

The structure and purpose of ICIA was thoroughly reviewed in the early 1960s. There was a desire to have ICIA certification procedures and minimum standards recognised at the national level in the US in the same way as CSGA procedures and standards were nationally recognised in Canada. Two separate committees were set up. The first examined the pros and cons of establishing a national US certification organisation. The second looked into the type of organisational structure that would be needed to accommodate a US national agency with the certification agency in Canada and in any other country that might be interested in joining.

It was decided not to establish a US national certification agency but to modify the existing organisation including changing the name to more accurately reflect its function. In 1968, ICIA became the Association of Official Seed Certifying Agencies under a new constitution (Hackleman and Scott 1990). AOSCA remained an international organisation, with Canadian and US agencies serving on the executive.

The US Federal Seed Act was changed to recognise AOSCA standards. An agreement between AOSCA and USDA provides for the monitoring of US agencies by designated AOSCA representatives to ensure conformity with minimum genetic certification standards.

Since the establishment of AOSCA, the separation between genetic certification and other aspects of seed quality has been more clearly stated. Seed that has been certified may remain in the same container for a considerable period of time. The certification tag can only attest to the genetic identity and purity but not the germination. In addition, weed seeds are of different significance in different areas and so the genetic certification is not the sole determinant of saleability. In some states there is a requirement for two tags, one for the genetic certification and the other for other quality aspects.

An AOSCA Advisory Committee was established in 1970 with representation from the major seed organisations in Canada and the US, federal research and regulatory organisations in both countries, together with researchers from the university and private sectors. The committee reviews proposed genetic standards before adoption by AOSCA and advises on policy. The committee acts as an invaluable clearing house for information for the seed sector in Canada and the US.

AOSCA–OECD Committee reviews the OECD certification schemes, in which Canada and the US participate, and recommends developments regarding existing and new schemes. Large quantities of seed are produced in both countries under these schemes and the committee assists the member agencies in complying with OECD requirements and with those of other organisations, particularly the European Union. AOSCA has now been recognised as an official participant in meetings of the OECD Seed Schemes.

AOSCA's work is carried out by a large number of committees that meet during the AOSCA annual meeting. There are commodity committees that advise on the particular rules for certification of species or groups of species. There are also general committees that cover issues relating to certification of all species. Committees may function by correspondence but the annual meeting provides the opportunity for interaction between certification specialists from across the US and Canada.

In recent years there has been interest from other countries in AOSCA. Representatives from Mexico have attended some annual meetings. More recently, New Zealand and Argentina have become members.

Legal Framework

AOSCA is incorporated under the state laws of Illinois as a not-for-profit corporation (Hackleman and Scott 1990).

Membership is open to national, provincial, state or other organisations officially designated to certify seed. Current members are Agriculture and Agri-Food Canada, Argentina, CSGA, New Zealand, and about 40 US state certifying agencies.

Achievements in the Technical Field

New cultivars are no longer lost because of contamination and naming is now standardised. Most seed planted is believed to be of improved cultivars, and yields and crop quality have improved immensely.

There are published uniform standards (AOSCA 1996a) and procedures (AOSCA 1996b) for maintaining genetic purity during seed multiplication. Uniform methods for field inspection have been adopted and there is standard terminology and labelling for different generations for all crops grown from seed in Canada and the US. In contrast, for seed potatoes, for which there is no equivalent organisation, the terminology and standards for different generations differ even between US states.

Seed produced in one area moves to another without loss of genetic certification status. Seed of new cultivars can be multiplied outside the area of origin of the cultivar and returned without genetic change.

A forum has been provided in which all participants in the seed sector in North America can review new developments and plan together how to adapt.

Possible Future Developments

Other countries may join AOSCA. In particular, there is movement of seed between North, Central and South America and several countries have expressed interest in participation in AOSCA.

Some seed companies will prefer their own systems to certification. In some species, most seed sold in the US is uncertified. In Canada, seed labelled with a cultivar name must be certified.

Despite recognition that certified seed is reliable, producers of self-pollinated crops will still save the progeny of certified seed for planting.

End users of particular crops are increasingly developing products for which only certain cultivars are useful. These users require some means of assurance that crops which they purchase are of the cultivar stated. AOSCA has played an active role in co-ordinating the development of cultivar identity preservation schemes which some certification agencies provide for crop purchasers. With the increasing sophistication of plant breeding, such schemes are likely to be of increasing importance and AOSCA can be expected to continue in their co-ordination.

Tests to verify cultivar identity are unlikely to replace seed certification in the near future because they do not prevent contamination occurring during production.

AOSCA will continue co-ordinating certification rules and procedures for its member agencies.

INTERNATIONAL SEED TESTING ASSOCIATION (ISTA)

A good test of the value of an organisation is to ask whether, if it did not exist, it would be necessary to invent it. In the case of the International Seed Testing Association (ISTA) the answer is 'yes – without doubt'. This conclusion is justified by one single activity of the Association, namely its continued development and publication of standardised protocols for seed testing appropriate for use in international trading. The ISTA International Rules for Seed Testing constitute an industry standard reference manual used throughout the world. In fact ISTA has a wide range of activities additional to the Rules and its role continues to expand as world trade increases.

The first formalised seed testing work was attributed to Professor Nobbe in Tharandt, Germany, in 1869, although doubtless since pre-history farmers had been aware of variations in seed performance. Indeed, the selection of better agronomic landraces from wild progenitors must have demanded skill in gathering, storing safely to retain viability, and recognising pure seed of the crop in question.

When seed trading developed on a commercial scale, quality control (quality assurance in today's parlance) must have been sadly lacking if one can judge by the UK Seeds Act of 1869.

The UK Seeds Act, 1869

> Whereas the practice of adulterating seeds, in fraud of Her Majesty's subjects, and to the great detriment of agriculture, requires to be repressed by more effectual laws than those which are now in force . . .
>
> Every person who, with intent to defraud or to enable another person to defraud, does any of the following things: that is to say,
>
> - Kills or causes to be killed any seeds; or
> - Dyes or causes to be dyed any seeds; or
> - Sells or causes to be sold any killed or dyed seeds, shall be punished as follows: that is to say,
> - For the first offence he shall be liable to a penalty not exceeding five pounds;
> - For the second and any subsequent offence he shall be liable to pay a penalty not exceeding fifty pounds.

The need for such an Act of Parliament indicates that poor husbandry of seed crops and downright fraud were rife! The development of a legal framework for the seed supply industry led inevitably to the need for standard test methods.

The Foundation of ISTA

Prior to the foundation of ISTA, a series of International Seed Testing Congresses had become established (Wold 1975). The first congress was held in Hamburg, Germany, in 1906. Nine countries were represented and there were 34 delegates. With a break caused by World War I, the next congress in Copenhagen in 1921 was attended by participants from 16 countries. At this meeting a European Seed Testing Association was founded. This European Association was short-lived, being supplanted by the International Seed Testing Association which was founded at a meeting held at the National Institute of Agricultural Botany, Cambridge, United Kingdom, in 1924. Present at the meeting were 42 representatives from 26 countries. The aims of the Association (Wold 1975) when founded were:

> To advance all questions connected with the testing and judgement of seeds by:
>
> - Comparative tests and other researches directed to achieving more accurate and uniform results than hitherto obtained.
> - The formulation of uniform methods and uniform terms in the analysis of seeds in international trade.
> - The organisation of international congresses attended by representatives of Official Seed Testing Stations for the purpose of mutual deliberation and information, the publication of treatises and reports on seed testing and mutual assistance in the training of technical officers.

Constitution of ISTA

In essence, ISTA is an independent self-governing organisation supported by national governments.

There are, however, some other sources of revenue, notably membership fees paid by other categories of members, and the sale of publications and international certificates. The latter are used by ISTA Accredited Laboratories for the purpose of reporting test results to be used in international trade.

The Constitution of ISTA in its current form was initially adopted at the Extra-Ordinary Meeting in Washington DC on 4 June 1971 and subsequently amended at the Ordinary Meetings in Warsaw 1974, Brisbane 1986 and Copenhagen 1995.

Objects of ISTA

- The primary purpose of the Association is to develop, adopt and publish standard procedures for sampling and testing seeds, and to promote uniform application of these procedures for evaluation of seeds moving in international trade.
- The secondary purposes of the Association are actively to promote research in all areas of seed science and technology, including sampling, testing, storing, processing, and distributing seeds, to encourage variety (cultivar) certification, to participate in conferences and training courses aimed at furthering these objectives, and to establish and maintain liaison with other organisations having common or related interests in seed.

The affairs of ISTA are administered from an Executive Office located in Zurich, Switzerland. An elected President serves for three years and others active in seed testing, quality assurance, and research serve on ISTA committees. Until June 1995, membership of ISTA was available only to '*officio*' laboratories and staff. For all practical purposes this meant government laboratories, but the trend to privatisation and the blurring of boundaries between public and private sector had resulted in the need for a new approach. At the ISTA Congress held in Copenhagen, Denmark, in 1995, the Association voted to extend membership to 'any person or laboratory engaged in seed science or seed testing and supporting the aims of the Association'.

Those persons nominated by national governments to participate in ISTA affairs are now known as Designated Members. Only Designated Members have a vote. The 'one country, one vote' criterion has been retained. A third category of membership, that of Accredited Laboratory, is available to government and independent non-government laboratories. But the latter must have government backing before accreditation is granted. ISTA does not recognise laboratories owned or operated by seed companies as falling into the 'independent' category.

Quality Assurance of Seed Testing

At the Copenhagen Congress, further decisions were taken in order to make more systematic and rigorous the accreditation of laboratories. Under the new scheme the accreditation and quality assurance programmes previously in place were amalgamated into an integrated system involving regular audit visits and a quality assurance scheme based on international standards. Full development and implementation of the system is expected by 1998. All existing laboratories will need to achieve re-accreditation within the next three to five years.

To comply with the new accreditation scheme, all laboratories must draw up Standard Operating Procedures, have a systematic staff training programme, and keep records of all tests and equipment calibrations performed. Evidence will also be needed of independence and freedom of action unconstrained by commercial interest in the sale or brokerage of seed. Accredited Laboratories must also take part in a referee test programme.

Membership of ISTA

At the time of writing (1997), 197 individuals and 64 countries were members of ISTA. The total of fully Accredited Laboratories, authorised to issue international certificates, was 97. The list of member countries is given below.

Argentina	India	Romania
Australia	Iran	Russia
Austria	Ireland, Rep. of	Slovakian Rep.
Bangladesh	Israel	Slovenia
Belgium	Italy	South Africa
Brazil	Ivory Coast	Spain
Bulgaria	Japan	Sri Lanka
Canada	Kenya	Sweden
Chile	Korea, Rep. of	Switzerland
Columbia	Latvia	Syria
Croatia	Lithuania	Taiwan
Cyprus	Luxemburg	Tanzania
Czech Republic	Macedonia	Thailand
Denmark	Morocco	Tunisia
Estonia	Nepal	Turkey
Egypt	Netherlands	United States
Finland	New Zealand	Uruguay
France	Norway	Yugoslavia
Germany	Pakistan	Zambia
Great Britain	Philippines	Zimbabwe
Greece	Poland	
Hungary	Portugal	

Benefits of ISTA Membership

The main benefits of ISTA membership are:

- free publications
- accreditation based on a quality management system
- audits
- reference test programme
- worldwide network of seed experts
- training at some ISTA stations
- workshops and symposia
- pool for communication
- triennial World Seed Congress.

The Future

Currently there exists a strong trend worldwide towards liberalisation and expansion of trade. Thus the membership of the World Trade Organisation (previously GATT) has risen from 92 countries in 1986 to 126 in 1996, with a further 30 countries, including China and Russia, wishing to join. Set against this backdrop, ISTA seems likely to continue as an international organisation of growing importance.

ASSOCIATION OF OFFICIAL SEED ANALYSTS (AOSA)

Objectives

Uniformity of methods of seed testing.

Historical background

The Association of Official Seed Analysts of North America, now AOSA, resulted from an informal 1908 meeting convened by the United States Department of Agriculture to discuss the condition of agricultural seed, the need to test such seed using uniform methods, and the development of a model seed law for the United States (US) (French 1958).

Those present at the meeting – representatives of the Canadian and US federal departments of agriculture and 16 states – decided to form an association of persons officially interested in seed testing in the US and Canada. The Association held annual meetings, and in 1910 adopted its first constitution. The early meetings were informal but in 1913 papers were read and these were then published in the Proceedings of the Association in 1914. Seed company officials and seed analysts attended these early meetings as observers.

From its inception, AOSA set up a committee to review the rules and procedures for seed testing that were in use. An extensive revision was adopted by the Association in 1917, and these were published as official rules for seed testing. The rules introduced definitions for terms used in purity and germination testing and more precisely defined germination requirements. Procedures were introduced for calculating allowable variation between replicates in purity and germination tests.

By 1919 there were 35 member states besides the Canadian and US federal departments (French 1919). Research on seed, particularly on test methods, was encouraged and papers presented at annual meetings were published in the Proceedings of AOSA, now the *Journal of Seed Technology*. Although research was reported on other aspects of seed quality, such as detection of disease and later the measurement of vigour, only purity and germination tests were standardised in such a way that they could be used for regulating seed quality.

Initially AOSA was concerned about state seed laws and set up a committee to develop recommendations. Although 39 states had some form of seed legislation by 1922, there were considerable differences and some were of questionable adequacy. By 1926, the AOSA legislation committee had developed a comprehensive Suggested Uniform State Seed Law. This led to the gradual strengthening of seed laws in most states and eventually to a Federal Seed Act in the US in 1939. Canada had had a federal seed law since 1905 but there are no provincial seed laws.

The implementation of the US federal seed legislation led to the establishment of formal seed inspection and testing programmes in the states. In some states the same person was the official seed analyst and the seed control official. The need for seed control officials in different states to consult with one another led to the development of regional seed control associations and eventually, in 1955, the Association of American Seed Control Officials (AASCO) was formally constituted (Midyette 1983).

For some years seed control officials had been reviewing the Suggested Uniform State Seed Law. In 1955 they recommended a procedure for future changes in federal legislation and in the Recommended Uniform State Seed Law (RUSSL). They recommended that there should be involvement of AOSA, the American Seed Trade Association and the International Crop Improvement Association in a joint legislative committee. Amendment of RUSSL is now the responsibility of AASCO. Members of AOSA continue to provide advice to AASCO regarding changes in RUSSL.

AOSA was represented at the establishment of the International Seed Testing Association (ISTA) in 1924 and was involved in developing the original International Rules for Seed Testing in the 1930s. When ISTA was reactivated in the 1950s, AOSA again played a very active role, particularly in recommending methods of testing.

ISTA changed its rules in 1971, limiting membership to seed testing stations. Organisations, including AOSA, were no longer eligible, but AOSA members have continued to play an active and prominent role in ISTA. In particular there have been continuous efforts to try to ensure compatibility between the rules of the two organisations.

Legal Framework

In its 1993 constitution, AOSA is incorporated as a not-for-profit corporation under the laws of the state of North Carolina (AOSA 1993).

Membership is open to government seed laboratories in the United States and Canada. Government laboratories in other countries and staff in non-member laboratories in universities, government organisations, and crop improvement agencies can participate as associate non-voting members.

Achievements in the Technical Field

There are now standardised rules for purity and germination testing in use in all US states and Canada; dispute over test results is very rare.

Seed laws based on RUSSL have been adopted in almost every state.

Test methods have been improved and internationally accepted.

Information has been published on seed quality including purity, germination, vigour, disease, and cultivar identification.

Concepts for accrediting analysts and laboratories have been developed. The Association offers a voluntary seed analyst certification programme to its member laboratories.

Possible Future Developments

Governments are cutting back laboratories and/or moving to full cost recovery of testing services. The use of private laboratory results for some 'official' purposes is increasingly accepted. There is a trend towards laboratory and analyst accreditation using International Standards Organisation concepts, with a potential combination of the laboratory accreditation process with analyst certification.

AOSA is considering further increasing its cooperation with the Society of Commercial Seed Technologists (SCST), which represents independent and seed company private analysts. Annual meetings of the two organisations are already normally held in conjunction with one another and SCST has representation on several AOSA technical committees. SCST has a process for training and certifying its analyst members, who are called registered seed technologists. They are familiar with AOSA, ISTA, and Canadian seed testing rules.

There are new information needs and capabilities. DNA technologies make it possible to determine genetic purity and to identify seed-borne disease more precisely.

There is a trend towards patenting testing protocols based on new technologies and restricting or requiring licence fees for use. This is a very different approach from the tradition of free exchange of information.

FÉDÉRATION INTERNATIONALE DU COMMERCE DES SEMENCES (FIS) – INTERNATIONAL SEED TRADE FEDERATION

Objectives

FIS is the global organisation acting on behalf of its members to promote the interests of the seed industry. FIS develops and facilitates the free movement of seeds with fair and reasonable regulations, to serve farmers, growers and consumers whilst protecting intellectual property. The Federation devotes its efforts to encouraging the use of modern technology in high quality seeds, in order to develop a sustainable agriculture for the production of food and industrial crops in a healthy environment.

The core business of FIS consists of:

- representing the interests of the members of the Federation;
- being receptive to the ideas and opinions of all members of the Federation;
- working to improve relations between members of the Federation and helping to solve problems which confront them;
- facilitating the marketing of seeds for sowing purposes and other propagating material by publishing rules for the trading of such items in international markets;
- facilitating the settlement of disputes between members through arbitration;
- stimulating the marketing of improved technology via high quality seed and plant propagating material for the benefit of agriculture and the consumer (the term 'agriculture' is used in the widest context to include farming, horticulture, amenity and forestry);
- encouraging and supporting the close liaison with, and education of, national and international bodies which influence the business environment in which seedsmen have to operate;
- encouraging and supporting the education and training of seedsmen throughout the world;
- recognising those who have made significant achievements to further the objectives of the Federation.

In relation with its sister organisation, ASSINSEL (Association Internationale des Sélectionneurs), FIS deals with other subjects such as the environment, biotechnology, intellectual property and genetic resources.

Historical Background

FIS was established in 1924 and has evolved to its present international status by steady development, which slowed only during the politically difficult periods of the 1930s and World War II.

The first International Seed Congress was held in London in 1924, followed by Bologna in 1928, Paris in 1929 and Budapest in 1930. These early congresses were successful in achieving the fundamental aims of establishing firm relationships within the great family of the seed industry and giving birth to the International Federation. The first Draft Rules for International Trading were submitted in Bologna in 1928. The first edition entered into force on 1 July 1929.

With the Paris congress in 1950, FIS was effectively reawakened. The Rules and Usages were revised and FIS activities were extended to include vegetable seed. In 1955 the Cereal Section was formed, followed by the Forest Tree Seed Group in 1964 and the Sugar Beet Seed Section in 1968. Countries continued to join FIS and, in 1970, the very important eighth edition of the FIS Rules and Usages was adopted. With this edition, the Canadian and US associations declared their commitment to the Rules.

A new step was reached in 1994, with the adoption of the 12th edition of the Rules and Usages, merging the Rules of the various Sections in one common text supplemented by small crop-specific annexes.

From an early stage FIS has stressed the importance of the technological aspects of the seed industry. In 1954 FIS was emphasising the importance of plant breeders' rights, and urging its members to press for government recognition of reciprocal protection of breeding products. At the 1958 FIS congress a resolution was passed, recognising breeders' rights and that licence fees could be charged when protected varieties are sold.

FIS has continued to evolve, and with each annual congress there is a better understanding between participants.

Legal Framework

FIS is a non-political, non-profit organisation. It is an Association according to Articles 60 *et seq.* of the Swiss Civil Code. FIS is registered in Nyon, Switzerland, where the seat of the Federation is located.

The Federation is composed of:

(a) Ordinary Members: the national seed trade associations;
(b) Associate Members: the individual companies active in the seed trade.

Members of national associations and Associate Members may participate legitimately in all open meetings of the Federation. Ordinary Members have voting rights.

The financial resources of the Federation are composed of the membership fees of the members as well as of the fees paid by the representatives of these members who participate in the congresses and other events of a professional nature organised by the Federation.

The governing bodies of FIS are the President's Council, the Executive Committee (composed of at least 10 members but not more than 20 members, each of them from different countries) and the General Assembly.

FIS comprises the following Sections: Herbage Seed; Vegetable Seed; Cereal Seed; Sugar and Fodder Beet Seed; Oil and Fiber Seed; and the Tree and Shrub Seed Group. The Federation now has the following Standing Committees: Membership and Dues; FIS/ISTA; Seed Treatment and Environment; and Arbitration Procedure Rules. If necessary, the Executive Committee can decide to form, cancel or combine Sections and Committees.

As at July 1996, FIS was composed of 50 Ordinary Members representing 40 countries and of 75 Associate Members from 39 countries. In all, FIS has members in 59 countries from the five continents.

Major Achievements

The Annual FIS/ASSINSEL World Congress

The development and health of our world is dependent on sustainable production of food and industrial crops which, in turn, is directly linked to seed quality and genetic value. Fundamental to this are the skills of plant breeders, the high level of financial input, the use of the most modern technology and the development of a sound global seed industry. To cope with all these demands and their complexity, the world seed industry needs a forum to discuss crucial issues, adopt positions and promote the interests of the seed industry. FIS achieves this by bringing together everybody involved in the breeding, multiplication and distribution of seed during the joint annual FIS/ASSINSEL congress. Industry representatives as well as representatives of other governmental and non-governmental organisations such as OECD (Organisation for Economic Co-operation and Development), FAO (Food and Agriculture Organisation of the United Nations), UPOV (International Union for the Protection of New Varieties of Plants), ISTA (International Seed Testing Association), WTO (World Trade Organisation), APSA (Asia and Pacific Seed Association), COSEMCO (Comité des Semences du marché Commun) and FELAS (Federación Latinoamericana de Asociaciones de Semillas) are invited to attend. Attendance exceeds in general 1000 participants and very often reaches more than 1500 participants (1550 participants in Amsterdam in 1996).

Adoption of Objectives and Motions

The General Assembly of FIS, the Executive Committee and the Sections adopt from time to time motions on topics of importance to the industry. During the past years, motions have been adopted on the following subjects: Plant Breeding; Official Variety Testing; Breeders' Rights; National Lists of Varieties; Seed Certification; Seed Testing; Seed-borne Diseases; and Seed Treatment. All these motions are widely circulated to interested circles and are available on request at FIS Secretariat.

International Seed Trade Federation Rules and Usages for the Trade in Seeds for Sowing Purposes

Since its formation in 1924, FIS has been concerned with the establishment of international trading rules as, at that time, no such rules existed. Seed was usually sold with specified purity

and germination with very low standard, leaving room for considerable difference of opinion. FIS has spent many years developing understandable and workable trading rules for its members. The first rules were adopted in 1929. The 12th edition of the Rules and Usages was adopted in 1994 and has been published in French, English, German and Italian. Today, nearly all international seed trading is done under FIS rules.

FIS Arbitration Procedure Rules for the International Seed Trade and Arbitration Facilities

FIS Rules and Usages for Trade end with an article providing for the settlement of all disputes arising from transactions concluded under FIS rules by arbitration. Although this may appear simple in its application, it is not. FIS established Arbitration Procedure Rules for the International Seed Trade as early as 1930. The fifth edition was adopted in 1996.

In order to create the necessary facilities, 'Arbitration Chambers' have been set up in most of the member countries of FIS. The Federation is proud of this organisation, which, while not unique in the world, is still one of the few really worldwide, specialised trade arbitration institutions.

The International Rules for Seed Testing and the International Seed Analysis Certificate

From its very start in 1924, FIS has adopted motions on the need to establish International Rules for Seed Testing and an International Seed Analysis Certificate. After long and fruitful discussions between ISTA and FIS, ISTA adopted International Rules for Seed Testing during its 1931 ordinary meeting in Wageningen. At the same time, an International Seed Analysis Certificate was approved and a specimen was included in the Rules.

Co-operation with Other International Organisations

Since 1924 FIS has developed very good links with several organisations having a role to play in the international seed movement. The most important for FIS are the following.

- ISTA: it has already been pointed out that the co-operation between FIS and ISTA dates back to the early 1920s, when FIS was founded. Since then FIS and ISTA have had regular formal and informal contacts, leading to solutions in general acceptable to both parties.
- OECD: the relations of FIS with the OECD and its forerunner OEEC (Organisation for European Economic Co-operation) date back to 1954. When OEEC established in 1958 a Scheme for the Varietal Certification of Herbage Seed Moving in International Trade, FIS was consulted and had the opportunity to make a number of suggestions. Since that time, FIS has always been consulted, and is still regularly consulted for the establishment of new schemes or their amendments.
- FIS also has efficient working contacts with FAO, UNEP (United Nations Environment Programme), WTO, and the Secretariat of the CBD (Convention on Biological Diversity), to quote the most important ones.
- Contact with UPOV is mainly through the FIS sister organisation ASSINSEL.

Future Development

FIS will continue to improve and update the results already achieved. In addition, the Executive Committee has adopted, in December 1995, five specific objectives for the coming years.

- To improve the free flow of information between FIS, its members and the seed companies to ensure that all are aware of FIS activities.
- To increase membership to be more representative of the industry at the global level.
- To widen and strengthen public relations activities towards the major organisations influencing the development of the international seed trade.

- To accelerate the process of arbitration and then to have a better follow-up of the decisions taken.
- To study a system of global product liability insurance and management of risks.

For each objective, a Working Group has been established to determine a detailed action plan. The progress made in achieving those objectives will be monitored at regular intervals.

ASIA AND PACIFIC SEED ASSOCIATION (APSA)

The aim of the Asia and Pacific Seed Association (APSA) is to improve production and trade in the Asia–Pacific region of quality seed and planting material of agricultural and horticultural crops. Functioning as a regional forum, the Association encourages collaboration among seed enterprises in the region, represents the interests of members to governments, compiles and disseminates information on technical, regulatory and market issues, assists in organising training and cultivar testing programmes, and strengthens links with regional and international organisations pursuing similar objectives.

The decision to establish APSA was taken during a regional consultation comprising representatives from the public and private seed sectors organised by the Food and Agriculture Organisation of the United Nations (FAO) in Bangkok in 1992. The Consultation established a Working Group of six seed specialists which conducted a comprehensive needs survey in the region before drafting the association constitution, work programme and budgets. The Working Group also prepared a project for funding by the Danish International Development Agency (DANIDA) to support development of the Association and planned the Foundation Meeting. The Foundation Meeting was held in September 1994 in Chiangmai, Thailand, and established APSA as an international non-profit and non-government organisation with its secretariat in the Royal Thai Government's Department of Agricultural

Extension in Bangkok. APSA is registered as a legal entity in Thailand.

The Association's membership consists of a wide range of enterprises and other organisations throughout the region, including national seed associations, government research and seed enterprises, private seed companies, regional research centres, and other national and regional organisations concerned with seed. Seed organisations outside the region are accepted as associate members. The Association is managed by a 16-member Executive Committee comprising representatives from the public and the private seed sectors. FAO has a permanent seat as an *ex-officio* member of the Committee. An annual general meeting of members reviews the Association's performance and formulates its work programmes and budgets. A Technical Committee meets regularly to formulate proposals for Association activities.

APSA membership benefits include a listing in its directory with detailed product information on each member, preferential fees for various services, copies of key FAO statistical and other bulletins, reports prepared or commissioned by the Association, eligibility to vote at the annual general meeting and election to the Executive and other committees, and APSA secretariat assistance.

Each year in September, APSA organises the Asian Seed Conference which is attended by about 500 seed sector representatives. The Asian Seed Conference highlights reports on critical technical and regulatory issues and comprehensive, country-specific seed industry analyses. During its first three annual conferences (1994, 1995 and 1996) the Association produced 12 substantial country reports and 17 technical papers on genetic engineering, hybrid technology, plant breeders' rights, and co-operation between public and private breeding programmes. The conference is also an invaluable opportunity for professional and business contacts.

APSA regularly organises subject-specific workshops to formulate recommendations concerning technical, legal and trade issues. A workshop in September 1996 formulated an APSA position paper on sensitive issues related to the introduction of plant breeders' rights legislation in the Asia–Pacific region, and a workshop in November 1996 developed improved strategies for supply of seed to Pacific island countries.

The Association publishes the *Asian Seed* magazine six times a year with articles on seed technology, regulations and trade.

The Association has also developed for its members a series of seed import/export statistics covering several countries in the region. In 1996 a project to strengthen this effort was initiated with French support through GNIS (Groupement National Interprofessionnel des Semences et Plants).

Study tours are being arranged to developed seed sectors in Asia, Europe and the USA with the purpose of facilitating transfer of technology and exploring business opportunities.

APSA, together with FAO, the International Maize and Wheat Improvement Center, and Kasetsart University in Thailand, participates in the Tropical Maize Network (TAMNET) and facilitates private seed companies entering their hybrids for testing in TAMNET's regional hybrid maize cultivar trials. The Association has contacts with other international agricultural research centres, ISTA (International Seed Testing Association), FIS (International Seed Trade Federation), and has observer status in sessions of UPOV (International Union for the Protection of New Varieties of Plants).

In the future APSA will consolidate the implementation of its existing work programme of meetings and publications with particular focus on the possibilities for the further strengthening of the collaboration between public and private seed sectors in the Asia–Pacific region. Efforts will be made to expand regional cultivar testing to other crops in addition to hybrid maize. Projects of relevance to the region's seed sectors will be formulated and presented to donors.

Part Two: Examples of National Legislation

CANADA

Objectives

There have always been two basic objectives in Canadian seed legislation:

(i) to exclude from the market seed not suitable for planting;
(ii) to inform buyers of the relative quality of seed offered for sale.

The factors considered in determining suitability for planting and for determining quality have been modified over time, but the basic objectives have remained.

Historical Background

Many of the organisational and legislative structures that are in place in Canada at the end of the 20th century were developed between 1899 and 1905. These include the seed law and the organisations for regulation, seed testing and certification.

A Seed Division was established in the federal Department of Agriculture in 1899. The division had an educational role with respect to production and use of high quality seed and an investigational role respecting the quality of seed sold and planted in Canada.

One aspect of the educational campaign led to the formation of an association of seed growers. The Canadian Seed Growers' Association (CSGA), founded in 1904, was assigned from its start the responsibility for preserving the pedigree records of seed crops. To undertake this work, it organised the inspection of seed crops in the field (Clayton 1990, 1995).

A government seed laboratory was set up in Ottawa in 1902 to investigate the quality of seed moving in trade. The surveys found some lots to be heavily contaminated with weed seeds. Other lots were found with relatively low levels of weed seeds. Investigations into the 'vitality' of field, root, and garden vegetable seeds found many lots to be of low vitality.

The publication of the results of the surveys on seed quality led inevitably to demands for legislation to restrict the sale of seed containing noxious weed seeds. A Seed Control Bill, introduced in Parliament in 1903, would have prohibited the sale of seed of most crops unless they were free from 12 specified weed seeds. In addition, there was provision for grading seeds of some crops based on standards for purity and vitality.

This first Bill met with organised opposition from seed merchants who considered that it unduly interfered with their business. The Bill was modified to take their concerns into account and the Seed Control Act went into effect in 1905. This Act was amended in 1911 with the support of the seed industry, merchants and seed growers to make the requirements more stringent.

The system that developed was one of grading seed. Higher grades allowed fewer weeds and required higher germination. Not all species were covered. Grading was done by the seed seller. Two principles, which still remain, had been established: number of weed seeds was considered

Encyclopaedia of Seed Production of World Crops.
Edited by A.F. Kelly and R.A.T. George. © 1998 John Wiley & Sons Ltd

rather than percentage by weight, and quality factors were considered in combination to establish the grade.

In 1923 the seed legislation was changed and seed control became a government function. Grading was either done by a government inspector or was based on a government test. All crop kinds were covered. Regional federal laboratories were established across Canada to facilitate testing.

Prior to 1923, CSGA arranged with various agencies and individuals to carry out crop inspection. In 1923 this was entirely taken on by the Canada Department of Agriculture, which still supplies the inspectors for crop inspection.

The most significant change introduced in the Seeds Act, 1923, was to restrict new cultivars permitted to be sold in Canada to those licensed by the Minister (Harvey 1995). Cultivars already on the market were exempted from the licensing requirement. Initially the basis for licensing was purely agronomic performance but in 1928 the Act was amended to provide for the exclusion from licensing of grain cultivars that were inferior in quality, i.e. unsuitable for bread-making or malting as the case might be. This excluded from the market cultivars that were unadapted to Canadian growing conditions or that were of limited use.

Originally licensing applied to cultivars of all crop kinds but in 1937 this was reduced to cereals, potatoes, forage crops and lawn or turf grasses. Quality as well as agronomic performance could now be considered as factors for licensing all these kinds.

In other federal jurisdictions seed is not always regulated at the national level or may be regulated at both federal and state levels as in the United States. A ruling in 1937, which declared that in Canada seed control is federal responsibility, has simplified the Canadian situation with one national law to which the provinces have contributed proposals.

In 1959 the Seeds Act was completely rewritten but the minister stated that there were no substantive changes. The essential basis for the regulation of seed in Canada had already been established. The main change was to introduce a short enabling Act with the detail to be established by regulation. The Act, however, has allowed a fundamental change from the arrangements under the 1923 Act when testing or grading was always done by government officials. Increasingly, under the 1959 Act, testing and grading have become the responsibility of accredited persons in the private sector.

The Act provided for seed to be sold by grade based on purity, germination, disease, quality and cultivar purity. The Act stipulated that the standards established by CSGA should be used for grades requiring cultivar purity. The role of CSGA in establishing standards and in determining cultivar purity were further clarified in an amendment to the Act in 1985.

The 1959 Seeds Act did not specify who was to grade seed. The initial regulations under this Act only provided for officials to grade pedigreed seed. Non-pedigreed seed was therefore graded by seed sellers. Since that time, progressive changes in regulations have provided for authorisations to persons meeting specified requirements to grade Certified Seed and to sell Certified Seed in bulk and this has gradually been extended to all classes of pedigreed seed.

The 1959 Act stated very precisely the kinds of seed that it covered. This Act was amended in 1977 to extend coverage to any plant of any species.

From 1923, sellers were to state the name of the cultivar if known; however, there were always concerns about the authenticity of non-pedigreed seed labelled with cultivar names, particularly of cross-pollinated species. From 1965, regulations were introduced to restrict the use of cultivar names to pedigreed seed. The first kinds covered were grasses and clovers but a similar restriction now applies to most crop kinds.

In restricting the cultivars that may be sold and the use of cultivar names, the emphasis in Canadian seed regulation is on pre-sale assurance of quality. This form of protection of seed buyers is considered more effective than providing penalties for persons selling seed failing to meet cultivar purity standards.

Quality Control Agency

The regulation of seed quality has always been the responsibility of the Canada Department of Agriculture (currently called Agriculture and Agrifood Canada (AAFC)). There is a small headquarters policy division, the Plant Products Division, based in Ottawa, and then regional staff throughout Canada who deliver the programme of inspection and enforcement.

Two departmental laboratories conduct testing for enforcement purposes but do not offer service testing for domestic purposes. To determine the grade of their seed, sellers use private laboratories which are accredited by the department.

A Canadian Seed Institute (CSI) has been established by the Canadian Seed Growers' Association, Canadian Seed Trade Association and the Commercial Seed Analysts' Association of Canada, with assistance from AAFC.

CSI is establishing quality systems standards, under ISO 9002, for establishments authorised to grade seed, for laboratories authorised to test seed and for importers authorised to import seed. The official authorisations will be granted by AAFC. Audit of those authorised will be by auditors accredited by CSI. Enforcement action will remain the responsibility of AAFC.

Cultivar Registration: Evaluation and Identification

Since 1989 the Seeds Act has specified that cultivars may only be sold if they have been registered in Canada, continuing the system started in 1923. The requirements for registration are detailed in the Seeds Regulations; only field crops are covered, and recently it was decided to exclude maize.

There is no single national organisation to arrange cultivar evaluation. Testing is organised by species and region through committees that are formally recognised under the Seeds Regulations. Recognition includes approval of test methods and of the system for deciding which cultivars shall be recommended for registration. In addition, the participation in the recommending

committee is examined to ensure that it represents a balance of those affected (e.g. agronomists, farmers, maltsters, millers, oilseed crushers, pathologists, plant breeders, seed growers or others, depending on the crop under consideration). The fact that a plant breeder is an interested party is taken into account, but it is also assumed that the breeder will be best able to suggest where the cultivar should be evaluated and what are its merits.

For most species there are laboratory testing requirements in addition to field trials over a specified period of years. These tests may include disease susceptibility and, depending on the crop, quality characteristics such as oil or protein content or milling and baking performance.

The most stringent requirements apply to spring wheat. By the time of registering a cultivar of hard red spring wheat suitable for the main production area in the prairies, there will have been field and laboratory tests over a period of years, at a large number of test sites. Field tests will have considered not only yield, maturity, straw strength and height, but susceptibility to insects and diseases. Laboratory tests will have measured resistance to known races of rust as well as the suitability of the grain for milling and bread-making. Detailed observations are made on such aspects as gluten strength and protein content.

In all the tests, comparisons are made with known standards. For registration a cultivar must be equal to or better than the standards in all important characteristics.

For cultivars of certain crops, e.g. potatoes and canola/rapeseed, the absence of certain toxins is a prerequisite for registration. The cultivar must have been tested by specified test methods and found to meet the appropriate standard.

The western Canadian grain handling system, particularly for wheat, has been based on the ability to distinguish cultivars with different quality characteristics by visual examination of their seed. Thus hard red wheat cultivars that are indistinguishable from cultivars suitable for bread-making will only be registered if they show the same suitability. This has excluded high yielding, low quality, indistinguishable cultivars from the Canadian market.

There are different market requirements in different areas of Canada. Low quality, high yielding cultivars of wheat may be unacceptable for the western area, where grain is graded for export by the Canadian Grain Commission and exported by the Canadian Wheat Board. The same cultivar, however, may be quite suitable for production and use, for animal feed, in eastern Canada. When a cultivar is of benefit in one region but is unacceptable in another region, it may be granted a regional registration. This is only done when it is considered necessary to keep the cultivar out of the region in which it is not registered.

At the time of cultivar registration, the breeder or applicant for registration must submit all the technical supporting data together with the recommendation of the committee that supported registration. In addition, the applicant must provide a botanical description of the cultivar and a reference sample. These are used in subsequent cultivar verification tests.

Quality Control of Seed Production

CSGA establishes the rules for certification (termed 'pedigreeing' in Canada), grants pedigreed status to seed lots, and maintains the pedigree records of all cultivars grown for seed in Canada.

For the entry of a cultivar into seed production, the first generation planted to be multiplied as pedigreed seed is termed Breeder Seed. Only plant breeders who have been formally recognised by CSGA can produce Breeder Seed. Recognition is based on academic qualifications and experience.

CSGA works closely with the federal agriculture department as new cultivars are registered, but CSGA also has a system for accepting new cultivars into seed production prior to registration. In addition, some cultivars are multiplied in Canada for return to their country of origin without ever being registered for sale for crop production. This can happen, for example, with foreign forage cultivars which will produce seed in Canada but may not produce acceptable quantities of forage.

A seed grower producing pedigreed seed becomes a member of CSGA. For each seed crop being produced, the grower must submit an application for field inspection together with the fees. CSGA passes the application to the appropriate regional official of AAFC, which undertakes the inspection as a service to CSGA.

After the inspection the inspector submits the report to CSGA where it is evaluated and a decision is made as to whether the crop meets the standards. Crops meeting the standards are granted crop certificates.

Quality Control of Seed after Harvest

Pedigreed seed must be labelled with an official tag, if sold in a container, or be accompanied by official forms, if sold in bulk. The label or form specifies the pedigreed status of the seed lot and provides information on the grade of the seed.

The information to be shown on labels and forms and the requirements to be allowed to use them are all established in the Seeds Regulations that are enforced by the federal Department of Agriculture. Grading and labelling of pedigreed seed are undertaken by persons authorised by the Department in accordance with the regulations. Authorisation is based on such requirements as access to cleaning facilities, technical competence, record keeping, access to seed testing, and record of performance.

Imported seed and all seed offered for sale, whether it is pedigreed or not, is subject to the requirements of the Seeds Regulations and may be subject to official inspection and testing. Seed that is inaccurately labelled may be required to be withdrawn from the market, and the seller may be subject to prosecution.

Requirements of Seller to give Information

The Seeds Regulations require certain basic information to be given by all seed sellers for all seed kinds. This is information identifying the seller and the seed kind and the quantity of seed per container (other than small packets).

For crop kinds, the seller must indicate the cultivar, if the seed is pedigreed, and the grade. The names of the grades and the standards that apply are set out in tables in the regulations. There are standards for maximum numbers of weed and other crop seeds, for the presence of certain disease organisms and minimum levels of germination. Certain weed species are prohibited and the presence of these species renders a seed lot unsaleable. The seller must provide information on the country of origin of imported seed.

Since a large proportion of cereal seed is sold in bulk on the basis of advertisements, there are regulations on seed advertising that are very similar to the labelling requirements.

Proportion of Controlled Seed on the Market

In theory, all seed sold in Canada is subject to the regulations and should be controlled. In practice, the situation is rather different.

For self-pollinated crops there is a tendency for farmers to save some or all of their seed requirements from their previous crop. A major factor in determining whether they will buy pedigreed seed is a change to a new cultivar.

In addition to the saving of seed, there is a long-established practice of sales from farmer to farmer. Again, this particularly applies to self-pollinated crops. Although most of these sales are probably contrary to the requirements of the regulations, it is difficult to control sales to willing buyers. Unfortunately, this provides unfair competition to pedigreed seed growers who must pay fees for crop inspection and membership in CSGA and royalties on seed sales.

In contrast, the use of controlled pedigreed seed is very high in cross-pollinated species and with hybrids where there is a risk of cultivar impurity with non-pedigreed seed. Similarly, there is a tendency to consider the risk of impurity and the consequent loss of market for the crop when buying canola/rapeseed.

Since farmer-to-farmer sales reduce the royalties received by plant breeders, under plant breeders' rights legislation, it could be expected that in the future there will be private enforcement action that will have the effect of increasing the proportion of controlled seed on the market. The Plant Breeders' Rights Act was only passed in 1990 and went into effect in 1991.

Possible Future Developments

The Canadian government is reducing its role in providing services to particular sectors such as the seed industry. The trend is towards privatisation and accreditation. When services are provided, they are to be at full cost recovery.

The seed industry in Canada is considered to be well developed and able to regulate its own activities. Increasingly the government's role will be only in setting standards and, in serious cases, enforcing regulations.

CHILE

Objective

From its very beginning, the national legislation applicable to seeds had as essential objectives to protect the farmer and to achieve a higher productivity of agriculture. Through the establishment of plant breeders' rights, it is intended to encourage the development in the country of improved cultivars and the introduction of foreign cultivars.

Starting in the 1980s, it has also been important to consider the reliability of the national seeds system within the context of international markets, through the establishment of clear rules applicable to the production of certified seeds and plants.

Thus, Law No. 1764 of 1977 set forth the basis for seed and plant certification, with a view to encouraging the production and trade of high quality seeds, and providing the buyer with additional support by guaranteeing the identity of cultivars and the purity of produced seeds and plants. This law recognises the existence of Common Seed, its production and trade being freely undertaken, i.e. being the direct responsibility of the producing companies, subject, however, to compliance with the packaging and labelling requirements, guaranteed genuineness and minimum quality standards set forth as mandatory by the official agency for marketing purposes. The agency further controls compliance with these provisions.

Plant breeders' rights are recognised in the country by Law No. 19342 of 1994 which covers all plant species. The breeder's consent is required for any trade in a protected cultivar.

Historical Background

The first official seed laboratory was set up in 1896. In 1942 rules covering the certification of fodder seeds were established and subsequently different regulations were established to define the trading standards applicable to seeds of other species.

The National Seeds Certification Programme was implemented in 1958, which mainly includes cereals, rice, corn (maize), alfalfa (lucerne), crimson clover, sunflower, beans, potato and beetroot.

At the same time, the central official laboratory was recognised by the International Seed Testing Association (ISTA).

In 1978, Chile was accepted by the Organisation for Economic Co-operation and Development (OECD) for the Schemes for Varietal Certification of Seeds Moving in International Trade. In 1980 the European Economic Community (EEC) accepted the equivalence of field inspections carried out by the national official body responsible for certification purposes and for the seeds produced by this process. As of this date, Chile was accepted as a third country supplying seeds to the Community.

Since 1977, Chile has recognised plant breeders' rights. During 1994 Law No. 19342 was promulgated, whereby the previous legislation was improved and made compatible with the principles laid down by the 1978 Act of UPOV (International Union for the Protection of New Varieties of Plants), of which the country is a member since January 1996.

Nature of the Quality Control Agency

The Seeds Department, Ministry of Agriculture, is entrusted with all official activities related to seeds. It is consequently in charge of issuing Certification Norms for Seeds and Plants, which are studied jointly with the producing companies, universities and plant breeders. It is also responsible for carrying out field inspections, which are undertaken by accredited personnel under the supervision of official inspectors.

The post-control certification is made by the Seeds Department in the different regions of the country. The quality analyses of Certified Seed are performed by the official laboratories of the Ministry of Agriculture, in accordance with ISTA Standards and Procedures.

The Common Seed produced under the companies' own responsibility can be analysed in the official laboratories or in private laboratories duly authorised by the Seeds Department.

Trade control is permanently exercised by the official body. Control samples are obtained and analysed in the official laboratories, in order to verify compliance with the minimum standards and to apply sanctions, as necessary.

The administrative and technical activities associated with plant breeders' rights are also performed by the Seeds Department, which is in charge of the Protected Varieties Registry.

Registry of Varieties: Identification and Evaluation

The study of the agronomic value and morphological characteristics of the candidate cultivars for inclusion in the national list is made directly by the interested parties themselves, beneficiaries or commercial companies, on the basis of the standards and procedures established by the official body. This body supervises the execution of these evaluation tests which last for at least two years. The results are then formalised and the approved cultivars are included in the national list.

For plant breeders' rights, the Seeds Department carries out the distinctness, uniformity and stability (DUS) tests throughout the country. The fundamental bases for cultivar identification are the UPOV descriptive guidelines.

Quality Control During Seed Production

The Chilean legislation establishes two categories of seeds: Certified and Common.

The production of Certified Seed is effected under the control of the Seeds Department and must comply with the certification standards established by the official body for each agricultural or fruit species.

These standards are in accord with the OECD and EEC standards. The production farms are registered with the Seeds Department, which carries out the field inspections, seed sampling, labelling and analyses.

The production of Common Seed is the exclusive responsibility of the producing company.

Post-Harvest Quality Control

The post-harvest quality control of Certified Seed is undertaken by the Seeds Department. For this purpose, the Seeds Department carries out the sampling and analysis of the seeds and authorises their marketing.

The post-harvest quality control of Common Seed is performed by the producing company or seed processor. These quality analyses may be effected in the official laboratories or in those of the companies, or in third parties' laboratories. In any case, the non-official laboratories must be authorised by the Seeds Department and are under its permanent control.

The seed sampling and analysis should be made under the legislation in force, according to ISTA guidelines and procedures.

Information which must be Provided by the Seed Dealer

Before seed of a particular cultivar may be marketed, the cultivar must be included in the national list.

The seed must be sold in a labelled container. The label must indicate the name of the producer, dealer, species, cultivar, and germination and purity percentage. In addition, it should indicate whether it corresponds to the Certified or Common Seed category.

Seed Marketing Patterns

Certification is not compulsory for seed marketing in the country and it is estimated that the use of this category of seeds reaches 30% for wheat, 25% for potatoes, 7% for maize and 30% for rice. Of the above, 50% of wheat, 30% of potatoes, 90% of maize and 60% of rice are sold as Common Seeds. The rest corresponds to saved seed through exchange among farmers and is not marketed through the formal channels.

Chile and the International Seeds Trade

In the last decade the country has become an important off-season multiplication centre, mainly for the countries in the northern hemisphere. Thanks to its latitude it permits the cultivation of the same species which are cultivated in those countries and principally hybrids of maize and sunflower, vegetables and flowers. In 1995 exports reached 50 000 t, of which 20 781 t were maize, 4372 sunflower, 15 100 t flowers and 9825 t vegetable seeds.

EUROPEAN UNION

Historical Background

The European Union of 15 member states has evolved from the original six member states that constituted the European Economic Community (EEC) established in 1954 under the Treaty of Rome. In 1964 the European Parliament expressed a favourable opinion in support of proposals by the Commission to introduce minimum quality requirements for the marketing of agricultural seeds in the EEC. It was recognised at the time that crop production occupied an important place in agriculture and the social fabric of rural life. The availability of high quality seed of improved varieties throughout the Community was seen as an important step in improving agricultural production. Council Directives were introduced in 1966 on the marketing of seed of the main agricultural species. Four Directives were issued (EC 1974) as follows:

– the marketing of cereal seed (66/402/EEC)
– the marketing of fodder plant seed (66/401/EEC)
– the marketing of seed of oil and fibre plants (69/208/EEC)
– the marketing of beet seed (69/400/EEC)

The Council Directives were described in terms of marketing rather than certification. This emphasises the objectives of supplying high quality seed to the farmer, and encouraging the unrestricted movement of seed between the member states of the Community. Seed certification provides agreed procedures for the production and control of seed quality to allow recognised minimum quality standards to be achieved throughout the Community.

The Council, Commission and the member states drew on three main sources of information in drafting the Directives: the experience of member states that had been operating official certification schemes; the recommendations of the Food and Agriculture Organisation of the United Nations (FAO) on the minimum standards for the certification of maize seed in European and Mediterranean countries; and the establishment of the scheme for the certification of herbage seed moving in international trade by the Organisation for Economic Co-operation and Development (OECD).

The original Directives allowed member states to restrict the marketing of seed to cultivars on their national lists. This restriction was an impediment to free marketing between member states. In 1970 the Council Directive on the Common Catalogue of Varieties of Agricultural Plant Species (70/457/EEC) was introduced (EC 1974). This directive laid down uniform rules for the acceptance of cultivars and required member states to compile national lists of cultivars accepted for certification and marketing in their territory. Procedures were also devised for producing a Common Catalogue of Varieties comprising all the cultivars on the national lists. This provided a basis to allow certified seed to be freely marketed throughout the Community.

The official control of seed quality was also extended in 1970 by the introduction of the Council Directive on the marketing of vegetable seed (70/458/EEC) (EC 1974). The Directive contains provisions for free marketing, the control of seed quality and the creation of national lists and a Common Catalogue.

The Directives have been amended from time to time in response to technical and political developments, and enlargement of the Community. The most recent legislative development has been the introduction of the Council Regulation on Community plant variety rights (EC 1994). This provides plant breeders with cultivar protection throughout the Community, and is based on the provisions of the 1991 International Union for the Protection of New Varieties and Plants Convention (UPOV 1991).

The Certifying Authority

Responsibility for the implementation of the requirements of the directives through statutory provisions lies with each member state. The routine operation of the statutory requirements is delegated to the certifying authority. The organisation of certification procedures in any one member state is likely to be influenced by national, legal and technical considerations. Some member states have more than one certifying authority with certification procedures co-ordinated at the national level. There are steps in the certification procedure that are required to be undertaken by officials of the certifying authority. These are known as official measures and are defined as measures taken:

(a) by state authorities;
(b) by any legal person, whether governed by public or private law, acting under the responsibility of the state; or
(c) in the case of ancillary activities, by any person duly sworn for that purpose.

Under (b) and (c) the persons should derive no private gain from such measures. As a result, some member states delegate some of the responsibilities normally associated with the certifying authority to regional authorities, or quasi-governmental organisations. In some cases the certifying authority itself may delegate specific tasks to persons other than its own employees.

Cultivar Registration

The main conditions for cultivars of agricultural crops to be accepted for addition to a national list are that the cultivars should be distinct, sufficiently uniform and stable (DUS), and have satisfactory value for cultivation and use (VCU). There are no requirements for VCU to be assessed for cultivars of amenity grasses, cultivars used solely as parents of hybrids and cultivars of vegetable crops, but similar provisions for DUS assessment apply.

Directives 70/457/EEC and 70/458/EEC (EC 1974) define DUS and VCU. The DUS requirements are similar to those defined in the UPOV Convention (UPOV 1991), except that a cultivar only has to be clearly distinguishable from any other cultivar known in the Community. This is not as wide as the common knowledge requirement of UPOV. VCU requires a new cultivar to show a clear improvement over other cultivars on a national list. Superior performance in one or more of the characteristics may be used to compensate for inferior performance in others.

Registration defines the cultivars accepted for certification and marketing by each member state. The national lists also contain the names and addresses of the maintainers of the cultivars. This is important in meeting the requirements of the certification procedures. A cultivar description is produced as a result of the DUS test and is used to determine identity and purity during seed certification.

Quality Control of Seed Production

Quality control during seed production is achieved mainly by:

- limiting the number of generations of certified seed
- placing the responsibility for the production of Basic Seed and earlier generations on the maintainer of the cultivars
- defining field standards for important factors, e.g. crop isolation and cultivar purity
- requiring at least a minimum number of crop inspections.

The categories of certified seed are Basic Seed and, in the case of self-pollinated crops, Certified Seed of the first generation and Certified Seed of the second generation. For cross-pollination crops only one generation of Certified Seed is permissible after Basic Seed.

Basic Seed is defined as seed that has been produced under the responsibility of the maintainer according to accepted practices for the maintenance of the cultivar. It may be divided into one or more generations referred to as Pre-Basic Seed, and is intended for the production of Certified Seed.

Certified Seed is defined as seed produced from Basic Seed and is intended only for crop production. For self-pollinated species, Certified Seed of the first generation is intended mainly for the production of Certified Seed of the second generation. The latter should only be used for crop production.

The standards to be met by certified seed are defined under two headings: standards to be met by the crop, and standards to be met by the seed. The crop standards include previous cropping of the seed production field, crop isolation distances, and condition of the crop including weediness and levels of disease present. For cross-pollinated species the levels of cultivar purity are usually defined as the number of off-types present in a unit area of the field. For self-pollinated species the minimum percentage cultivar purity to be achieved by the seed is usually prescribed. It is accepted that this is mainly examined during crop inspection.

Provisions for the production of Basic Seed and Certified Seed are made in the vegetable plant directive, although most vegetable seed is marketed as Standard Seed. This category of seed only requires official checking in post-control tests that prescribed standards have been met.

Quality Control of the Seed after Harvest

The identity of the seed must be maintained from harvesting and during cleaning and processing. Once these operations are completed the seed should be retained in discrete seed lots not exceeding the maximum sizes prescribed in the rules of the International Seed Testing Association (ISTA 1993). The seed lots are also required to be sampled and sealed in accordance with ISTA rules. An officially drawn sample from each seed lot is sent to an official seed testing laboratory. Analyses for percentage purity, number of specified weed and other crop seeds present, and percentage germination, are carried out using methods prescribed in ISTA rules. In most cases there are prescribed minimum standards that must be met for the seed to be certified (EC 1974).

Most member states grow control plots as a means of monitoring the effectiveness of the certification procedures, although this is not a directive requirement. However, there is an exception for those member states using authorised crop inspectors who are employees of private companies; in this case control plots should be grown in accordance with the rules of the OECD Seed Schemes (OECD 1995c). This arrangement was introduced as an experimental system (EC 1989), but is likely to become permanent following amendments to the Directives.

Information Required at the Point of Sale

All certified seed lots must carry an official label. The minimum size and colour of the label and the information it should give are prescribed for each category of seed (EC 1974). The quality of the seed is indicated by reference to the category, and information such as percentage germination is not required. Sellers may give such information but, if it is to be included on the official label, there must be a clear division between 'official' and 'additional' information.

There is no defined period over which a seed lot remains certified provided that the containers remain labelled and sealed; once certified it is the responsibility of the seller to ensure that the certification standards are met during seed marketing.

Proportion of Controlled Seed on the Market

It is a requirement of the Directives that only certified seed, or Standard Seed of vegetable plants, may be placed on the market. However, the marketing of seed for cleaning and processing, and seeds for experimental, plant breeding and testing purposes is not subject to this requirement. Farmers may save seed for use on their own farms, but for protected cultivars this may require the permission of the breeder and may be subject to the payment of a royalty (EC 1994).

The Future

The removal of checks by customs officers at national boundaries within the Community, as a result of the development of the single European market, will lead to amendments to the Directives to remove permissive actions which member states have been able to use voluntarily. Such actions are now considered as potential barriers to trade.

The development of legislation to control the release into the market of genetically modified organisms (EC 1990) will affect the marketing of seeds. Genetically modified plant cultivars will require the consent of all member states before release onto the open market. Before a consent to release is given, the breeder must show that the cultivar will not present an unacceptable risk to the environment, or to human or animal health. It is likely that for genetically modified cultivars, consent to release will become a condition for addition to a national list.

A major development is likely to be the introduction of quality control and accreditation systems that will permit some activities reserved for officials to be delegated to seed companies and other private enterprises. The first steps in the process are represented by the temporary authorisation of crop inspectors (EC 1989), and similar arrangements in the OECD certification schemes.

GHANA

Objectives

Seed legislation in Ghana is aimed at ensuring the optimisation of crop production through the production and supply to farmers of high quality disease-free seeds and planting materials. The legislation is also aimed at preventing the in-flow of undesirable seeds from external sources and makes provision for policy development, co-ordination and regulation to ensure that the national seed industry develops in tune with the needs of the country's agriculture and farmers.

Historical Background

Ghana's seed programme development dates from 1959 when the first efforts in the establishment of the Seed Multiplication Unit (SMU) were made. Between 1959 and the 1970s the SMU stood as an arm of the Ministry of Food and Agriculture which led to the creation of improved seed awareness, seed production of the main cereal and legume crops, and a countrywide seed supply network which often involved close collaboration with agricultural extension.

Prior to 1959 the production of improved seed was on a plot scale, spasmodic and mainly limited to the supply of small quantities of seed of new cultivars from research stations to interested or neighbouring farmers. The Crops Research Institute and the Agricultural Faculty of the University of Ghana were particularly active in this respect.

The cultivar development efforts of the Crops Research Institute grew rapidly from the late 1960s in response to the increasing farmer demand for improved seeds. The rapid expansion in rice (*Oryza sativa* L.) production in the northern parts of Ghana and farmer satisfaction with the yields of the first locally bred maize (*Zea mays* L.) cultivars also accelerated cultivar development.

The rapid growth in the national seed programme under the SMU and the financial and administrative burden it imposed led to government efforts at privatisation in the late 1970s. When it became obvious that the then private sector did not find it attractive to invest in the seed industry, the government, in 1979, following a series of studies, converted the SMU into the Ghana Seed Company, a private limited liability company wholly owned by government.

The Ghana Seed Company succeeded in operating a nationwide seed grower programme and established seed marketing outlets throughout Ghana. It installed five processing and storage plants in the main growing areas of the important crops – maize, rice, groundnut (*Arachis hypogea* L.), sorghum (*Sorghum bicolor* (L.) Moench.) and cowpeas (*Vigna ungiculata* (L.) Walp.) – as well as small quantities of a range of local vegetable seeds, in all producing about 10% of total seed demand, although for rice the supply level peaked at 40%.

Although a Seeds (Certification and Standards) Decree had been passed in 1972 and Seeds

(Certification Standards) Regulations in 1973 during the time of the SMU, the two legislative arrangements were not seriously implemented since an envisaged independent seed certification agency could not be established. This left the Ghana Seed Company and its seed growers to adopt internal quality control, although the provisions of the seed law and regulations were adhered to as far as possible. Additionally, the National Seed Committee and the Variety Release Committee, which were provided for under the seed law, were established and did make some contribution in the early years of their existence.

Following the government's adoption of a free market style economy, the Ghana Seed Company was closed down in 1989 and the seed industry declared privatised. This action resulted in the development of autonomous seed growers who were provided with custom services using the facilities created by the Ghana Seed Company. In order to ensure the success of the privatisation drive, two agencies were created in the Ministry of Food and Agriculture in 1990 to recruit, organise, train and assist the private sector in production, processing and marketing. The two agencies are the National Seed Service (NSS), responsible for organising and training seed growers and sales merchants, and the Ghana Seed Inspection Unit (GSIU) responsible for seed quality control and certification as well as training of personnel in the seed industry.

In 1991 a Task Force of the Ministry of Food and Agriculture reviewed the Seeds (Certification and Standards) Decree of 1972 and made recommendations on the rules and regulations pertaining to the operations of the seed industry in view of its new structure. The Task Force subsequently proposed an amended Seed Law and Seed Regulations which differ from the earlier legislation in the sense that they recognise and mandate the newly created GSIU and NSS to carry out their responsibilities with the full force of law.

The highlights of the Seed Law are:

(i)   restrictions on the importation of seeds into the country;

(ii) a register to be kept of the names of importers, growers and cleaners of seed;

(iii) a National Seed Committee to be set up, with the Deputy Minister responsible for crops as Chairman, with functions to include:
  – formulation of policies for the development of production
  – processing and marketing
  – monitoring the supply of seeds for purposes of seed security in Ghana
  – fixing fees for the certification and testing of seeds;

(iv) the appointment of Sub-Committees by the National Seed Committee; and

(v)   prohibition of publication of false advertisements.

The areas covered by the Regulations are:

(i)   certification of Breeder, Foundation, Registered and Certified Seed and provision of tags;

(ii) obligation of growers to apply to the certifying authority for approval before undertaking the production of Foundation, Registered or Certified Seed;

(iii) obligation of seed cleaners to ensure that before they accept any seeds for cleaning there should be a certificate for them issued by the certifying authority;

(iv) roguing of fields; and

(v)   obligation of those who want to have a cultivar released to provide the required information.

The Regulations have a schedule which contains rules for certifying seed of maize, groundnut, rice and sorghum.

Although the amended legislation has been taken through the administrative processes before enactment, it has not yet (1996) been passed as an Act of Parliament. Nevertheless, since the GSIU and the entire seed industry of Ghana view the proposed legislation as an educational tool, a vehicle for the orderly growth of the seed industry and a vital protection mechanism for the country's agriculture, the provisions of the new legislation are already being implemented with the full support of the entire seed industry in place of the earlier Decree and Regulations. However, in spite of laudable efforts, less than 5% of the scheduled crops were certified in 1996.

Cultivar Registration, Evaluation and Identification

The Seed Law calls for the establishment of a Seeds Register in which shall be recorded:

(i) seeds and cultivars produced in the country or imported into the country;
(ii) the names, addresses, principal places of business and other particulars of seed dealers; and
(iii) such other information as the Minister of Food and Agriculture, on the advice of the Committee, considers necessary to request.

The establishment of the Seeds Register brings into play the Variety Release and Registration Sub-Committee which shall recommend, for the approval of the National Seed Committee, cultivars to be entered in or withdrawn from the register.

The Sub-Committee liaises with the NSS, the Department of Agricultural Extension, the GSIU, the Crops Research Institute and the Grains and Legumes Development Board and considers all available cultivar data and trial performance records before making its recommendations.

In line with this, the Regulations provide as follows.

(i) A cultivar of an existing or new crop submitted by a plant breeder to be entered in the seed certification programme must be submitted to the National Seed Committee with a complete description of the cultivar for identification and recommended regions for adoption.
(ii) Cultivars of existing or new crops shall be tested for yield, survival, disease reaction and

other important characteristics in comparison with standard commercial cultivars of the relevant crop using experimental techniques which ensure valid measurement of differences and their significance.

(iii) Each performance test shall consist of not less than three replicates of each candidate.
(iv) Each performance test shall include for comparison not less than three competitive cultivars of the relevant crop or selections thereof.
(v) The result of such performance test shall be reviewed each year by the Committee.
(vi) A report on such performance test shall be lodged with the certifying authority at the end of every year.
(vii) No seed of a cultivar of a new or existing crop shall be approved for certification under these Regulations unless the cultivar has been shown to be superior in quality to existing commercial cultivars of the relevant crop and is at least satisfactory in other major requirements.

Quality Control of Seed Production

The three main institutions working in the generational seed multiplication system are:

(i) the Crops Research Institute, which generates Breeder Seed of approved cultivars;
(ii) the Ghana Grains and Legumes Board, which multiplies the Breeder Seed to attain Foundation Seed; and
(iii) private autonomous seed growers which undertake the last stage of multiplication to achieve the Certified Seed class which eventually is planted by farmers.

Throughout the system the standards set by the Regulations are enforced by the GSIU.

Seed producers engaged in the production of Foundation, Registered or Certified Seed have to apply to the GSIU for approval to commence production. In issuing this approval, factors which the GSIU considers include: the cropping history of the proposed site; isolation; source and type of initial seed; and adequacy of farm equipment for planting, cultivation and harvesting.

The Regulations provide for the inspection of land intended for seed production in order to ensure that it has not been used for the production of any other cultivar of the same crop or any crop of the same cultivar in respect of which no certification has been granted for a period of time specified in the standards established for the relevant crop. The presence of volunteer plants of the same crop as well as weeds, the seeds of which it is impracticable to separate from the crop seed, also disqualify a field from being used for seed production.

The Regulations also provide for isolation of seed fields to avoid undue contamination from foreign pollen.

Quality Control of Seed after Harvest

Because many private seed producers have little experience, both the GSIU and the NSS assist seed producers in the correct methods of harvesting, processing and storage. In addition, the facilities established by the former Ghana Seed Company are utilised to render technical services to seed growers in processing and storage.

The most important tests and control procedures following harvest are sampling of seed batches for analyses to determine conformity to seed standards prescribed in the Regulations for purity, germination and moisture content. Following satisfactory laboratory testing, which largely conforms with International Seed Testing Association rules, tagging and sealing are carried out. The tag, which is coloured differently for the three classes of seed, carries the following information:

- name of crop
- cultivar
- code number of the grower
- purity analysis
- percentage germination
- date of certification

- batch number
- treatment of seed (i.e. conspicuous label marked 'POISON' and the name of the chemical used)
- date of expiry
- certification number of the inspector.

The GSIU recognises the need to exercise flexibility in seed certification control as the seed industry is currently (1996) in a developmental stage. Accordingly, standards are modestly set and seed health testing is not mandatory, although it is to be introduced soon.

Requirements for Sellers to give Information

Private sector participation in seed marketing has only recently become significant and the present stage of seed control in marketing involves more education than enforcement. Farmers mostly save their own seed and generally renew their seed from the formal seed sector when a new cultivar is released (hybrid seed usage is non-existent except for some imported vegetable seed). The informal seed sector and farmers' own saved seed make up nearly 95% of seed of cereals and legumes and nearly 100% of local vegetable seed and vegetative planting material. As private sector seed distribution picks up, enforcement of the provisions of seed legislation will be tightened.

The Seed Law requires that the names, addresses, principal places of business and other particulars of seed dealers be provided and recorded. The Law prohibits false advertisements such as describing seed as certified when it has not been certified. Seed dealers are required to be approved and registered to distribute seed and for this registration the dealer is required to conform to prescribed requirements in the Law and Regulations. The Seeds Register contains the relevant dealer and cultivar information.

In line with the enforcement of seed distribution control measures, seed inspectors have wide powers to enter seed facilities for the purpose of checking adherence to the provisions of the legislation. Sample-taking and examination of documents in respect of seed to which the Law applies are some aspects of seed control measures.

The Law guards against the importation of prescribed 'restricted' seed into the country without an official permit, and stipulates a maximum fine and imprisonment term for obstruction of seed inspectors in the execution of their duties or refusal to comply with a legitimate request by an inspector.

Seed dealers who have failed to comply with any of the stipulated conditions could have their registration suspended or cancelled.

Possible Future Developments

The pace of privatisation following the closure of the Ghana Seed Company has not been as brisk as required and the formal seed industry has declined. There are few seed grower groups, installed seed plant capacity has been grossly under-used, and seed distributors are few in number and largely confined to imported vegetable seed.

Current efforts aimed at enhancing private sector participation will need to be strengthened. An incentive package for privatisation will need to be set up and the GSIU and NSS properly equipped and strengthened to provide the necessary support, especially with respect to laboratory facilities as well as trained and equipped manpower.

At present only a few crops, namely maize (open-pollinated), rice, sorghum and groundnut, have standards for certification. It is expected that standards will soon be developed for cassava (*Manihot esculenta* Crantz.), soyabean (*Glycine max* (L.) Merr.), cotton (*Gossypium* spp.) and cowpea in view of their increasing importance in the country's agriculture. Additionally, the need has been felt for establishing minimum standards for the certification of local cultivars of vegetable crops following their purification.

Incentives for public sector plant breeders in the short term and some form of plant breeders' rights in the long term are also possible future developments.

MOROCCO

Objectives

The concern for organising on a rational basis the production and diffusion of cultivars and species well adapted to the local conditions, led the Moroccan legislative to promulgate a number of laws and regulations, which had the objective to ensure the quality of seed and save the rights of farmers and users, and to develop adequate supplies of Foundation and Certified Seed.

Historical Background

Legislation and Regulations

The organisation of seed production in Morocco began in 1927. Centres for the multiplication of improved cereal seeds were established as early as 1928 by a Ministerial Decree. The function of these centres is to multiply the cereal cultivars that are best suited to local conditions and to the requirements of the domestic and foreign markets. This decree was replaced by two decrees in 1932 and in 1936.

In 1937, a decree was promulgated by the Director of The Economics Business, relative to the production and marketing of wheat. Each producer was authorised to grow only a single cultivar of the same species. He had to follow the recommended cultural conditions. In return, the producer would be granted a subsidy based on specific weight, specific purity, germination capacity and cultivar purity of the seed. The seed growers were free to sell the seeds.

From 1939 onwards, marketing of seed wheat (seed meeting specific standards and packed in sealed bags with labels showing the name of the producer, the place of production, the cultivar, and the technical character of the lot) may be done only by co-operative bodies or specially approved dealers who are required to keep records of their purchases and sales.

The Government checks for seed quality. When analysis shows that the seed does not fulfil the conditions, the dealer is no longer authorised to sell seeds for one year. The supply price of wheat seeds is free from control.

In 1940, 1960, 1968 and 1969, further regulations were added to control the production, inspection, marketing and importation of seeds and propagating material.

Today, the production of seed and propagating material is governed by Law No. 1.69.169 of 25 July 1969, modified and completed by Law No. 1.76.472 of 19 September 1977 regulating the production, control and marketing of seed and propagating material. This Law was made effective through the following decrees:

- a decree regulating the conditions for registration of species and cultivars in the Official Catalogue of Species and Cultivars that can be cultivated in Morocco;
- a decree concerning the composition and responsibilities of the National Committee for the Improvement of Seeds and Propagating Material;
- a joint decree of the Minister of Agriculture and the Minister of Finance fixing the rate and modalities of fees for the control of seed production;
- decrees approving technical regulations related to production, control, processing and certification;
- an inter-ministerial decree fixing the rate and modalities of fees for registration in the Official Catalogue;
- decrees allowing companies to market seeds and plants;
- a Law of Plant Breeders' Rights was promulgated in 1996, based on the 1991 Convention of the International Union for the Protection of New Varieties of Plants (UPOV).

Nature of Current Quality Control Agency

The seed testing and certification is carried out by the Service of Seed Control and Certification, Ministry of Agriculture (Government department).

Cultivar Registration: Evaluation and Identification

Only cultivars included in the Official Catalogue or on the provisional lists can be commercialised in Morocco.

The Official Catalogue was created by decree of the Ministry of Agriculture and Agarian Reform on 22 September 1977; it was modified and replaced by a decree which is currently being gazetted. According to this decree, only seeds and propagating materials of cultivars which are registered in the Official Catalogue or on the provisional lists of cultivars eligible for certification can be marketed in Morocco.

The Official Catalogue consists of two lists:

(i) the A-list of species and cultivars whose seed or propagating materials can be certified and marketed in Morocco and abroad;
(ii) the B-list of species and cultivars whose seed or propagating materials can be multiplied in Morocco exclusively for export purposes.

The provisional list includes new cultivars belonging to species for which the Official Catalogue does not exist yet. A breeder cannot submit more than three cultivars per species per year for registration in the Catalogue. The experiments for inscription in the Catalogue comprise the following.

(a) For inclusion in the A-list:
   - a test carried out in one or two locations to determine distinctness, uniformity and stability (DUS) of the cultivar;
   - an agronomical and technological value test (value for cultivation and use) carried out in the main zones of production for which the cultivar has been developed.
(b) For inclusion in the B-list:
   - a DUS test is carried out only.

Results of the tests are submitted to the National Committee for the Improvement of Seeds and Propagating Material, which will discuss the results of the tests carried out. Based on the results and the economic value of the cultivar to Moroccan agriculture, the Committee will propose to the Minister of Agriculture whether or not to register the cultivar in the Official Catalogue. The registration of a cultivar in the Catalogue is authorised by a decree of the Minister of Agriculture. The decree also indicates the duration of the registration and the conditions for a second registration. The cultivars included in the Official Catalogue are periodically reviewed to eliminate those which are no longer of interest and those for which no request for a second registration has been made.

(c) For inclusion in the provisional list:
  – tests should be conducted under the supervision of the breeder or his representative and results should be submitted to the Directorate of Plant Protection, Technical Control and Fraud Restraining (DPVCTRF). The breeder or the applicant will have to provide a short description and a reference sample of seed of the cultivar.

Quality Control of Seed Production

The control of seed production is governed by Law No. 1.69.169 of 25 July 1969 and made effective through technical regulations approved by decrees of the Minister of Agriculture and Agrarian Reform. The control is carried out at all stages of production, processing, storage and marketing, and is exercised for seed of all generations: each seed lot can be traced back to its origin.

The Service of Seed Control and Certification certifies seed of major agricultural crops (cereals, food legumes and forage crops, potatoes, etc.) through field inspection, laboratory seed testing and growing post-control plots.

Field inspectors inspect Pre-Basic, Basic and Certified Seed fields on one or more occasions depending on species and classes of seed to be produced.

Quality Control of Seed after Harvest

Testing of Seed in the Laboratory

The laboratory seed testing is carried out on samples (taken by the official inspectors) from cleaned seed lots of approved seed fields. Sampling is carried out according to the methods of the International Seed Testing Association (ISTA). For each species a deadline for sampling is fixed, by which date the seed grower should have declared the quantities available for certification.

Post-control Plots

Post-control is carried out after seeds have been certified and allows a final judgement of the quality of seeds. The tests are carried out according to the Organisation for Economic Co-operation and Development (OECD) schemes.

Requirements for Seller to give Information

Seeds are marketed through organisations approved by decrees of the Minister of Agriculture. The number of companies has reached 192. Approval is granted to the companies, which:

- can commercialise seeds or propagating materials of cultivars which can be produced locally or imported and are registered in the Official Catalogue or on the provisional lists;
- have a professional qualification in seeds and/or propagating materials;
- have facilities allowing the proper production and storage of seeds and/or propagating materials.

The decree allowing a company to market seed indicates the species that can be handled by the company.

Before 1976, the marketing of cereal seed was handled by SCAM (Moroccan Co-operative Agricultural Society) and CMA (Moroccan Agricultural Co-operatives). In 1976, SONACOS (National Company for Seed Marketing) was created and replaced SCAM and CMA and has the sole right to market seed of winter cereals (durum wheat, bread wheat and barley) and to import seed of sugar beet and potatoes.

Seed of maize, forage species and vegetables are marketed by a number of approved private enterprises.

In 1989, a decision of the Minister of Agriculture authorised companies that have access to cereal cultivars registered in the Official Catalogue to market Certified Seeds of these species. Each company is required to keep records of their purchases and sales for which a special book is made which contains the following information:

- lots received
- quantity of the lots
- the quantity processed for each cultivar, category, lot
- the certified amount by cultivar, category, lot
- the quantity by cultivar, category, lot of seed sold.

Proportion of Controlled Seed on the Market and Possible Future Developments

Seed Production

Cereal seed production represents in general 93% of the area intended for seed production and more than 98% of the total amount of produced seed.

For cereal Certified Seed, the area intended for seed production varied from 13 489 ha in 1961 to 68 368 ha in 1986, and the production of seed from 6390 t in 1961 to 92 508 t in 1986. In 1995/96, the area was 43 390 ha and the amount of the seed produced 46 080 t.

Before 1984 the National Institute of Agricultural Research (INRA) was responsible for the production of Pre-Basic and Basic Seed. Since 1984, INRA has started to reduce the production of Basic Seed to concentrate on the production of Pre-Basic Seed. Average production between 1980 and 1984 was 1433 t, i.e. 96% of the national production of Pre-Basic and Basic Seed. In 1991, INRA produced only 3% of the national production of these two categories of seed.

Participation in International Organisations

Morocco participates in the activities of the following international organisations:

(i)   International Seed Testing Association. Morocco has been a member of ISTA since 1964; it participates in various technical committees.

(ii) International Union for the Protection of New Varieties of Plants. Morocco has been an observer at UPOV since 1978.

(iii) Organisation for Economic Co-operation and Development. Morocco has participated since January 1989 in OECD schemes for the varietal certification of seeds of cereals, forages, oil crops and maize, and for the control of seed of legumes for international trade.

(iv) European Union (EU). Since April 1992 Morocco has benefited from the equivalence from the EU regarding the certification of seeds of some species of cereals, food and forage legumes, and oil crops.

(v)   West Asia–North Africa (WANA) Seed Network. Morocco is a member of the WANA Seed Network and a member of the Steering Committee. Morocco's responsibilities in the network are:
- to conduct referee tests at regular intervals which verify the way seed analyses are carried out by laboratories in member countries;
- to prepare and keep up to date a catalogue of cultivars commercialised in member countries;
- to harmonise methods of seed certification.

PAKISTAN

In Pakistan, the Seed Act 1976, a national legislation, is the source of rules and regulations pertaining to cultivar registration and release, seed certification and quality control, marketing,

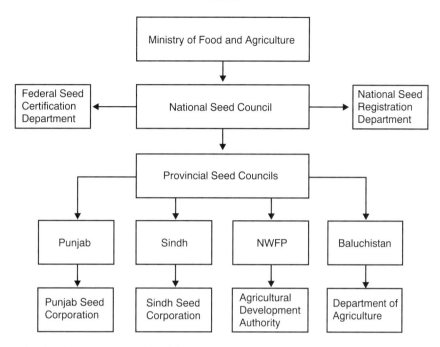

FIGURE 2.1   Institutional arrangements in Pakistan

import and export of seed. This legislation has provided a base for setting up the seed industry and its institutional framework for various aspects of seed development at federal and provincial levels (Figure 2.1). At the federal level, under the auspices of the Ministry of Food, Agriculture and Livestock, the required arrangements are made with a particular aim of giving an overall direction and providing a regulatory framework to the seed industry. This includes the following components:

- National Seed Council
- Federal Seed Certification Department
- National Seed Registration Department.

At provincial level, four Provincial Seed Councils have been established. The Punjab Seed Corporation in the Punjab Province, the Sindh Seed Corporation in Sindh Province, and the Agricultural Development Authority in NWFP (North West Frontier Province) look after seed multiplication, production and distribution of seed of various crops. In the fourth

province, Baluchistan, the seed programmes continue to be looked after by the Department of Agriculture. These public sector agencies produce and distribute seed of wheat, cotton, rice and maize.

The seed industry since 1976 has experienced a number of changes, from an earlier complete monopoly of the public sector to allowing participation of the private sector in the 1980s. The private sector, which remained hesitant for a while to take up seed ventures, has now become a viable partner of the public sector in seed supply in the country at a stage when seed business has been declared a seed industry. The private sector is of two types: formal (private registered seed companies), and informal (where about 90% of the seed flows from farmer to farmer or from unauthorised sources like commission agents, retailers and shopkeepers). The formal private seed sector, which comprises 142 seed companies (138 national and four multinational companies) is beginning to organise seed multiplication, production and distribution on scientific lines. There is no restriction on the private sector, which can distribute seed of any crops. Therefore, the field for seed production is wide open for the private sector. Additionally, the free market economy concept is under adaptation and the government is vigorously encouraging privatisation. These circumstances offer favourable opportunities for the organised private sector in seed production and distribution. In the changing scenario, the demand for quality seed has increased very significantly and a market has emerged to operate the Seed Act to benefit the end users of seed, i.e. the farmers.

Objectives

The Seed Act 1976 (Anon 1976a) explicitly aims to provide arrangements for controlling and regulating quality of seed of various cultivars of crops, thus ensuring a national mechanism for cultivar registration, approval and release, seed quality control in the market, seed import and export.

Cultivar, Registration and Release

Seed rules and regulations in vogue for cultivar registration and release do not discriminate between cultivars from the public and private sectors submitted for the purpose. It is mandatory, before a cultivar is registered and released, for the breeder to submit a seed sample to the National Seed Registration Department (NSRD) to determine its distinctiveness, uniformity and stability (DUS) characteristics. Simultaneously, the seed samples are also provided to the Variety Evaluation Committee (VEC) to evaluate a candidate cultivar for value for cultivation and use (VCU). A cultivar that meets the requirements of VCU and DUS is submitted to the Federal Seed Registration Committee (FSRC) at national level for registration. The Provincial Seed Council then considers candidate cultivars for approval and release. If the cultivar is for the whole of the country, it is then approved and released by the National Seed Council. The federal government finally notifies the crop cultivars which are eligible for certified seed production.

Seed Certification and Quality Control

Seed certification is compulsory for notified crop cultivars and is implemented by the Federal Seed Certification Department, a department attached to the Ministry of Food and Agriculture. For the production of certified seed, only cultivars registered and approved by either the National Seed Council or the Provincial Seed Councils are accepted in the system. The genetic purity of the crop and the physical purity of the seed lots are evaluated according to the prescribed minimum seed certification standards (Ahmad and Gilani 1993).

The following classes of seed are recognised in the system:

1. Pre-Basic Seed (identified by a white tag with a violet diagonal line);
2. Basic Seed (white tag);
3. Certified Seed (blue tag);
4. Certified Seed – II (red tag).

In the system there is provision for another class of seed known as 'Approved Seed'. This class of seed is produced from certified seed in isolated and quarantined areas under the supervision of seed producing agencies and is tested by the Federal Seed Certification Department (Figure 2.2). This class of seed is not designated by an official tag. However, seed agencies can use their own tags for this class of seed.

The quality of seed of unnotified cultivars being offered for sale in the local market is monitored through Truth-in-Labelling (Seed) Rules, 1991, framed under Seed Act, 1976. These rules envisage labelling of all packets/containers of seeds, specifying the following information:

– lot no.
– crop/species
– variety
– pure seed percentage

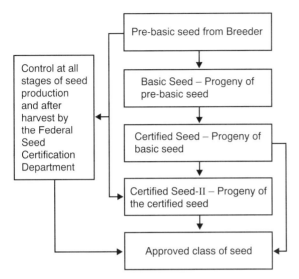

FIGURE 2.2 Production and control of various classes of seed in Pakistan

- germination percentage
- other crop seeds percentage
- weed seeds percentage
- inert matter percentage
- month/year of production
- date of expiry.

Import and Export of Seed

Quarantine laws are enforced in the country through the Pakistan Plant Quarantine Act 1976 (Anon 1976b), which lays down that 'No person is allowed to import any plant material which may be a source or medium of infestation or infection by diseases and plant pests destructive to agriculture or medium for the introduction of noxious weeds, except under a valid import permit obtained prior to such importation'. Under this Act for export of plant material, the requirements are: 'Any person intending to export plant-material shall get such material inspected for any injurious insects and plant diseases'. The Act also prescribes that 'No commercial fumigator shall engage in, perform or offer to perform any fumigation unless registered with the Department'.

Keeping in view the peculiar nature of seed, its imports and exports are regulated under 'Truth-in-Labelling (Seed) Rules 1991'. For the import and export of seed, the following requirements need to be fulfilled.

All imported/exported seed shall bear a label with the following information:

- lot no.
- crop/species
- variety
- quantity
- purity percentage
- germination percentage
- other seeds percentage
- month/year of production
- date of expiry.

Under the rules:

(a) Import shall be allowed only of seeds of those cultivars which have been approved by the Agricultural Research Institutes and entered and notified by the federal government in the National Register for seed and crop production in Pakistan.

(b) The National Seed Registration Department shall ensure supply of a copy of the National Seed Register and update it periodically.

(c) A small packet of seed not exceeding 10 kg is allowed to be imported for experimental purposes subject to approval by the Ministry of Food and Agriculture.

(d) The importer shall inform the Federal Seed Certification Department, on the prescribed application form, of the probable date of arrival of the seed and other propagating material at the port of entry and shall on arrival of the seed and propagating material notify the Federal Seed Certification Department for drawing of samples for testing, etc.

Offences and Penalties

The Seed Act prescribes the following penalty: Whoever:

(i) contravenes any provision of this Act or any rule: or

(ii) prevents a Seed Certification Officer or a Seed Inspector from taking a sample or inspecting seed under this Act; or

(iii) prevents any official from exercising any power conferred on him by or under this Act, shall be punishable:

(a) for the first offence, with a fine not exceeding one thousand rupees;

(b) where the offence continues after conviction, with a further fine of one hundred rupees for each day during which the offence continues; and

(c) for a subsequent offence, with imprisonment for a term which may extend to six months, or with a fine, or with both.

If any person is convicted of an offence punishable under this Act in respect of any notified cultivar or species of seed, the court convicting him shall further direct that the seed shall be forfeited to the federal government.

In the changing scenario of privatisation of the seed industry in particular, and the future world trade order in general, steps are being taken to introduce plant breeders' rights.

ZIMBABWE

Summary

Zimbabwe has a relatively sophisticated seed industry which is backed by a sound agricultural policy and a national seed legislation developed by the Ministry of Agriculture in recognition of its agricultural based economy. Formal research and development of improved cultivars formed the basis for the implementation of national legislation through which the seed industry was organised and regulated to produce high quality disease- and pest-free certified seed.

The Seeds Act, Chapter 133, was promulgated in 1965 and enabled by the Seeds (Certification Scheme) Notice and the Seeds Regulations in 1971 which provide for the production and marketing of all kinds of crop species. The Plant Breeders' Rights Act, Act 53/73, was promulgated in 1973 followed by the Plant Breeders' Rights Regulations in 1974, which provide for the patenting of plant material. In addition, the Plant Pests and Diseases Act and its enabling Regulations of 1982, the Tobacco Marketing and Levies Act No. 32 of 1977, and the Cotton Marketing and Control Act of 1969 are employed in plant pest and disease control and the use of superior cultivars for commercial production.

Historical Background

Cultivars grown at the beginning of the century were either landraces of indigenous species or introductions of exotic species. Indigenous farmers used their skills, currently acknowledged as indigenous technology, in the selection and genetic stabilisation of landraces, and in the increase, maintenance and storage of seed of indigenous crop species. Exotic species were introduced by the Department of Agriculture at the turn of the century. Settler farmers obtained small quantities of seed from the department and bulked and maintained the seed on their own farms. Crops at that time were either self-pollinated or open-pollinated as hybrid technology was unknown.

Research into cultivar improvement has been one of the most important factors which has contributed to increased crop yields. In the pre-independence era, formal agricultural research was conducted by government, parastatals, academic and private organisations, while the post-independence era brought in international, multinational and non-governmental organisations, further strengthening the research base.

Organised seed production was introduced in 1940 by the Department of Agriculture when the Southern Rhodesia (Zimbabwe) Seed Maize Association was founded by a small group of farmers. The objective at that time was to produce certified open-pollinated maize seed. Supervision and advice on the best methods of seed production were provided by the Department of Agriculture. The formation of the Association has gone down in history as the foundation for organised seed production in Zimbabwe.

Since then, associations for other crop species have been established to undertake seed production. The most significant and commendable feature is the policy to develop a privatised seed industry rather than the state-controlled and subsidised seed programmes seen in most African countries. Further exclusive rights of government-bred cultivars were granted to associations, except for cotton where seed production was handled through its marketing parastatal, the Cotton Marketing Board and its successor, the Cotton Company of Zimbabwe. Thus the planting seed needs of the country were satisfied through this well established seed production base followed by a lucrative export market which generated foreign currency. At present, production has grown to more than 100 000 ha to include maize, sorghum, millets, sunflowers, groundnuts, wheat, barley, oats, soyabeans, cotton, potatoes, pasture grasses, legumes, tobacco, dry beans, vegetables and flowers.

Quality Control Agency

Seed Services, a unit under the Research Services Division of the Department of Research and Specialist Services, is the official organ for the quality control of seeds and is the designated Certifying Authority. Since 1971 this exclusive but specialised organisation has full statutory powers to administer the Seeds Act empowered through the Seeds Regulations and the Seeds (Certification Scheme) Notice. The Plant Breeders' Rights Act is currently administered from Seed Services but this may not continue in the future.

Cultivar Registration Evaluation and Identification

In common with procedures followed worldwide, new cultivars are compulsorily tested for distinctness, uniformity and stability. The Variety Release Committee, with the responsibility of debating and scrutinising new cultivars, is made up of scientists, plant breeders, representatives from the agro-industry, the Certifying Authority and the Registrar of Plant Breeders' Rights. The plant breeder submits to this committee information regarding the origin and breeding procedures of the cultivar, performance data in support for its commercial value for cultivation, description of its morphological characters and any other relevant data. The new cultivars essentially should have some characters superior to the existing cultivars. An official declaration accompanied by an original sample of the new cultivar is submitted to the Plant Breeders' Rights Office for record keeping and reference. After it has been approved for release, the cultivar is listed on the national list of cultivars. Only cultivars of certain prescribed kinds of plants are afforded protection in terms of the Plant Breeders' Rights Act. Initially protection is offered for a period of 20 years, which may be extended for a further period of five years if the registrar considers it to be desirable.

Quality Control of Seed Production

Seed is produced, processed and marketed by seed producers' associations, privately controlled seed companies and seed multiplication schemes operated by commodity boards through commodity seed producers' associations. These activities essentially were limited to the large-scale commercial sector. Of significance in the last decade is the emergence of the small-scale sector engagement in seed production. Seed production is contracted to large numbers of small units in the communal, the resettlement and the small-scale commercial farming areas.

In line with international practice, Zimbabwe has adopted the three-generation system, namely Breeder (Pre-Basic) Seed, Foundation (Basic) Seed and Certified Seed. Seed qualities strictly adhered to in the certification scheme are cultivar purity, analytical purity, freedom from foreign seeds, germination, freedom from seed-borne diseases and freedom from live insects. The provisions of the seed certification scheme apply to maize, wheat, barley, oats, beans, cowpeas, velvet beans, soyabeans, groundnuts, sorghum, sunflowers, tobacco, cotton, pasture grasses, pasture legumes and seed potatoes. The production, processing and marketing of certified seed is undertaken by the following certifying agencies recognised by the Certifying Authority.

(1)   The Zimbabwe Seed Maize Association is the oldest association. The commercialisation of government-bred cultivars was implemented through a legally binding agreement between government, Seed Maize Association and the Commercial Farmers' Union. The agreement binds government to supply the Breeder Seed and the other two signatories to produce and satisfy the seed demands of the nation and maintain a 20% reserve stock as seed security for the ensuing year. The adherence to standards of the scheme is policed by a team of inspectors gazetted by the Minister of Agriculture and the seed is tested at the seed testing laboratory licensed by the Certifying Authority. The production levels and pricing are controlled annually by these signatories.

(2)   The Zimbabwe Crop Seeds Association was established in 1957 to produce seed of groundnuts, sorghum, soyabeans, wheat, barley and sunflowers through a bipartite agreement with government. For the supply of Breeder Seed, the Association holds a 30% reserve stock in terms of the agreement. The production levels and pricing are set annually by the two signatories in consultation with relevant organisations.

The above two associations amalgamated their interests as the Seed Co-operative Company of Zimbabwe in 1982 which further consolidated by becoming a public company in 1996, quoted on the Zimbabwe Stock Exchange.

(3)   The Zimbabwe Tobacco Seed Association was established in 1957 to produce seed of permitted cultivars of flue-cured, burley and oriental tobacco and now operates under the Tobacco Marketing and Levies Act of 1977. Members handle with ease the self-pollinated and hybrid seed production. Certified Seed may only be produced from Breeder Seed. Seed price is set by the Association in consultation with the Zimbabwe Tobacco Association.

(4)   The Seed Potato Association was established in 1957. It is unique in that seed is produced through a controlled scheme in collaboration with the national potato breeder. The breeder produces virus- and bacterial blight-free Breeder Seed annually by screening each plant at an isolated site 2000 m above sea level. The Foundation Seed is bulked in a specially designated area at an even higher altitude under the Plant Pests and Disease Act, which requires all fields to be subjected to rotation with specified grasses. Certified Seed is produced at lower altitudes. The arrangement between government and the Association to produce seed is informal. The Association is required to satisfy seed demands of the nation.

(5)   The Zimbabwe Pasture Growers' Association was established in 1953, making it the first organisation in the world to attempt a certification scheme in subtropical grasses. The organisation produces seed for export through contract with other countries. Seed Services ensures that field standards adhere to the scheme and the seed is tested according to International Seed Testing Procedures for the issuance of an Orange International Certificate.

(6)   The Cotton Company of Zimbabwe has undertaken certified seed production since 1970. Government-bred cultivars which may be commercialised are exclusively released to its certifying agency under the Cotton Marketing and Control Act, 1969. Breeder Seed is released and a team of inspectors gazetted by the Minister of Agriculture oversee the certification process. Seed producers are paid premiums for seed.

Private Production

Several international, multinational and wholly or partially owned Zimbabwe companies and non-governmental organisations are engaged in seed production adhering to the schemes for the various crops. Cultivars are bred and released locally. Some companies are actively engaged in vegetable and flower seed production for the domestic and export markets. The Forestry Commission is entirely responsible for the seed production of indigenous and exotic trees for the domestic and export markets.

Quality Control of Seed after Harvest

The Seeds Regulations laid down all the procedures for sampling of seed, purity analysis, the determination of prohibited weed seed, weed seed and other crop seed, the procedure for germination testing and the purpose of tolerances, of prescribed and non-prescribed kinds of plants. The certification scheme laid down the minimum standards for three categories of seed of prescribed kinds of plants.

Requirements for Sellers of Seed

The Seeds Regulations outline the procedures and standards required for registration as sellers of seed, keeping of records by sellers of seed, requirements relating to seed sold, containers of seed, seed imported, seed exported, seizure of seed and sampling of seed for information to be declared on a seed analysis certificate. The Seeds Act further cautions seed sellers against false advertisements of seed.

Proportion of Controlled Seed on the Market and Possible Future Development

On-farm seed retention is not practised for highly commercial crops, e.g. tobacco, cotton and hybrid maize, because of strict cultivars control by commodity and farming organisations and finance institutions which require the use of certified seed of recommended cultivars. Farmers of all sectors are well informed on the poor performance of retained seed of hybrids. It is estimated that commercial farmers retain 30% of their annual wheat and soyabean seed requirements whilst small-holder farmers retain over 90% of their seed requirements of crops such as cowpeas, beans, sunflower and small grains.

The supply and demand needs of the local and export markets of certified seed of highly commercial crops are met through carefully planned production, though shortages have been experienced for crops of indigenous interest, e.g. groundnuts, cowpeas, beans, sunflower and small grains. With the advent of small-holder farmer participation in seed production, the problem has been partially alleviated as seed of some cultivars of both known and unknown origin is available on the market and at household levels.

There is widespread concern and discontent from the public and private sector plant breeders, non-governmental organisations, farmers and private sector seed trade over the current regulatory framework. The regulatory body is seriously constrained due to lack of capacity to monitor small-holder seed production. It would be appropriate for government and the regulatory body to acknowledge the formal and informal sectors that have evolved for the commercial and less commercial crops. A pragmatic approach would be to find the right level of certification for crops engaged in by the small-holder farmer from amongst the range of certification levels available.

REFERENCES

Ahmad, S.I. and Gilani, S.M.M. (1993). *Minimum Crop and Seed Certification Standards*. Federal Seed Certification Department, Islamabad.

Anon. (1976a). *Seed Act 1976*. Federal Seed Certification Department, Ministry of Food, Agriculture and Livestock, Islamabad.

Anon. (1976b). *Plant Quarantine Act 1976*. Ministry of Food, Agriculture and Livestock, Department of Plant Protection, Plant Quarantine Division, Islamabad.

Anon. (1980). *International Code of Nomenclature for Cultivated Plants*. The International Bureau for Plant Taxonomy and Nomenclature, Utrecht.

AOSA (1993). Association of Official Seed Analysts Inc. Bylaws. *Newsletter of the Association of Official Seed Analysts*, **63**(3), 93–105.

AOSCA (1996a). *Genetic and Crop Standards*. Mississippi State University, Mississippi.

AOSCA (1996b). *Operational Procedures*. Mississippi State University, Mississippi.

Clayton, O. (1990). The history of the Canadian Seed Growers' Association. *Plant Varieties and Seeds*, **3**, 127–138.

Clayton, O. (1995). The Canadian Seed Growers' Association. In *Harvest of Gold* (Eds A.E. Slinkard and D.R. Knott). University of Saskatchewan, Saskatoon, 335–343.

EC (1974). Information and Notices. *Official Journal of the European Communities*, **17**, No. C66, 2–76.

EC (1989). Information and Notices. *Official Journal of the European Communities*, **L286**, 24–26.

EC (1990). Information and Notices. *Official Journal of the European Communities*, **L117**, 15–27.

EC (1994). Information and Notices. *Official Journal of the European Communities*, **L227**, 1–30.

FAO (1981). *Report on the FAO/SIDA Technical Conference on Improved Seed Production*. AGP/SIDP 81/76. FAO, Rome.

FAO (1982). *Seeds. Proceedings FAO/SIDA Technical Conference in Improved Seed Production*. FAO, Rome.

FAO (1986). *FAO Variety Description Forms and Variety Passports*. AGP/SIDP/86/3, FAO, Rome.

FAO (1987). *Report on the Expert Consultation on Interstate Movement of Seeds*. AGP/SIDP/87/6, FAO, Rome.

FAO (1993). *Quality Declared Seed*. FAO Plant Production and Protection Paper 117, FAO, Rome.

French, G.T. (1919). Organisation, Development of the Association of Official Seed Analysts of North America. *Proceedings of the Association of Official Analysts*, **11**, 15–20.

French, G.T. (1958). Notes on AOSA Highlights. In *Fifty Years of Seed Testing (1908–1958)*. Association of Official Seed Analysts, 18–20.

Hackleman, J.C. (1961). *History of the International Crop Improvement Association 1919–1961*. Chambers Printing Company, Clemson, South Carolina.

Hackleman, J.C. and Scott, W.O. (1990). *A History of Seed Certification in the United States and Canada*. Association of Official Seed Certifying Agencies, Raleigh, North Carolina.

Harvey, B. (1995). The western grain registration committees and their role in cultivar development. In *Harvest of Gold* (Eds A.E. Slinkard and D.R. Knott). University of Saskatchewan, Saskatoon, 326–334.

ISTA (1993). International Rules for Seed Testing. *Seed Science and Technology*, **21**, 288pp.

Midyette, J.W. (1983). RUSSL – A History of its Development. *Proceedings of the Association of American Seed Control Officials*, 1983, 81–88.

OECD (1995a). *List of Cultivars Eligible for Certification* (Annual publication). OECD, Paris.

OECD (1995b). *OECD Guidelines to Methods Used in Control Plot Tests and Field Inspection*. OECD, Paris.

OECD (1995c). *Up-to-date version of the Seed Schemes as of 15 June 1995*. Organisation for Economic Co-operation and Development, Paris.

OECD (1996a). *OECD Scheme for the Varietal Certification of Herbage and Oil Seed Moving in International Trade*. OECD, Paris.

OECD (1996b). *OECD Scheme for the Varietal Certification of Cereal Seed Moving in International Trade*. OECD, Paris.

OECD (1996c). *OECD Scheme for the Varietal Certification of Sugar Beet and Fodder Beet Seed Moving in International Trade*. OECD, Paris.

OECD (1996d). *OECD Scheme for the Varietal Certification of Vegetable Seed Moving in International Trade*. OECD, Paris.

OECD (1996e). *OECD Scheme for the Varietal Certification of Seed of Subterranean Clover and Similar Species Moving in International Trade*. OECD, Paris.

OECD (1996f). *OECD Scheme for the Varietal Certification of Maize and Sorghum Seed Moving in International Trade*. OECD, Paris.

UPOV (1991). *International Convention for the Protection of New Varieties of Plants* of December 2, 1961, as revised at Geneva on November 10, 1972, on October 23, 1978, and on March 19, 1991. International Union for the Protection of New Varieties of Plants, Geneva.

Wold, A. (1975). Background and history of the International Seed Testing Association. *Seed Science and Technology*, **3**, 33–35.

Further Reading

Anon. (1995). *ISTA News Bulletin (Special Edition)*. International Seed Testing Association, Zurich.

Anon. (1996). International Rules for Seed Testing. *Seed Science and Technology*, **24**, Supplement 1–335.

Decree-Law No. 1764 (1977). Establishes standards for research, production and marketing of seeds. Ministry of Agriculture, Servicio Agricola y Ganadero, Santiago, Chile.

FIS (1974). *A 50 Year Old Family 1924–1974*. FIS, Nyon, 144 pp.

FIS (1994). *International Seed Trade Federation Rules and Usages for the Trade in Seeds for Sowing Purposes*, 12th edition. FIS, Nyon, 121 pp.

FIS (1996). *FIS Strategic Plan*. FIS, Nyon, 48 pp.

FIS (1996). *International Seed Trade Federation*

Arbitration Procedure Rules for the Trade in Seeds for Sowing Purposes, 5th edition. FIS, Nyon, 42 pp.

Instituto Nacional de Estadisticas (1995). *Agricultural and Livestock Statistics*. Santiago, Chile.

Law No. 19342 (1994). Regulates the right of plant variety beneficiaries. Official Bulletin of Seed Department, Servicio Agricola y Ganadero, Santiago, Chile.

Leenders, H.H. (1967). The function of the International Seed Trade Federation in the international seed trade. *Proceedings of the International Seed Testing Association*, **31**(2), 245–257.

Plant Variety Protection. UPOV Publication No. 438 (E), Gazette and Newsletter of the International Union for the Protection of New Varieties of Plants, Geneva.

SAG (1994). *General and Specific Seed Certification Standards*. Seeds Department, Servicio Agricola y Ganadero, Santiago, Chile.

Servicio Nacional de Aduanas (1995). *Monthly Reports on Export Statistics*. Santiago, Chile.

TRIPS Agreement (1995). Agreement on Trade-Related Aspects of Intellectual Property Rights Article 27.3(b), World Trade Organization, Geneva.

3

Technical Aspects of Quality Control

DEFINITIONS OF QUALITY: COMPONENTS OF A SYSTEM

The term 'seed quality' embraces several individual components which can each be separately defined and assessed, but which collectively provide an overall indication of the value and usefulness of a given seed lot. The individual components of seed quality are:

- genetic quality
- viability
- physical quality
- seed health and
- moisture content.

Genetic Quality

Farmers and growers generally obtain seed of a particular cultivar assuming that the resulting crop will have certain attributes such as total crop yield, season of maturity, relative uniformity of crop, suitability for a particular market outlet, or known resistance to specific pests and pathogens. In the case of some of the vegetable species the grower is looking for additional potential from the cultivar which would include more subtle characters such as flavour, size grade of individual fruit and shelf life.

The farmer can only confirm the seed's suitability for a particular purpose after it has been sown, whereas the seed technologist has the facilities to check genetic quality prior to the seed being marketed. Therefore there is a need to be able to provide evidence that a particular cultivar is of value for cultivation and use and also that when marketed, the seed lot is of the cultivar stated.

Genetic purity of a seed lot is often referred to as 'cultivar purity', or 'varietal purity' in circles where the term 'variety' is used in preference to 'cultivar'.

Viability

It is of great interest and importance to the farmer and grower to know that the majority of the seeds in a seed lot are capable of germination and of subsequently producing a seedling of the chosen cultivar. In practice it is appreciated that not each seed will succeed in this way, but it is important that there be positive and reliable information regarding the germination potential and that the stated information has been obtained and is

Encyclopaedia of Seed Production of World Crops.
Edited by A.F. Kelly and R.A.T. George. © 1998 John Wiley & Sons Ltd

expressed in a protocol which is in line with the prevailing legislation.

However high the potential germination may be, there is no real guarantee that the seed will germinate or the seedlings survive if sown in adverse field or soil conditions. The term 'seedling vigour', usually referred to simply as 'vigour', can provide an assessment of the seed lot's ability to withstand inclement germination and seedling establishment conditions. Although vigour tests are not always available, results of such tests can be of use to the farmer in providing a guide on points such as suitability for early sowing.

Physical Quality

The physical analysis of a statistically valid sample taken from a seed lot will provide information on materials which are present but which are not seed of the cultivar. During physical analysis (a purity test), the components of a seed lot can be classed into clearly defined categories. These include pure seed (i.e. the stated species), the number of seeds of other species (subdivided into weed and other crops), and contaminants. The contaminants in a seed lot can be subdivided into one of several definitions according to their origin, namely:

(i)   material derived from seeds;
(ii) material derived from other parts of the plant;
(iii) material derived from living organisms (but not plants);
(iv) material not derived from living organisms (e.g. soil).

Therefore the physical purity of a seed lot will significantly influence the monetary value; in addition, the presence of weed seeds, especially if noxious or parasitic species, can be potentially very harmful to farming environments.

Seed Health

Many of the pathogens which are of economic importance are transmitted from one generation to the next via the seed. Thus even small levels of infection in the seed crop will act as a source of initial infection in the next generation. Some nematode pests are also seed-borne.

Ideally the seed-borne pathogens and pests should be controlled during seed production, but in practice this is not always totally effective.

There are laboratory tests which are specific to determining the presence of individual pathogens or nematodes on seeds, and their application during seed testing can provide vital information as to which seed treatments should be applied before the seed is sown.

Seed which is healthy at the outset of storage can become infected with storage pests and diseases which can reduce the seed's germination and storage potential in addition to significantly reducing its value.

Moisture Content

The moisture content of seed is normally expressed as the percentage moisture of the seed lot as determined by an oven drying process specified by the International Seed Testing Association (ISTA 1996a, b). Other relatively quick electronic methods are available but usually require calibration of the instrument, especially when changing from one species to another. Information on seed moisture content has two distinct applications.

Firstly, it may be commercially important for purchasers of large quantities of seed to know the percentage of moisture so as to be able to determine the 'farm-gate' value of the consignment. Although this concept is not applied so frequently in the present seed industry trading systems, it is still applied on occasions.

The second use of moisture content data is the critical application for the safe storage of seed in order to gain the maximum potential storage life for the seed lot. The seed's moisture content is at equilibrium with the atmosphere in which it is stored, therefore regular checks should be made on stored seed, especially in climates with fluctuating ambient relative humidities. An exception to this is seed stored in vapour-proof containers or moisture-controlled environments, but in these cases it is also important to have accurate information on the seed lot's percentage moisture content when it is being lowered to the safe level at the start of storage.

GENETIC QUALITY

An unequivocal genetic identity is one of the foundations of quality seed. After the breeder has developed a cultivar with suitable characteristics for commercial use it is essential that no change is allowed to interrupt the genetic quality and detract from the expectations of the grower of the crop or the fulfilment of end-user requirements. In all of the steps of the seed production process it is important to recognise that a set of genes is being maintained.

Plant reproductive systems vary widely in the rate at which genetic identity may naturally be eroded. Foremost here is the mode of fertilisation in which we distinguish two main types: self-fertilisation and cross-fertilisation. These phenomena control the rate at which new genes may be introduced as well as providing the mechanism by which genetic balance is maintained. Many species of plants, however, do not fall clearly within either of these groups. The successful seed producer must understand the reproductive system of his crop in order to preserve cultivar identity.

Plant Reproductive Systems

Self-fertilised Species

In self-fertilised, or *autogamous* species, male and female gametes produced on the same individual plant fuse. A species is generally considered autogamous if at least 90% of the progeny are the result of self-pollination (Mayo 1987). Ideally, a self-fertilised species will also express *cleistogamy*,

TABLE 3.1 Examples of self-fertilised crop species

Barley	Peanut	Subterranean clover
Lettuce	Peas	Tobacco
Linseed	Rice	Tomato
Oats	Soyabean	Wheat

in which fertilisation takes place within the unopened floret. Many of the crop species classified as self-fertilising (Table 3.1) will also cross-fertilise at a low level depending on genotype and environment. For example, in wheat, a self-fertiliser, the range of out-crossing has been placed at 0–4% (Heyne and Smith 1967). Significant cross-fertilisation may also occur between plants within a crop of nominally self-fertilised plants. If the crop is of a pure line such crossing is referred to as *geitonogamy* and will not, of course, lead to a breakdown in genetic purity. If a cultivar has been based on more than one homozygous line, an erosion of the trueness to type of the cultivar will take place as the lines cross, leading to segregation.

In most instances, the breeding method used for a self-fertilising species is to produce a pure line. Thus, each cultivar will consist of a single homozygous genotype and it will be the aim of the seed producer to maintain the cultivar in this condition, taking suitable steps to prevent the introduction of new genotypes. Such an introduction may arise from without by cross-fertilisation, or from within by mutation.

Even in the presence of foreign pollen, the majority of seed from crop species in this reproductive group is formed as the result of self-fertilisation. However, a number of species, although often self-compatible, will readily cross-fertilise if pollen from other cultivars is available. This appears to happen because the plants concerned are 'open flowering'. For the practical seed producer, these plants generally require similar treatment to that given to allogamous species as far as the prevention of cross-fertilisation is concerned. Table 3.2 gives examples of species in this partially out-crossing category.

A breeding strategy occasionally adopted for those species able to cross-fertilise is to encourage

TABLE 3.2 Examples of self-fertilised crop species prone to out-crossing

Cotton	Parsnip
Faba beans	Sorghum
Oilseed rape	Triticale

TABLE 3.3 Examples of cross-fertilised crop species

Brussels sprouts	Maize	Runner beans
Cabbage	Mangel	Rye
Carrot	Meadow fescue	Ryegrass species
Cucumber	Onion	Sugar beet
Hemp	Parsley	Sunflower
Kale	Radish	Timothy
Lucerne	Red clover	White clover

TABLE 3.4 Incompatibility in crop plants

Incompatibility system	Controlling genotype	Example species
Gametophytic	Pollen	Rye
Sporophytic	Pollen parent	Brussels sprouts

cross-fertilisation by incorporating genes for male sterility. The objective is to facilitate heterosis. Sorghum has been handled in this way (House 1985).

Cross-fertilised Species

A plant of a true cross-fertilising, or *allogamous*, species requires pollen from another genotype for fertilisation to take place. Examples of cross-fertilised crop species are given in Table 3.3. There is variation in the degree of toleration of inbreeding. In principle, however, a cross-fertilised cultivar is essentially a collection, or population, of genotypes which intercross between themselves. In this way, the individuals of this population share a common gene pool in a continuous process, making it possible to maintain a population of constant identity over many generations provided steps are taken to prevent the addition or removal of genotypes. The genetic basis of genetic stability in an open-pollinated cultivar is *Hardy–Weinberg equilibrium*, by which it can be demonstrated mathematically that a population of *infinite* size is at equilibrium provided (a) mating within the population is random, and (b) selection, migration and mutation are all at zero (Allard 1960). In absolute terms, these conditions will rarely, if ever, be achieved but, at a practical level, it is possible to maintain a cross-

fertilising cultivar on the basis of Hardy–Weinberg equilibrium.

The mechanism ensuring cross-fertilisation is usually referred to as *incompatibility*. This is common in the plant kingdom and may result from structural, physiological or genetic factors. The form most often conferring inability to self-fertilise in crop species arises where the style will not accept pollen carrying identical alleles at a locus controlling compatibility. There are two systems: either the male parent or the pollen genotype determines incompatibility. Both use many alleles, with the sporophytic being the most complex (Table 3.4).

Cross-fertilising species have given rise to several types of cultivar (see below). All of these, apart from hybrids, are effectively open-pollinated populations and the breeder's objective is to maintain the frequency of alleles representing the cultivar. Selection methods (such as mass selection and recurrent selection) control the breadth of genotypes incorporated and dictate the extent to which the breeder can determine the breeding value which component genotypes confer to the cultivar.

Types of Cultivars

Pure-line Cultivars

A pure-line cultivar consists of a single genotype in which all loci are homozygous. The traditional method of producing a pure line is by repeated selfing over seven or more generations running concurrently with the process of selecting a line with the desired performance characters. Heterozygous loci will normally be reduced in this way by 50% at each generation. However, the establishment of a pure stock will often be supported by careful checking of progenies and the removal, prior to pollination, of any suspected variant plants – an operation of particular importance as the cultivar approaches release (Hanson 1973). Such a procedure is referred to as the *pedigree method.* An alternative to pure-line breeding is the *bulk method* in which segregants are multiplied together for several generations in a bulk until homozygosity has been reached. Although mass selection may be carried out, the selection of lines will be carried out later.

The conventional methods of breeding a pure-line cultivar require several years to complete the process. Techniques have been developed to accelerate the breeding of pure lines. Two of these deserve particular mention, namely single-seed descent (SSD) and the use of dihaploids (DH). As the name implies, with SSD, the breeder carries forward to plant for the next generation one seed from each plant. With SSD, the principal difference from conventional breeding lies in the delay of the selection phase until homozygosity is close. SSD is usually carried out in growth cabinets to enable several generations to be grown in a single year (Choo et al. 1985). In terms of seed production and the establishment of a pure stock, SSD is similar to the conventional system in that it requires several generations of selfing to achieve homozygosity. On the other hand, DH relies on the production and identification of haploid plants having a single set of chromosomes (Snape et al. 1986). The plants are then treated with colchicine to produce an exact copy of the haploid alleles to give a dihaploid plant which is fully homozygous at all loci. Dihaploids may be produced at F2 or, to allow some segregation to take place, the process may be carried out at a later generation. DH may also be used to achieve full homozygosity in advanced line material bred by conventional methods.

The success of seed production of a pure-line cultivar, regardless of breeding method, depends on the production of a pure stock of Foundation Seed. Not only is this critical for a new cultivar, but each successive multiplication of Foundation Seed must represent the cultivar. Genetic shift will occur if the individual plants used to maintain the cultivar deviate from trueness to type, and careful checking of these plants is essential.

Inbred Line

In the strict sense, if a line is self-fertilising and homozygous it is an inbred line. The term is, however, generally used only for inbred lines of species which would otherwise normally consist of open-pollinated cultivars. Inbreds are developed by repeated self-fertilisation and selection which may lead to problems of low fertility and reduce vigour in normally cross-fertilised species, although material with improved performance is available in a number of crops, including maize.

Multiline Cultivars

Pure-line cultivars facilitate well targeted management of crop agronomy and are desirable for the end-user, particularly where uniformity of produce is required. However, a frequent objection has been raised on the grounds that monoculture increases the risk of disease epidemics (Marshall and Brown 1973) and may also be less stable in the absence of multiple alleles giving a possible 'buffering' effect against environmental stresses. The multiline approach seeks to provide multiple alleles so that, should one allele conferring resistance to a disease break down, another will be present to provide a degree of compensation and to limit the opportunity for pathogens to spread from plant to plant. Strictly, the multiline cultivar is developed from a single pure line from which multiple lines have been bred containing a series of alleles offering different resistance factors to the pathogen in question.

For the seed producer, the component lines should be maintained as separate cultivars and then blended in the correct proportions prior to sale to the commercial grower. Clearly, accurate stock control will be required to prevent confusion of the true identity of each line from the initial production of the Foundation Seed to the final cultivar sold to the grower.

Considerable breeding effort will be required to produce a multiline cultivar. A simpler approach has used a blend of conventionally bred cultivars (Priestley et al. 1988). Whilst this might achieve a similar result in terms of the management of disease resistance, the resulting blend will be mixed for many characters of importance to grower and end-user.

Open-pollinated Populations

An open-pollinated cultivar consists of a uniform and stable interpollinating population at genetic equilibrium. Such a cultivar should be genetically heterogeneous and heterozygous, with gene frequencies at a constant level of expression. There are a variety of procedures by which an open-pollinated cultivar will be developed. Mass selection of plants on the basis of a visual assessment is fairly effective for characters readily discernible by eye. Seed from plants identified as desirable will be bulked to form the next generation. The disadvantage is that selection, practised in this way, only controls the maternal contribution to the progeny and, whilst it may increase the proportion of favourable genotypes, the results will be limited. An attempt to reduce variation from the desired performance by limiting the number of genotypes is likely to lead to inbreeding depression, as will an attempt to develop a cultivar of an allogamous species from a single plant. Moreover, those characters which are strongly influenced by the environment, such as yield, are difficult to assess. One of the simpler forms of *recurrent selection*, with one of the forms of progeny testing, may be used to develop an improved cultivar over a number of generations.

Composite Cultivars

A composite cultivar is constructed by compositing a number of separately selected lines (Allard 1960). These lines will, in general, have been assessed individually in progeny tests and will be bulked together to form the composite cultivar which will then be multiplied by random mating prior to release. It will be important to maintain the genetic identity of the composite, in order to prevent significant drift from the original type caused by inbreeding, selection or other forces. This may include periodic re-establishment of the composite from the original lines. The number of generations of multiplication allowable for an open-pollinated population will vary with species.

Synthetic Cultivars

A synthetic cultivar is constructed by intercrossing a number of defined and more or less inbred genotypes which have been selected for general combining ability in all possible hybrid combinations (Becker 1989). The synthetic is established by random mating of the lines in such a way that all possible combinations have an equal probability of occurring. The number of generations for which a synthetic may be multiplied as an open-pollinated population will depend on the species. Essentially, the objective behind the breeding of a synthetic is to utilise heterosis in those species where it is not possible cost-effectively to harness the phenomenon in an F1 hybrid. The key question affecting the success of this strategy is the number of random mating lines incorporated into the synthetic. It can be shown that the reduction in vigour from the F1 to F2 (or more correctly, Syn 2) declines as the number of lines increases. Following Syn 2, the synthetic should retain its performance provided it has reached equilibrium. This requires that mating is random and that the other conditions of Hardy–Weinberg are met. In practice, periodic reconstruction of the synthetic will be necessary using the original 'lines'.

Clonal Cultivars

A clone is a group of individuals descended from a common ancestor by mitosis. Clonal cultivars are principally of importance in horticulture (see remarks on somatic embryogenesis below). Although not forming part of the subject of seed production, clones require similar treatment in certain respects to ensure that the quality and identity of reproductive material is maintained.

Apomictic Cultivars

Strictly, *apomixis* includes all forms of asexual reproduction. However, the term is most frequently used to describe the production of seed without normal meiosis and fertilisation (Asker 1980). The plants so produced are known as *apomicts*. Considerable work on the induction of apomixis in a range of species has taken place over the past half century. Apomictic seed of *Poa* spp. has been produced, but in many species there are problems in obtaining stable cultivars.

Hybrid Cultivars

Seed of a commercial hybrid cultivar is the result of a large-scale crossing programme involving selected parents. Unless a self-incompatibility mechanism is used, it will be necessary to render the seed parent male sterile in order to force it to cross with the intended pollen parent, and several methods for achieving this exist (Table 3.5).

Types of Hybrid Cross

A number of different types of hybrids have been devised to meet particular problems in producing the parental seed and to procure male sterility. The level of homozygosity and homogeneity of the parental seed controls the genetic structure and visual uniformity of the hybrid. In cases where it is not possible to produce a hybrid from two pure lines, it may be necessary to select the parental materials to avoid conspicuous segregation in commercial crops.

Types of Pollination Control

Chemical hybridising agents (CHA) Since 1950, many chemical compounds have been tested with cereals and other crops such as cotton (*Gossipium hirsutum*) for potential selectively to induce male sterility (McRae 1985). The problems encountered include inadequate male sterility and a reduction in female fertility resulting in low yields of poor quality seed. It should be noted that formulations may function differently; for example, those used on wheat include CHAs which prevent pollen

TABLE 3.5 Types of hybrid cultivars

F1 hybrid	The first generation of a cross of two defined parental cultivars or lines. If these are pure inbreds, the F1 will be uniform.
F2 hybrid	The progeny derived by selfing, or intercrossing, an F1 hybrid.
Single cross hybrid	The F1 of a cross of two inbred lines.
Three-way cross hybrid	The F1 of a cross of an inbred line and a single cross hybrid. Normally, the inbred line will be male.
Double cross hybrid	The F1 of a cross of two single cross hybrids.
Triple cross hybrid	The hybrid derived from the cross of two three-way hybrids.
Top cross hybrid	The F1 of a cross of an inbred line or single cross hybrid and an open-pollinated cultivar.
Intercultivar hybrid	The F1 of a cross of two open-pollinated cultivars.

production and others which, although allowing pollen development, render it non-functional. A number of breeders continue to use CHAs on wheat. There has also been limited application to other species.

Mechanical emasculation This is obviously only feasible for commercial seed production in special circumstances. Monoecious species with suitable floral morphology may be mechanically emasculated; these include maize (*Zea mays*) and cucumber (*Cucumis sativus*). The detasselling of maize is usually carried out by hand, sometimes with the aid of a personnel transporter. Where the multiplication rate of a species is high, the labour input for emasculating perfect flowers may be justified, as in the tomato (*Lycopersicon lyropersicon*).

Self-incompatibility (SI) Many species of plants express self-incompatibility which prevents self-fertilisation in fully fertile plants. SI is common in the *Brassica* spp. where it has received considerable study and application (de Nettancourt 1993; Ockenden and Smith 1993). A series of alleles conferring self-incompatibility has been identified and lines homozygous for S allele have been developed. Maintenance of an SI line may be achieved by circumventing self-incompatibility by one of a number of methods including bud pollination, the pollination of young flowers, exposure to high temperatures and the application of chemicals. The choice of method will largely depend on the species involved. In the hybrid seed production field, pollination by pollen carrying another S allele will produce hybrid seed in which self-compatibility is dominant.

Cytoplasmic male sterility (CMS) There are now more than 140 species in which cytoplasmic male sterility has been found (Hanson and Conde 1985; Kaul 1988). In a number of these, pollen abortion is associated with the presence in a species of alien cytoplasm. It may also be caused in certain species, including maize, by a mutation in mitochondrial DNA (Levings and Dewey 1988). Generally, the sterilising cytoplasm must be incorporated by backcrossing with the recurrent parent acting as the female. The major advantage of CMS is that, in most plants, the cytoplasm is not carried by the pollen and is therefore maternally inherited, making maintenance of a male sterile line a relatively simple procedure. The use of CMS suffered a setback when its use in maize became implicated in an outbreak of southern corn blight (*Heminthosporium maydis*).

A CMS line is only able to produce seed if a pollinator is present. As the cytoplasm from the CMS line will be present in the hybrid, a nuclear restorer gene, or genes, must be present if seed is to be obtained. For the maintenance of the CMS line, it will be pollinated by a line carrying the same nuclear genotype without restorer genes but having normal cytoplasm; this is known as the 'maintainer'. In certain species affected by inbreeding depression, synthetic restorers may be used.

Two circumstances may, however, arise where the male hybrid parent will not be a restorer. If seed is not the end product of the crop, there will be no need for restoration and there may be commercial reasons for wishing to prevent multiplication of the seed. A further reason is found in species affected by inbreeding depression where inbred lines exhibit lower fertility. In order to reduce the cost of seed production for these, double or three-way crosses may be produced. Thus, if an F1 hybrid is to be used as a female parent, obviously it must retain male sterility from the original seed parent. Conversely, if destined to act as a male parent, it must carry restorers. In hybrids where a satisfactory restoration system has presented difficulties, a pollinator has been blended with the hybrid seed for sowing for commercial crop production. Clearly, a species managed in this way must be able to cross-fertilise readily in the field if reliable commercial crop production is to be achieved.

Nuclear male sterility (NMS) Male sterility, most frequently controlled by a single recessive allele, has been located in at least 35 species (Driscoll 1986; Kaul 1988). The difficulty with NMS is the complex line maintenance involved. The female parent must carry a nuclear gene for male sterility, often as a recessive. Normally, it is not possible for the locus coding for male sterility to be homozygous since maintenance would not be achieved. Heterozygosity at this locus allows multiplication but gives rise to male fertile segregants in the female parent and these will require roguing. To facilitate this, a marker gene should be associated with the plants showing fertility. Proposals have been made for circumventing roguing by genetic means; an example is *certation*, or differential pollen tube growth, as a means of eliminating gametes conferring male fertility to wheat, a crop where roguing would be too expensive.

Male parent maintenance, on the other hand, is simpler under the NMS system than the CMS. No restorer genes are needed since any normal cultivar will carry dominant alleles at the locus coding for male sterility and restoration will therefore be automatic. Maintenance of the male parent should therefore follow the practice for the normal breeding system of the species.

TABLE 3.6 Classification of impurities in cultivars of self-fertilising cereals

Cause	Typical percentage	Species most affected
Cross-fertilisation	1–5	Winter barley, triticale
Aneuploidy	1–5	Bread wheat, triticale
Point mutation	<1	Bread wheat
Residual heterozygosity	<1	All species

The Management of Genetic Quality

The reproductive systems of plants are susceptible to a number of genetic problems which result in a loss of trueness to cultivar type. Of particular interest are the self-fertilised cereals where genetic problems can lead to conspicuous off-types (Table 3.6).

Foreign Pollen

There are many reports of unexpected cross-fertilisation interfering with pure seed production. Although pollen dehiscence in the self-fertilising cereals normally occurs within the floret, some pollen is generally discharged from the plant, thereby facilitating cross-fertilisation. There are differences between the self-fertilising cereals in proclivity to out-cross. Weather patterns give rise to seasonal variation in flowering behaviour. Clearly, any out-crossed progeny will be F1 hybrids and, in later generations, these will show segregation. Trueness to cultivar will thereby be lost and the contaminants will often be very conspicuous.

Winter barley in Europe has demonstrated a strong proclivity to cross-fertilisation. Giles (1989) investigated causes of out-crossing in this cereal and concluded that earliness of sowing may affect timing of growth stages involved in reproduction, making the plant more vulnerable to weather. It is possible that a proportion of florets may thus be male sterile. Pollen from another cultivar, if available, is then able to effect fertilisation with the consequence that cultivar purity is affected.

The production of pure stocks requires avoidance of cross-fertilisation. The measures adopted may include the selection of cultivars showing a lower tendency to cross-fertilise. However, isolation is the proven method of avoiding contamination from foreign pollen. Prescribed isolation distances reflect common experience with the species. In the UK, self-fertilising species with a tendency to out-cross are isolated by 20–100 m, depending on crop. In a cross-fertilising species – sugar beet (*Beta vulgaris*) – Dark (1971) found the outside margins of crops to be most seriously affected by contaminating pollen. The removal of the outside margin of a crop at harvest may be a useful precaution.

Aneuploidy

In a number of crops, incorrect cell division at meiosis leads to plants having lost or gained whole or part chromosomes. The state of having an incorrect chromosome complement is termed *aneuploidy*, in contrast to *euploidy* which refers to the normal condition (Riley and Kimber 1961; Worland and Law 1985). Wheat, oats and triticale are affected by irregularity in chromosome complement. These may give rise to plants having an abnormal phenotype. For example, plants of semi-dwarf wheat lacking the chromosome carrying the dwarfing genes *Rht1* or *Rht2* will be taller than the normal cultivar. Plants lacking the chromosome carrying a speltoid suppressor gene will produce 'speltoid ears' having some of the characteristics of *Triticum spelta* and not conforming to the cultivar. The extent of the problem varies with cultivar, but not all aneuploids will be detected by morphological examination. Experience has shown that aneuploids arise spontaneously and then multiply in seed stocks.

Point Mutation

Point mutations occasionally lead to the introduction of variants into a cultivar. The 'fatuoid oat' may be caused in this way. This variant arises within most oat cultivars in small numbers. It exhibits the characteristics of the cultivated oat

(*Avena sativa*) but also expresses certain seed features of wild oats (*Avena fatua*), causing confusion at crop inspection. The plant breeder should rogue mutants from all species, especially from the early generations.

Residual Heterozygosity

Wheat cultivars are often based on a single F5 plant which will be multiplied and given further selection in ear/plant rows prior to the first official trials for which F7 or F8 seed will be submitted. Most segregation will have ceased by this stage and this will be assisted if every ear-row showing variants is completely removed prior to anthesis. The breeder is, however, under considerable pressure at the critical time of the year as a result of having many lines to handle, and there is a possibility that less obvious variants will escape detection.

Tests for Genetic Quality

The most difficult stage for establishing pure stocks is immediately after the initial breeding phase when it is particularly important to recognise and eliminate residual heterozygosity. However, the cultivar must also be maintained true to type for life without genetic shift. To achieve this in the self-fertilising cereals, the maintainer will start each cycle of multiplication with an adequate number of carefully monitored ear-rows. Similar principles may be applied to other self-fertilising crops.

The *ear-row test* is a technique for identifying genetic problems. For this, the progeny of single ears are grown separately in rows of approximately one metre. Within rows the resulting plants have a known filial relationship and cannot be derived from mechanical contamination. The plant-row is based on a similar principle but involves more material. In both cases, a mixed row will indicate residual heterozygosity whereas single variant plants suggest either mutation or out-crossing.

The Management of Identity and Purity in Hybrid Cultivars

There are three aspects to ensuring the cultivar identity and purity of hybrids. As a first step, the parent stocks must be verified in pre-control according to the certification system for inbreds or normal seed production of the species. In effect, it may be possible to post-control parental cultivar purity in the hybrid seed production field. Secondly, the effectiveness of male sterility can be assessed, where appropriate, by means of ear-bagging. Thirdly, the hybrid seed may be examined by growing-on techniques, but since these will cause a delay, a suitable chemotaxonomic test may be preferable. Electrophoretic tests have been used to confirm the identity of hybrid maize and wheat (Smith and Wych 1986).

Future Developments

Transgenic Technology

The foremost application of genetic engineering in plant breeding has been the introduction of genes to a species by non-sexual methods to confer new characters. Considerable effort is being made to incorporate resistance to biotic and abiotic stresses. An example of the former is corn borer resistance in maize (Evola 1996), and of the latter, oxidative stress resistance in higher plants (Hérouart et al. 1994). Transgenic approaches are also being used to facilitate new uses for the end product as illustrated by laurate oilseed rape (Sovero 1996). Distinct progress has been made in the development of a system for the production of hybrids (Mariani et al. 1990).

Ultimately, transgenic technology may affect the seed producer in a number of ways. For example, new cultivars may be introduced more quickly but may also be subject to exacting control procedures by the authorities responsible for safeguarding the environment. It should not be overlooked that wide crossing by conventional means has long been practised, with the result that a number of important crop species contain genes from other genera.

In the maintenance of genetic quality, the primary concern will be the effect of any changes in reproductive system, intentional or otherwise. It is possible to envisage circumstances where a transgenic might be more prone to out-crossing. It might also be considered a greater hazard to other crops if ability to synthesise a new product, such as an industrial oil in oilseed rape, is conferred by the pollen to crops destined for human consumption.

Non-transgenic Technology

Of the many possible ways non-transgenic innovation in cultivar breeding and development might affect seed production, there are three deserving particular mention because each has the potential to change the way in which genetic systems are used by the plant breeder.

Hybrids   Following the successful introduction of maize hybrids, there has been a steady increase in the number of agricultural and horticultural species for which F1 hybrids are available (Mayo 1987). Crops with significant areas grown from hybrid seed now include: Brussels sprout, rice, rye, sorghum, sugar beet and sunflowers. It appears that F1 hybrids may soon dominate oilseed rape (*Brassica napus*). The hybrid route is, however, unlikely to suit every crop: the decision is likely to be based on an assessment of the cost-effectiveness and commercial desirability of hybrids, taking into account the costs of development and production. One of the foremost factors here will be the nature of the reproductive system of the crop and the ease with which that system allows genetic potential to be fixed in a non-hybrid cultivar and harnessed in seed production.

Asexual propagation   Under this heading, a number of techniques could be mentioned. Micropropagation by tissue culture is, of course, a well established procedure for a number of horticultural crops. Techniques for asexual reproduction are principally of interest in those species where it is difficult to fix desirable genotypes, perhaps showing heterotic yield increase, in true

breeding cultivars. In these circumstances, propagation of a clone would be highly advantageous provided production and handling of the units of propagation could be accomplished at a cost not too far in excess of conventional seed. Clearly, for the lower value, large-scale field crops, there is a requirement for an efficient method of producing propagules suitable for handling as synthetic 'seed'. The only system that currently appears to offer a potential to meet these requirements is *somatic embryogenesis* (Bornman 1993). Essentially, this term describes processes by which embryos may be formed from somatic cells of plants, i.e. cells not produced by gametic fusion. Although considerable work has been carried out on somatic embryogenesis, a commercial system has yet to be developed.

Gene identification   In addition to providing the means for manipulating genetic material in the production of transgenic cultivars, the revolution in gene technology has also affected non-transgenic breeding. The ability to identify genes using molecular tests can assist the breeder by making it possible to determine the presence of a sought-after gene in a segregant much more swiftly than by conventional growing-on techniques. This topic is covered elsewhere in this volume; in this section we simply remark that molecular identification has the potential to increase the precision of plant breeding.

GENERATION CONTROL: SEED CLASSES

Seed production on a commercial scale requires forward planning. A commercial quantity of seed is only achieved after a series of multiplication steps. Starting from a maintenance programme (by the breeder or a person or organisation authorised as the 'maintainer' for a cultivar), it is necessary to multiply the seed through several generations. In most cases this involves several years as it is generally possible to grow only one crop per year except in some favoured, mostly tropical, areas. Through all of these steps it is necessary to exercise a system of quality control to ensure that the essential qualities of the cultivar are maintained.

Each time that a cultivar goes through a cycle of multiplication there is the possibility that it may lose some of its essential qualities or that they may be eroded. There are several reasons for this. First, there is the possibility that the seed production field may be contaminated with seed of another cultivar which produces plants within the seed crop, so causing mixture at harvest; or, in handling the seed used to sow the seed crop or the seed harvested from the crop, some admixture may occur. Second, there is the possibility of a genetic change to some of the plants in the seed crop caused by mutation which will produce aberrant seed. And third, there is the possibility that some of the plants in the seed crop may be fertilised by pollen from outside the crop or from aberrant plants within the crop, although this possibility is less likely in the self-fertilised species. However, even in species classified as self-fertilising there can be a small proportion of cross-fertilisation; for example, Smith and Whittington (1991) state that out-pollination in wheat may occur with a frequency of 4% and Bickelmann (1993) found 1.1 to 8.7% of out-crossing in oats.

The longer the sequence of multiplication, the greater the possibility that one or all of these factors may occur to erode the essential qualities of the cultivar. Because cross-fertilised species show a greater plant-to-plant variation than self-fertilised, they are more likely to show shift in the characteristics of cultivars. When cultivars of some cross-fertilised species are multiplied outside their 'area of origin' there is evidence that a genetic shift may occur so that the whole character of the cultivar changes as more generations are produced in the new area. For instance, Kelly and Boyd (1966) reported 'shift' in some grass cultivars from the UK when grown for seed in the USA. In general these early results indicated that growing seed for one generation in a different environment does not have a significant effect, but that in later generations the effect can be cumulative.

These considerations have led to the current system of seed classes which is based on the number of generations that the seed is removed from the maintenance programme.

The most generally used terminology is that which has been devised for use in the Organisation for Economic Co-operation and Development (OECD) Seed Schemes. The following definitions are taken from the OECD Herbage and Oil Seed Scheme (1995a):

1. *Parental Material.* The smallest unit used by the maintainer to maintain his cultivar from which all seed of the cultivar is derived through one or more generations.
2. *Pre-Basic Seed.* Seed of generations preceding Basic Seed, which may be at any generation between Parental Material and Basic Seed.
3. *Basic Seed*
   (i) *Bred cultivars.* Seed which has been produced under the responsibility of the maintainer according to the generally accepted practices for the maintenance of the cultivar and is intended for the production of Certified Seed. For hybrid cultivars it includes seed sown to produce the pollen-parent plants as well as seed sown to produce the seed-parent plants. Basic Seed must conform to the appropriate conditions in the Scheme and the fulfilment of these conditions must be confirmed by an official examination.
   (ii) *Local cultivars.* Seed which has been produced under official supervision from material officially admitted for the purpose of the local cultivar on one or more farms situated in an adequately defined region of origin and is intended for the production of Certified Seed. It must conform to the appropriate conditions in the Scheme and the fulfilment of these conditions must be confirmed by an official examination.
4. *Certified Seed*
   (i) *Cultivars other than Hybrid.* Seed which is of direct descent from either Basic Seed or Certified Seed of a cultivar and is intended for the production of either

Certified Seed or of crops for purposes other than seed production. It must conform to the appropriate conditions in the Scheme and the fulfilment of these conditions must be confirmed by an official examination.

The first generation from Basic Seed is known as *Certified Seed 1st Generation*.

Further generations are known as *Certified Seed 2nd Generation*, etc. the appropriate generation being designated.

(ii) *Hybrid cultivars.* Seed which is the first generation of hybridisation of Basic Seed of a cultivar and is intended for the production of crops for purposes other than seed production. It must conform to the appropriate conditions of the Scheme and the fulfilment of these conditions must be confirmed by an official examination. In the production of a multiple-cross hybrid, Certified Seed may on occasion be used to produce pollen-parent or seed-parent plants. The Designated Authority may reclassify it as Basic Seed for this purpose only.

These definitions have remained with very little change since the schemes were first formulated, the main amendment being to substitute the word 'maintainer' for 'breeder'. This is to take account of the fact that in modern trading conditions it is not always the breeder who maintains a cultivar, but may be someone authorised to do so. There may be different maintainers authorised in different countries.

In North America a somewhat different terminology is used. The handbook of the Association of Official Seed Certifying Agencies (AOSCA 1971) gives equivalences between the two systems (Table 3.7).

Table 3.7 was prepared before the introduction of the term 'Parental Material' in the OECD Scheme. The difference between the two terminologies is that whereas in the OECD Schemes once Basic Seed has been reached it can be one

TABLE 3.7 Comparison between seed classes in North America and those in the OECD Seed Schemes (adapted from AOSCA 1971)

OECD seed classes	AOSCA seed classes
Pre-Basic Seed (can include several generations)	Breeder Seed
Basic Seed	Foundation Seed (can include several generations) or Registered Seed
Certified Seed 1st Generation	Certified Seed
Certified Seed 2nd (3rd etc.) Generation	None

generation only, in the AOSCA Scheme there can be more than one generation within Foundation Seed followed by a single generation as Registered Seed. It might be argued, therefore, that Registered Seed is the equivalent of Basic Seed and that all that gocs before is Pre-Basic; alternatively, Registered Seed might be regarded as the equivalent of Certified Seed 1st Generation. The important point is that both schemes provide some flexibility within the progression through the generations so that it is possible to build up marketable quantities of seed while maintaining control over the number of generations. This is necessary because different quantities of seed will be required depending on the popularity of different cultivars.

While certification schemes cover most of the agricultural crops, a different approach is often adopted for vegetables. This is because the quantities of seed of individual cultivars required are generally relatively small and the seed crops often widely dispersed, so that to apply the same system of official seed certification would not be economic. However, the same general principles apply and the same terminology is used up to Certified Seed 1st Generation. Thereafter another category known as *Standard Seed* has been introduced in the OECD Scheme for Vegetable Seed (OECD 1995b). The definition of Standard Seed is: 'seed which is declared by the supplier as being true to the cultivar and of satisfactory varietal purity. It must conform to the appropriate conditions of the Scheme'.

Although the seed classes defined above are taken from officially recognised international seed certification schemes, the terminology is also sometimes adapted for use for seed produced under the control of a private company when an effective quality control system is operated by the company. From the point of view of the purchaser of the seed it is important to know how far the seed on offer is removed from the breeder's or maintainer's maintenance programme as this will have a direct bearing on the value of the seed.

In addition to the need for generation control it is necessary to define the different standards of quality which apply to the different classes. Thus the earlier generations are usually given a higher standard of cultivar purity than the later. For such qualities as germination or freedom from seed of weeds or other species, however, there is normally no difference in the standards adopted for the earlier and the later generations. Qualities other than cultivar purity are generally specific to the crop species. In most quality control schemes there is provision for a lowering of germination standards in a year when harvest conditions are difficult; this applies particularly to the earlier generations – Pre-Basic Seed, Basic Seed – where it is considered best to make use of the seed available of a valuable cultivar rather than to delay production by a year.

All of these aspects of quality control are addressed internationally by the appropriate organisations described in Chapter 2 so that there is agreement between countries on the meanings of the terms used and also agreement on the methods to be used in testing to determine whether or not particular standards have been met for a particular seed lot. Thus seed can be traded between countries using easily understood descriptions of quality. The extent to which these descriptions have to be used for any seed offered for sale in a country will depend on the national legislation or on the accepted practices of commerce as described in Chapter 2.

CULTIVAR IDENTIFICATION: GENERAL DISCUSSION

The whole purpose of plant breeding and the production of improved cultivars would be negated if it were not possible to identify the cultivar at least at certain key points during seed multiplication. It is essential to know that seed offered for sale is of the cultivar it purports to be, even though it may not always be possible to determine the purity of a seed lot as regards cultivar by reference to easily observed characteristics of the seed or seedlings. There are now some methods for testing samples of seed or other plant parts in a laboratory, which will be described later in this chapter, and these are proving very useful as additional tools for the identification of seed lots as to cultivar. Traditionally, however, the seed producer has relied on knowledge of the origin of the seed used to sow the seed crop, and on checks made during the growing, harvesting and subsequent preparation for sale, to ensure that the essential characteristics of the cultivar are maintained and that no admixture has occurred.

To identify a seed lot with a cultivar it is first necessary to understand the meaning of the word 'cultivar'. There has to be some definition which makes it possible to state that one particular group of plants is a cultivar or that another group of plants is not.

Definitions of Cultivar

A *cultivar* is defined in the International Code for the Nomenclature of Cultivated Plants (Anon 1980) in Articles 10, 11 and 12 as follows.

> Article 10. The international term 'cultivar' denotes an assemblage of cultivated plants which is clearly distinguished by any characters (morphological, physiological, cytological, chemical or others) and which, when reproduced (sexually or asexually) retains its distinguishing characters. The cultivar is the lowest category under which names are recognized in this Code.

The term is derived from *culti*vated *vari*ety, or their etymological equivalents in other languages.

Note 1. Mode of origin is irrelevant when considering whether two populations belong to the same or to different cultivars.

Note 2. The concept of cultivar is essentially different from the concept of botanical variety, *varietas*. The latter is a category below that of species. Names of botanical varieties are always in Latin form and are governed by the Botanical Code. Rules for the formation of cultivar names are set out in the present Code.

Article 11. Cultivars differ in their modes of reproduction. The following are examples of categories that can be distinguished:

a. A cultivar consisting of one clone or several closely similar clones.

b. A cultivar consisting of one or more similar lines of normally self-fertilizing individuals or inbred lines of normally cross-fertilizing individuals.

c. A cultivar consisting of cross-fertilized individuals which may show genetical differences but having one or more characters by which it can be differentiated from other cultivars.

d. A cultivar consisting of an assemblage of individuals reconstituted on each occasion by crossing. This includes single-crosses, double-crosses, three-way crosses, top-crosses and intervarietal (intercultivar) hybrids.

e. A cultivar consisting of one clone or several closely similar clones which have a habit of growth which is clearly distinguishable from the normal habit and which is retained by appropriate methods of propagation. [*Authors' note: this category is confined to certain types of trees and is not referred to further in this book.*]

Article 12. The practice of designating a selection of a cultivar as a strain or equivalent term is not adopted in this Code. Any such selection showing sufficient differences from the parent cultivar to render it worthy of a name is to be regarded as a distinct cultivar.

The Code states unambiguously that to be designated as a new cultivar the assemblage of plants of which it consists must be 'clearly distinguishable' from other such assemblages. There is no reference to cultural value as a criterion for deciding whether or not a particular assemblage of plants can be designated as a new cultivar. Characteristics which give such a value can be used to distinguish cultivars one from another, but are not exclusive; other characteristics are given equal weight and are usually more easily observed.

However, it is still a matter of subjective judgement as to what constitutes a clear difference. Obviously this is a question which is of vital importance to the development of plant breeders' rights and the International Convention for the Protection of New Varieties of Plants (Anon 1991) also defines a 'variety' (cultivar). Article 1 (vi) reads:

. . . 'variety' means a plant grouping within a single botanical taxon of the lowest known rank, which grouping, irrespective of whether the conditions for the grant of a breeder's right are fully met, can be

- defined by the expression of the characteristics resulting from a given genotype or combination of genotypes,
distinguished from any other plant grouping by the expression of at least one of the said characteristics and
- considered as a unit with regard to its suitability for being propagated unchanged.

The Convention goes on in Articles 7, 8 and 9 to define Distinctness, Uniformity and Stability. To be distinct a cultivar must be 'clearly distinguishable from any other variety whose existence is a matter of common knowledge'. A cultivar is to be considered uniform if 'subject to the variation that may be expected from the particular features of its propagation, it is sufficiently uniform in its relevant characteristics'. And a cultivar is stable if 'its relevant characteristics remain unchanged after repeated propagation or, in the case of a particular cycle of propagation, at the end of each cycle'.

The Use of Cultivar Definitions in Practice

Taking these definitions, it is evident that a cultivar must be defined by some clear characteristics which have to be uniformly expressed by all plants of the cultivar on each occasion that the cultivar is grown, subject only to any expression of natural variation which is inherent in the mode of reproduction of the plants. Characteristics which can be determined by a relatively quick test in a laboratory or glasshouse on seed or seedlings can give a result in a relatively short time which can be used immediately, but most easily observed morphological characteristics occur as the plants mature. The Convention definition in Article 1 (vi) introduces the idea of 'genotype'; the characteristics of a cultivar must result from a 'given genotype or combination of genotypes'. In practice this means that, except when laboratory tests are available which will define the genotype *per se*, the characteristics are expressed phenotypically. It is possible that some characteristics which are used for cultivar definition will be influenced by the environment in which the cultivar is growing and it is necessary to use, as far as possible, those which are least influenced by environment.

There is, therefore, a difficulty in the fact that plants must be grown to maturity in order to describe fully all possible characteristics when using those which are directly observable either on seed or on growing plants. Whereas, for purposes of initial determination as to whether or not a new potential assemblage of plants conforms to the criteria of a cultivar, it is possible to make observations over one or more entire growing seasons, for quality control during seed production this is not possible in the seed growing fields. During seed production (except for the early generations which are under the direct control of the breeder or maintainer of the cultivar), seed crops are normally dispersed and it is not possible to keep them under day-to-day observation; it is normal to rely on one or two spot-checks at times when it is expected that some key characteristics are visible. These checks are known as 'field or seed crop inspections' in a quality control system, and their number and timing are usually specified in a seed certification scheme.

An alternative method is to grow small samples of the seed used to sow the seed crops in control plots; for earlier generations this can provide information in parallel with the field inspection but obtained throughout the growing season. The seed samples are representative of production in the previous growing season. For the final generation in a multiplication cycle, however, the results are normally available too late to prevent sale of a substandard seed lot (unless the seed is stored for one growing season for the plot results to be obtained) and act only as a 'post-mortem' on or reference to the previous season's quality control work.

Both of these control measures provide information only on the state of the seed crop; they are used to predict what the harvested seed will be like but do not guarantee its cultivar purity. Apart from laboratory tests on samples of the harvested seed, the only way to determine cultivar purity of seed before it is sown is to store it for a growing season while control plot tests of samples are conducted. These quality control measures are illustrated in Figure 3.1.

Kinds of Characteristics

The International Union for the Protection of New Varieties of Plants (UPOV) has published many 'Guidelines for the Conduct of Tests for Distinctness, Homogeneity and Stability' which describe the way in which cultivars can be defined by reference to mainly morphological characteristics (there are some features in some of the guidelines which are not morphological, e.g. resistance to disease or ploidy). Whole plants or any part of a plant can be described and used as a means of distinguishing between cultivars. For example, the whole plant may be described in terms of growth habit, or by reference to size; stems and branches may differ in number or in length; leaves may vary in colour, in anthocyanin colouration, or in features of the leaf surface such as blistering, waviness or indentation of the margins; flowers have numerous parts which provide distinguishing features. The following lists the main kinds of characteristic which are easily observable for use in discriminating between cultivars; their usefulness varies between different crop species and between different

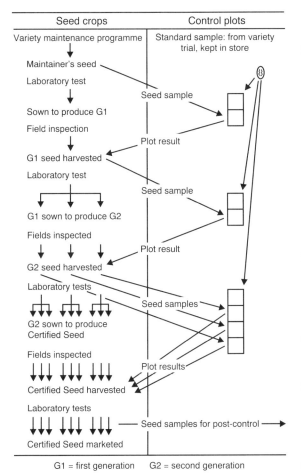

Seed crops	Control plots
Variety maintenance programme	Standard sample: from variety trial, kept in store

Maintainer's seed

Laboratory test

Seed sample

Sown to produce G1

Field inspection

Plot result

G1 seed harvested

Laboratory test

Seed sample

G1 sown to produce G2

Fields inspected

Plot result

G2 seed harvested

Laboratory tests

Seed samples

G2 sown to produce
Certified Seed

Fields inspected

Plot results

Certified Seed harvested

Laboratory tests

Seed samples for post-control →

Certified Seed marketed

G1 = first generation   G2 = second generation

FIGURE 3.1   Quality control system

groups of cultivars within a species. Appropriate details are given in Section II, which deals with individual species.

1. *Shape.* Plants and plant parts differ in shape – seeds, stems, leaves and flowers or parts of these organs may all have distinct shapes which can vary between cultivars. The shape or habit of growth of the whole plant is also often a useful characteristic. Shape may be described either from observation or from measurement of certain key parameters (e.g. ratio of length to width). The comparatively new method of 'image analysis', using cameras and analysing digitalised data from

the image recorded, is described later in this chapter.

2. *Colour.* Flowers and parts of flowers vary in colour. There are also differences in the green colouration of stems and leaves and some cultivars show pigmentation (e.g. from anthocyanin colouration) while others do not. When determining colour by observation it is often advisable to use a colour chart such as that published by the Royal Horticultural Society of England (RHS Colour Chart). New tools have recently been developed for measuring colour with greater accuracy than can be obtained from observation, and are described later in this chapter.

3. *Hairiness.* Plant parts may differ from glabrous to pubescent and the extent or nature of pubescence can differ between cultivars. For example, hairs may be sparse or thick and also soft or stiff.

4. *Waxiness.* Stems or leaves may vary between cultivars in the extent to which they show waxiness or glaucosity.

5. *Dimensions.* Plant height and plant diameter are often useful means of comparison. Length and width of plant parts such as leaves are also useful characteristics and may be used as an indicator of shape (see above).

6. *Size.* Some plant parts may be large or small in different cultivars.

7. *Number.* Certain cultivars or groups of cultivars may show differences in the number of some plant parts. For example, the number of seeds per pod or the number of tillers or stem branches are useful for some species.

Most of these criteria can be observed directly or measured when the plants are growing; in some instances, part of the plant has to be removed to enable the observation to be made (e.g. observation of shape or number of parts of flowers or counting of seeds in a pod).

Additionally, there are three frequently used characteristics which are not morphological.

(a) *Maturity*: the time taken from sowing to flowering (emergence of the inflorescence) or to maturation of the seed can be used as a criterion for identification.

(b) *Disease or pest resistance*: resistance or susceptibility to or tolerance of diseases or pests are characteristics of the cultivar. There are specific tests to identify cultivar reaction to particular pathogens. For some fungal diseases, different races or pathotypes of the disease react differently when inoculated onto different cultivars.

(c) *Ploidy*: in some species, different groups of cultivars differ in ploidy (e.g. diploid and tetraploid cultivars of ryegrass – *Lolium* spp.).

Classification

The usual way to describe a cultivar using these characteristics is to compare it with other similar cultivars. To do this it is first necessary to group cultivars in conveniently sized classes.

Classification is a tool which is used to enable the easy identification of one object from a large array of similar objects. The characteristics which are used to classify objects are assembled in a hierarchical arrangement which is structured so as to give the best classification for the objects to be identified. The objective is to have each class as small as possible so as to limit the number of individual comparisons which need to be made between pairs of cultivars. The order in which the characteristics are listed is thus important in achieving the most efficient classification. Each crop therefore has its own classification hierarchy.

Using the Characteristics

The characteristics used for describing cultivars fall into two categories. Some are *qualitative*, that is they are expressed as discrete states. The number of possible states is not limited; for example, ploidy or disease reaction can be described in discrete states for each cultivar and colours often occur in discrete states in flowers or seed coats. These characteristics are least influenced by environment.

UPOV (Anon 1979) recommends that these characteristics be described by a continuous numbering system, from 1 up to however many states it is desired to allocate. Absence or presence of a particular feature would be shown as: 1 – absent and 9 – present.

The second type of characteristic is *quantitative*. These show a continuous variation from one extreme to another and may be more or less influenced by the environment in which the plants are growing. For example, height can vary between cultivars from very short to very tall. Generally these characteristics are less exact than those which are qualitative and it is necessary to arrive at a consensus as to what constitutes a clear difference. All such characteristics can be measured so that statistically significant differences can be identified; however, measurement is very time-consuming and therefore they are observed and recorded on a scale. UPOV (Anon 1979) recommends the use of a 1 to 9 scale, with 1 being the least and 9 the greatest expression of the characteristic, so giving 5 as the mid-point. On such a scale the number of points which are required to separate two cultivars is arrived at from experience, often based on previous measurements to determine a significant difference (generally at the 1% level of probability).

Qualitative characteristics are ideal for distinguishing purposes, but generally occur much less frequently than quantitative. Thus they are usually used at an early stage in a classification and final comparisons are made using quantitative characteristics.

For purposes of quality control, the characteristics are used to confirm the cultivar growing in a seed crop. This may be either by growing control plots from seed samples taken from the seed used to sow the crop or by inspecting the seed crop at appropriate times.

Control Plots

The following are the main points to be considered in setting up a system of control plots.

1. The plots should be sited on uniform land with an easily worked soil so as to ensure timely sowing. The field must be free from all volunteer plants of the same or similar species and also from weeds. The soil should not be too fertile as it is necessary to avoid excessive growth which might conceal some cultivar characteristics.

2. There should be a collection of authentic seed samples of the cultivars being grown; often it is possible to retain a portion of the first seed used during initial tests of the cultivar and to store it under conditions which will maintain viability for several years. This sample will provide a living description of the cultivar and is used to check subsequent samples obtained from the maintainer. These subsequent samples are used as a standard of comparison in the control plots. They are generally known as *Standard Samples*.

3. In addition to the Standard Samples there should be available written descriptions of the cultivars.

4. When characteristics are to be assessed by observation, the plots can be arranged systematically with all those of one cultivar together and standard samples at appropriate intervals. Any sample which differs from the rest will then be obvious.

5. If characteristics are to measured, it will be necessary to adopt an arrangement that will provide data which can be subjected to appropriate statistical analysis. A difference from the Standard Sample which is significant at the 1% level of probability is usually sufficient to reject any seed crops grown from the seed lot represented by the plot.

6. If standards of purity as to cultivar are specified in the quality control system it will be necessary to have a size of plot which will give the possibility to assess whether or not the standard has been met. As an approximate guide, the plot size should be sufficient to provide $4n$ plants, where the standard of purity is 1 in n. Thus a purity standard of 1 in 1000 would require a plot of at least 4000 plants to give a reasonably accurate assess-

ment of whether or not the standard had been maintained. A more detailed account of these calculations is given in Kelly (1988).

7. Observations should be by trained observers. The number of off-types in a plot is counted as also is the total number of plants in the plot; this enables the cultivar purity of the plot to be assessed.

Seed Crop Inspection

Seed crop or field inspection is carried out at a time when meaningful observation can be made of some of the more important characteristics of the cultivar. This is dependent on the crop species, but the most useful time for many species is at flowering time when the plants are fully grown and show some easily observed characteristics. For some other species, however, earlier or later inspection is desirable. The best time for inspection is noted for each species in Section II. Seed crop inspection enables a check to be made on the work of the seed grower but it does not substitute for the grower's responsibilities.

The following are the main points about seed crop inspection.

1. The person making the inspection must be properly trained both in the identification of cultivars and in the inspection technique. It is generally necessary to follow a specified method for making an inspection.

2. All previous details about the seed lot used to sow the crop to be inspected should be studied before making the inspection. This will include any results available from control plots.

3. The inspection can only make use of easily observed characteristics of the cultivar. Those which take a long time to determine, or which require parts of the plants to be examined in detail or measured, cannot be used because the time available for each inspection will not permit it. This may mean that the inspector's report is in negative form: 'nothing was observed which indicated that the crop was not of the cultivar stated'.

4. The inspector will be required to see as much as possible in the time available. On entering the field the first task is to check that the plants generally conform to the description of the cultivar at that stage of growth. Subsequently a predetermined route is followed and the purity of the stock is assessed by reference to a suitable sampling procedure; the number of sample areas examined in detail will be a compromise between a desire for statistically significant data (using the guide of $4n$ described above) and the need to inspect several seed crops in the time available when the crops are at the correct stage of development. Kelly (1988) gives further details of field inspection.

These quality control measures are used in both public and private seed growing systems (see, for example, the OECD Seed Schemes) but they form only part of the whole concept. Equally important is the system of generation control described earlier, so that each seed lot offered for sale can be traced back through one or more generations to the Basic Seed produced under the control of the breeder or maintainer of the cultivar.

CULTIVAR IDENTIFICATION: REVIEW OF NEW METHODS

Introduction

The ability to describe, distinguish between and identify cultivars (varieties) of agriculturally important crop species is of benefit to all sectors of the seed industry. Cultivar identification as a means of consumer protection and quality control extends throughout the entire seed trade and allied industries, from plant breeders to food consumers. For instance, many countries already have, or are introducing, schemes for plant breeders' (or variety) rights (PBR, PVR). These reward breeders financially for their efforts and offer protection for their newly bred cultivars, but in turn require that these are distinct from others and also uniform and stable in the expression of their characteristics (the so-called DUS criteria). Again, seed certification, which forms a link between cultivar registration and seed production, assures the quality of seed marketed to farmers by setting rigid standards for identity and purity. Finally, the ultimate consumers of the harvested produce need to know what they are buying, particularly if there is a premium paid for certain cultivars or if the produce is to be used for large-scale processing (e.g. mechanised bread-making, malting, oil crushing). The contribution of cultivar identification to seed quality and its importance to the different industry sectors have been reviewed (Cooke 1995a, b).

There are various approaches that can be taken to cultivar identification, ranging from morphological examination to an assessment of DNA polymorphisms (Cooke 1995b, c). The choice of method depends to a large degree on the specific identification requirements, which will vary from one situation to the next. Thus the priorities of a grain miller, who needs to be able to identify wheats suitable for processing into particular types of flour product, differ greatly from those of someone involved in cultivar registration work, where a detailed description of a new cultivar is required, along with data that will distinguish it from all others. In addition, there are differing interpretations of the concept of cultivar identification, i.e. (i) identification *in sensu stricto* – what cultivar is this? (ii) verification of identity – does this sample conform to the description of the authentic reference sample? (iii) distinctness – is this cultivar different from all others of this species? (iv) assessment of purity – is there more than one cultivar in this sample? (v) description – can data be obtained that can be used to describe cultivars and hence assist in their characterisation and classification? The requirements of these various concepts obviously differ and hence the approaches taken will inevitably differ as well. Also, different types of crops and cultivars offer varying possibilities for analysis. Thus a wide range of solutions to the question of cultivar identification can be envisaged.

Cultivar identification has traditionally been carried out by what is arguably a classical taxonomic approach, based on the observation and recording of a range of morphological characters or descriptors. In practice, this must be seen as largely successful and still forms the basis, for instance, of most current DUS testing procedures. However, it can be expensive and time-consuming, since large areas of land are needed along with highly skilled and trained personnel, making what are often subjective decisions. Many of the morphological descriptors used are continuous characters, the expression of which is altered by environmental factors. Hence replication is needed, which adds to the time and cost of the operation. Again, in some species the number of descriptors is limited or is not sufficient for discrimination between an increasing number of cultivars. There are thus good grounds for finding alternative procedures to augment this morphologically based approach. The phenotype of a cultivar results from the interaction between its genotype and the environment. In general, it is an increasingly desirable objective to reduce or eliminate the environmental influence, thus enabling the genotype of a cultivar to be observed more directly.

The rest of this section is concerned with new methods which are being applied to cultivar identification. Such techniques are of three types: (i) the use of computerised image analysis systems to capture and process morphological information; (ii) the use of biochemical methods to analyse components of plants, primarily proteins and enzymes; (iii) the use of various DNA profiling techniques. Rather than reviewing currently available methods comprehensively, which has been done previously (Cooke 1988, 1995c, d; Smith and Smith 1992; Morell et al. 1995; Wrigley 1995), attention is focused on new developments and possibilities.

Cultivar Identification using Image Analysis

Conventional cultivar identification relies to a large extent on the observation and recording of a multitude of morphological characters. This can

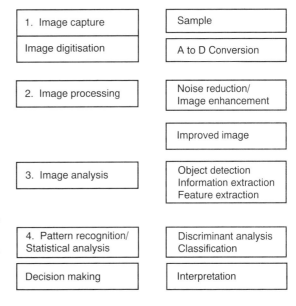

FIGURE 3.2 Summary of the features of a typical image analysis system that could be used for applications in cultivar and seed identification

be a subjective process which is tiring and tedious for skilled operatives to perform and by its nature is imprecise. In principle, an objective, quantitative and automated method for the routine measurement of these same characters would be highly advantageous. The use of computerised methods of image analysis (IA) – machine, robot or computer vision – is now enabling such systems to be developed. IA techniques have a number of other attractions, including speed, the potential to measure new characters and the electronic storage and transmission of data. Also, IA is largely non-destructive and can be used 'remotely', in that photographs, negatives, photocopies or other representations can be analysed as readily as living biological specimens.

Given these features, it is not surprising that a great deal of interest has been expressed in the use of IA for cultivar identification. The basic features of a typical image analysis system that might be used in cultivar and seeds work are shown in Figure 3.2 (see also Cooke 1995b, c) and have been well described previously (Sapirstein 1995). Briefly, the features comprise the following.

(i) *Image capture and digitisation* – the samples to be measured are placed in front of an image acquisition device (TV or video camera). The output from the camera is transmitted to a computer, which must contain an analogue to digital (A to D) converter. This converts the signal into a numerical form, which can then be stored. Images are divided into a grid of pixels, each of which is given a number indicating the intensity of the image in that particular square.

(ii) *Image processing* – the stored image is modified by changing the pixel intensity numbers. Many modifications can be made in this way, including enhancing the contrast of the image, removing extraneous background features or noise, or filling in gaps or holes.

(iii) *Image analysis* – extraction of information from the processed image.

(iv) *Pattern recognition* – the development of statistical and other tools to sort and compare images.

(v) *Decision-making* – the development of computer software to interpret the data produced from the above operations.

Using systems based on these features, various authors have reported attempts to discriminate between and identify cultivars. Much of this work has been directed towards the classification of cereal grains, primarily wheat, in the quality control context, because of the economic potential that a successful machine in this area would have. The results have been mixed, but generally encouraging, bearing in mind that this is a relatively young science with only about 10 years of experience to its name. Early work was based mainly on the statistical analysis of grain/seed shape and size features and successful discrimination between cultivars in different commercial grades varied from 15 to 96% (Sapirstein 1995). The use of other descriptors of grain shape, including colour images and textual features, and other pattern recognition systems has improved the success rates (e.g. Barker et al. 1992).

The work most directly related to wheat cultivar identification *per se* is probably that of Keefe (1992), who reported the development and use of a 'dedicated wheat grain analyser'. This utilised a charge-coupled device (CCD) camera to view individual wheat seeds. To make 33 separate measurements from 50 individual seeds and process the resultant data took five minutes. From these measurements, 69 shape parameters were derived, from which an identity could be assigned to an unknown sample. Although this approach was generally successful, problems became apparent because of the genetic relatedness of cultivars and the heterogeneity of seed characters within cultivars, which made it impossible to assign any given seed unequivocally. Others have reported similar problems (Sapirstein 1995).

This places a limitation on IA technology and it is thus possible that IA will be most useful with regard to cultivar taxonomy and characterisation rather than identification in the strict sense. In particular, the potential for providing better and more objective descriptions and new characters for DUS testing and certification is considerable. An interesting example is provided by Draper and Keefe (1989) who investigated onion bulb shapes. IA was performed on 35 mm photographic negatives of bulbs. A model for bulb shape was derived which treated the bulb as a 16-sided polygon. Points were defined on the bulb circumference and selected empirically to give the best visual fit to the recorded image. This work demonstrated clearly that not only was IA capable of reproducing measurements already utilised in registration work in an objective and cost-effective fashion, but it could also generate novel characters which were useful for improving discrimination. Similar conclusions have been drawn by other workers, for instance van de Vooren and van der Heijden (1993) who used IA to measure pod size and shape in *Phaseolus* beans. More recent work in this area has begun addressing rather more complex problems. For example, work at NIAB (National Institute of Agricultural Botany) (Warren 1997) is developing a system for the recognition and description of leaf shape in chrysanthemum cultivars. Leaf images are captured using a desk-top document scanner and are then analysed in order to record automatically characteristics that are currently recorded manually for DUS testing purposes. Again, the system will enable potentially new aspects of shape to be utilised and the data to be stored and searched electronically. Other groups are looking at the applications of colour imaging and novel approaches to pattern recognition, such as adaptive neural networks (Chtioui et al. 1996). Given the improvements in imaging technology and software, as well as in affordable computer processing power, it is likely that further applications of IA in cultivar description can be anticipated in the near future.

Cultivar Identification using Biochemical Techniques

Biochemical methods have several attractions as tools for taxonomy, including the reduction of environmental interaction. Chemotaxonomists generally recognise two groups of useful compounds: (i) episemantic, or secondary compounds (e.g. anthocyanins and other flavonoids, fatty acids), and (ii) semantides, or sense-carrying compounds (proteins, nucleic acids).

A good example of the analysis of episemantic compounds in relation to cultivar testing is the phenol test in wheat, which assesses tyrosinase activity in seeds by a simple colour reaction and is used both for seed certification and DUS testing purposes, primarily for uniformity assessment and verification of identity. Another colour test, the vanillin test which detects tannins in field beans, can be used for DUS and classification purposes. Further examples of such rapid and simple tests are given in Table 3.8 and in van der Burg and van Zwol (1991). There is also a handbook published by the International Seed Testing Association (Payne 1993) on *Rapid Chemical Identification Techniques*.

The analysis of colour as a means of cultivar discrimination is well established in ornamental and other species and various attempts have been made to adopt an instrumental approach to this

TABLE 3.8 Some of the simple and rapid tests that can be used for species and cultivar identification

Type of test	Species
Phenol	Rice, oats, barley, wheats, ryegrass, *Poa pratensis*, *Vicia striata*, triticale
Fluorescence +/– chemical treatment	Ryegrass (roots), oats (seed), peas (seed), soyabean (roots), fescue (roots), mustard (seeds)
Chromosome counting (root tips)	Sugar beet, ryegrass spp., clovers
NaOH/KOH	Wheats, sorghum, rice, red rice
$CuSO_4$-NH_3	Sweet clovers
HCl	Oats, barley (pearled), mustard
Iodine	Wheat, lupins
Vanillin	Beans, barley

problem, using spectrocolorimeters. This would provide an absolute quantitative evaluation of colour and hence remove the need for colour charts and comparisons (Biolley and Jay 1993). Various chromatographic methods, primarily gas liquid chromatography (GLC) and high performance liquid chromatography (HPLC), have also been used to analyse compounds such as anthocyanins and hence distinguish between cultivars, particularly of flowering and ornamental species (Cooke 1995c). Fatty acids, glucosinolates and other compounds in oilseed crops have likewise been shown to be indicators of identity (e.g. Mailer et al. 1993). However, the analysis of semantides has been much more widely undertaken and is of particular value for cultivars and seeds.

Protein Analysis and Cultivar Identification

Proteins are direct gene products and can be used as markers for the genes that encode them. The close relationship between protein synthesis and the primary genetic information (DNA) reduces, or even eliminates, any environmental effects. Thus a comparison of protein compositions in different organisms can be considered to be a comparison of the underlying genetic variation. Hence those analytical methods which reveal and compare protein compositions are ideally suited for differentiating between cultivars, which are collections of genetic material with differing expressions.

It should thus not be surprising that the two most common separation methods in biochemistry – HPLC and gel electrophoresis – have been used extensively for identification work. The precise way in which such techniques are utilised varies according to the species in question. Two main approaches have been recognised, known as: (i) the direct (multilocus) approach, in which polymorphic proteins that are genetically encoded at multiple loci (e.g. cereal seed storage proteins) are analysed; and (ii) the indirect (single locus) approach, involving the examination of proteins which, although polymorphic, are derived from a single locus (isozymes) (Cooke 1995c, d). The optimal way in which HPLC or electrophoresis can be used for identification purposes is determined by the mode of reproduction (and hence the genetic structure) of the cultivar in question. For instance, self-pollinating crops and those which are vegetatively or clonally propagated or reproduce asexually are usually best analysed using a multilocus approach, whilst it is normally more productive to utilise single locus analysis when dealing with cross-pollinating species.

HPLC analysis of proteins for cultivar discrimination and identification purposes has been widely used, especially for cereal species such as wheat, barley, rice, maize and oats, and has been well reviewed previously (Cooke 1995c; Bietz and Huebner 1995). The profiles produced generally enable cultivars to be distinguished using both qualitative (absence/presence of peaks at certain places) and quantitative (height/area of specific peaks) differences. Again, HPLC separations are usually rapid (completed in an hour or less), the process can be automated to a substantial degree, and the data produced are in a format ideally suited for electronic storage and/or manipulation.

However, despite these considerable advantages, HPLC has not found substantial use outside the research laboratory. Perhaps the main cause of this is the poor stability of the columns most widely used, which inevitably leads to a lack of reproducibility. This has severe consequences for identification work, especially where computerised pattern-matching procedures are employed, and could lead to incorrect identification decisions being made. The conventional bonded-phase columns most commonly employed for reversed phase HPLC of proteins are unstable over relatively short periods of time when used at low pH in the presence of strong acids, which are precisely the conditions utilised in most protein separations. Although new silane-modified silica columns, reputedly far more stable, have been shown to be suitable for separation of wheat seed proteins, it remains to be seen whether this will lead to a resurgence of interest in HPLC for cultivar identification.

Thus whilst HPLC in principle is extremely useful for identification of crop cultivars, by far the most successful techniques, which have made the largest impact in this area, are the various types of gel electrophoresis.

Electrophoresis and Cultivar Identification

The use of gel electrophoresis to analyse seed and seedling proteins and/or isozymes and subsequently distinguish between and identify cultivars is now well established and has been reviewed on several occasions (Cooke 1988, 1995d; Smith and Smith 1992; Lookhart and Wrigley 1995). In general, electrophoresis methods are easy to perform and do not require particularly extensive laboratory facilities, making them attractive for more routine applications, as well as for research.

Much of our knowledge of the uses of electrophoresis for cultivar identification has come from studies of self-pollinated and vegetatively propagated species. As this includes many of the most important crops in world agriculture, such as rice, wheat, barley and potatoes, this is perhaps not so surprising. Generally, a direct comparative approach has been taken, using a range of electrophoretic techniques to analyse storage proteins. As an example, Cooke (1988) listed 27 published methods for wheat cultivar identification using electrophoresis and 16 for barley. Various organisations have sought to rationalise this situation by adopting standard methods for the more important crops. Prominent amongst these have been the International Association of Cereal Science and Technology (ICC) and the International Seed Testing Association (ISTA), and both have standard reference methods for the identification of wheat cultivars by electrophoresis. The methods are similar, involving the use of polyacrylamide gel electrophoresis (PAGE) at acid pH to separate the alcohol-soluble wheat seed storage proteins known as gliadins (Figure 3.3). ISTA also has a variation of this method which can be used for the verification of cultivar identity in barley (White and Cooke 1992), and has been active in promoting the adoption of standard reference methods for other crops. Details can be found in the ISTA handbook of *Electrophoresis Testing* (Cooke 1992) and in Lookhart and Wrigley (1995).

These methods can also be used, with modifications, in other cereals such as oats, rice, maize, triticale and sorghum. The criterion for difference between cultivars is usually taken as the presence or absence of particular protein bands at specific places on the gel (see Figure 3.3). Impressive levels of discrimination can be achieved. For example, Cooke (1995e) reported that in a collection of 706 barley cultivars from 15 different countries, 105 distinct groups could be recognised on the basis of the storage protein (hordein) allele composition.

Other electrophoresis methods have also proved to be extremely useful for identification of cultivars of self-pollinating crops. PAGE in the presence of sodium dodecyl sulphate (SDS-PAGE) is particularly suitable for legume species such as peas and beans, but has also been widely applied to the analysis of hordeins (Figure 3.4) and the glutenin proteins of wheat (Cooke 1995f). Isoelectric focusing (IEF) of seed and other proteins can be useful in some instances (Cooke 1988, 1995d; Smith and Smith 1992). In many of these cases, catalogues of protein profiles have been produced by various authors to facilitate identification and discrimination between cultivars.

Cultivars of self-pollinating crops usually consist of a single homozygous line and hence are uniform, both phenotypically and genotypically.

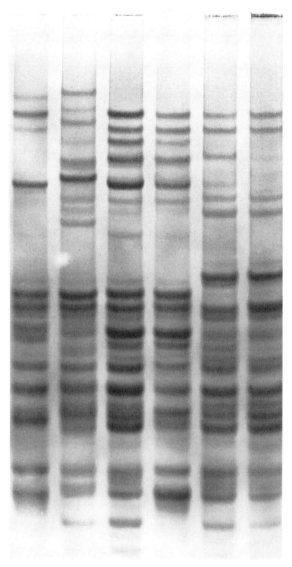

FIGURE 3.3 The use of acid PAGE of gliadins to distinguish between cultivars of wheat. Each track represents the gliadins extracted from a single seed of a different wheat cultivar, analysed according to the ISTA standard reference method. Note the polymorphism of gliadins and that different cultivars have characteristic profiles (unpublished data of Dida and Cooke)

However, it is usually the case that a proportion of cultivars consists of two or more electrophoretically distinguishable lines, sometimes called biotypes (Cooke 1988, 1995e, f). This is a

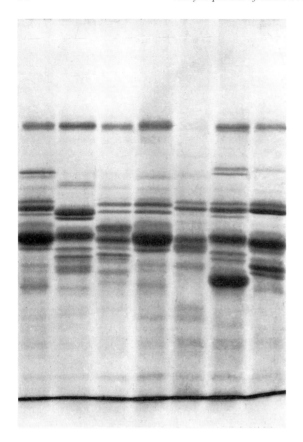

FIGURE 3.4 The use of SDS-PAGE of hordeins to distinguish between cultivars of barley. Each track represents the hordeins extracted from a single seed of a different barley cultivar. Note the polymorphism of hordeins and that different cultivars have characteristic profiles (unpublished data of White and Cooke)

Braunschweig. Over 15 000 potato cultivars, botanical varieties, species, subspecies and wild types have now been examined by electrophoresis (Huaman and Stegemann 1989). The most generally applied method is PAGE at pH 7.9 or 8.9 of the soluble tuber proteins or esterases expressed from potato sap. This method allows almost all of the cultivars examined to be distinguished from one another. Similar techniques can be successfully used for other root crops such as sweet potato (Stegemann et al. 1992).

Rather less extensive work has been carried out on cross-pollinating crops, largely because these species can present special problems from the electrophoretic identification point of view. Cultivars of such species are populations with individual plants expressing a range of phenotypic characters and containing different combinations of homozygous and heterozygous genes, including those encoding proteins and enzymes. This variation is maintained in equilibrium over generations. Two ways of tackling the problem of biochemical cultivar identification in these species have been devised: (i) analysis of a bulked or pooled extract of seeds or plants, to obtain an overall cultivar protein profile – a direct comparative approach, examining seed storage proteins for instance, can then be taken; (ii) analysis of single locus proteins or isozymes from individual seeds or plants from a cultivar, and determination of the variability within and between cultivars statistically.

Both approaches have been successfully used in cross-pollinating species. For instance in ryegrass, analysis by SDS-PAGE of the seed storage protein composition of bulked (approximately 2 g) seed samples of cultivars provides a high level of discrimination between species and cultivars (Cooke 1995d; Gilliland 1989). ISTA (Cooke 1992) has adopted SDS-PAGE as a standard reference method for verifying the identity of commercial seed lots of ryegrass cultivars and species. Most studies of single locus proteins in ryegrass have used starch gel electrophoresis to analyse isozymes from young leaf enzymes, with phosphoglucoisomerase (PGI) being particularly well researched (Nielsen 1985; Greneche et al. 1991). Extracts of leaves from 50–100 individual

consequence of the lack of selection for protein homogeneity and, whilst it does not detract from the utility of electrophoresis for identification purposes, it is important for investigators to analyse individual seeds or plants and hence be aware of any non-uniformity.

Several vegetatively propagated or asexually reproducing crops have been successfully investigated electrophoretically, including bluegrass, bananas, roses, strawberries, pears and other fruits (Cooke 1988, 1995d; Gilliland 1989). Undoubtedly, though, the most thoroughly researched is the potato and a major part of this work has been carried out by the group in

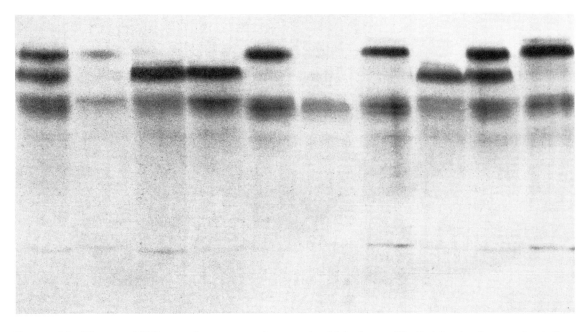

FIGURE 3.5   The use of IEF to analyse esterases from seeds of faba beans. The first five tracks are extracts from single seeds of the cultivar Troy and the second five tracks are from the cultivar Banner. To distinguish between cultivars, the frequency of occurrence of each enzyme phenotype within each cultivar would be measured from a sample of 50–100 seeds. The frequencies would then be compared for significance using a suitable statistical test (unpublished data of Goodrich and Cooke)

plants are taken and subjected to electrophoresis, and the gels are then stained for PGI activity. The frequency of occurrence of each isozyme phenotype (which can be directly related to genotype) in a cultivar is determined and these frequencies in different cultivars are then compared using a chi-squared or other suitable analysis.

Thus cultivars are discriminated from one another on the basis of differences in the frequency of occurrence of specific alleles. This method, even when using only a single enzyme, can be very powerful. For example, 70% of a collection of 149 perennial and Italian ryegrass varieties could be distinguished ($P<0.05$) on the basis of PGI allele frequencies (Gilliland 1989). By combining the data from more isozymes and varying the conditions of analysis, greater levels of discrimination are achievable (Greneche et al. 1991; Loos and Degenaars 1993). Again, catalogues of the allelic compositions of collections of cultivars have been published as aids to identification.

Several other cross-pollinating species have been investigated using either the bulked approach or the analysis of individuals, and various electrophoretic techniques have been employed (Figure 3.5). Thus PAGE, SDS-PAGE and/or IEF of a range of proteins and enzymes have been successfully used to identify cultivars of crops such as rye, Faba beans, oilseed rape and vegetable brassicas, sugar beet, alfalfa, fescue, timothy, lovegrass and various horticultural species, vegetables and fruits (see, for example, Cooke 1988, 1995c; Smith and Smith 1992; Konishi 1995).

There can thus be no doubt that the analysis of proteins and enzymes by gel electrophoresis is a very successful and widely used approach to the question of cultivar identification in many crop species. Electrophoresis methods reveal variability at a number of loci encoding storage and other proteins. If this variability exists, then it follows that there must also be variation in the genetic material (DNA) itself. In fact it is likely that there will be more DNA variability, since not all differences occur in expressed regions of DNA. Thus the analysis of DNA polymorphisms would provide an extremely powerful tool for cultivar identification. Relatively recent advances in molecular biology make such analyses a possibility.

Cultivar Identification using DNA Profiling Techniques

Detailed consideration of the various types of DNA profiling methods available is beyond the scope of this section and is available elsewhere (Fowler and Kijas 1994; Smith 1995; Morell et al. 1995; Staub et al. 1996). However, it is useful to consider the different approaches briefly, since they each have features of interest for identification work. Profiling methods can be divided into essentially two types: probe-based technologies and amplification technologies.

Probe-based DNA Profiling

Restriction fragment length polymorphisms (RFLPs) were the first profiling technique to be widely used for revealing DNA sequence variations in a diverse range of organisms, including cultivars (Ainsworth and Sharp 1989). RFLP analysis generally utilises single-, low- or multicopy genomic of cDNA clones as probes (see Figure 3.6) following cutting (restriction) of the target DNA with a specific enzyme. This is a very effective way of revealing differences between cultivars, and the large number of restriction enzyme/probe combinations available make this a powerful approach. A no doubt incomplete survey of the literature published between 1993 and early 1997 (Cooke, unpublished) has revealed reports of RFLPs in over 50 agricultural and horticultural species, ranging from the major crops, such as wheat, barley, rice and maize, to fruits (apples, grapes), vegetables (brassicas, cucumber) and trees (oil palm, willow).

RFLPs can be a very discriminating means of profiling. For instance, Smith and Smith (1991)

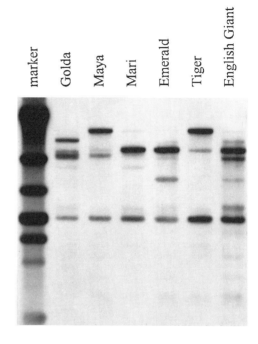

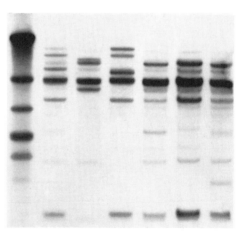

FIGURE 3.6   An example of RFLP analysis in cultivars of oilseed rape. DNA was extracted from leaves of the cultivars and restricted using the enzyme *Eco*RV. The fragments were separated by electrophoresis on an agarose gel and transferred by blotting onto a nylon membrane filter. The filter was probed either with the low-copy-number genomic clone pN107 (top) or the higher-copy-number clone pN180 (bottom), both of which were radioactively labelled and then subjected to autoradiography. Note the differing polymorphisms revealed by the use of the different probes. These RFLPs could clearly be used to distinguish between the cultivars (unpublished data of Lee, Reeves and Cooke)

reported that an array of enzyme/probe combinations readily distinguished between 78 US maize hybrids and inbred lines. Again, Lee et al (1996a) showed that by using a single restriction and two multicopy probes, 99.2% of the pair-wise comparisons within a set of 62 *Brassica napus* cultivars were distinguishable. Gorg et al. (1992) reported that they were able to identify uniquely 122 out of 134 potato cultivars using a marker known as GP-35 as a probe.

Some of the probes used for DNA 'fingerprinting' in humans and other animals can also reveal useful polymorphisms in plants and have been used for identification purposes. Thus the M13 repeat probe has been shown by Nybom et al. (1990) to distinguish between cultivars of species such as apples, blackberries and raspberries. Other workers have used synthetic repetitive oligonucleotides as probes. The (GATA)-type probes have proven to be particularly effective and appear to be applicable to a range of plant species, including oilseed rape (Cooke 1995c; Lee et al. 1996a).

There are some disadvantages to RFLPs, primarily the need for relatively large amounts of good quality DNA and the fact that most probe-based work has used radioactively (^{32}P) labelled probes followed by autoradiography to reveal the DNA profiles. This widespread use of radioactivity undoubtedly places a limitation on the more routine use of RFLPs in many cultivar identification situations. Although there are various fluorescent and chemiluminescent labels that can be used as alternatives, none of these seems to be as reliable or as sensitive as ^{32}P.

Amplification-based DNA Profiling

Technologies based on the amplification of DNA sequences via the polymerase chain reaction (PCR) offer some advantages over RFLPs for cultivar identification. PCR-based methods are quicker, need relatively little target DNA, avoid the need for the use of radioactivity and can be automated. The reports (Welsh and McClelland 1990; Williams et al. 1990) that arbitrarily chosen oligonucleotide primers can act as templates for

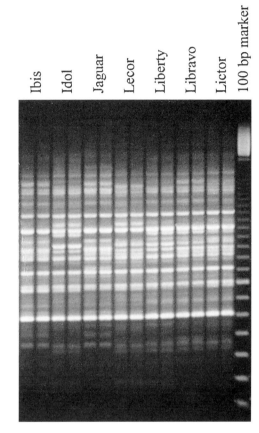

FIGURE 3.7 An example of RAPD analysis in cultivars of oilseed rape. DNA was extracted from leaves of oilseed rape cultivars and subjected to PCR using the arbitrary primer 33 (CGGTAGCCGC). The products were separated by electrophoresis on an agarose gel and stained with ethidium bromide. Note the polymorphisms between the cultivars (loaded in pairs) which could clearly be used to distinguish between them (data of Lee, Reeves and Cooke)

the amplification of several fragments of genomic DNA have led to a new generation of techniques (Caetano-Anolles 1996), the most widely used of which is random amplified polymorphic DNA (RAPD). The size of the amplified fragments depends on the primer sequence used in the PCR and on the DNA being analysed. Different primers give rise to different amplified bands, and polymorphisms at the priming sites result in the disappearance of an amplified band (Figure 3.7). Thus RAPDs will detect polymorphisms distributed throughout the genome, with a primer amplifying several bands, each of which probably originates from a different locus.

Although a comparatively newly reported technique, there has been huge interest in the use of RAPDs for cultivar discrimination and identification and there are now reports of the use of RAPDs in well over 75 different plant species. These include all of the widely grown cereal, oilseed and root crops, but also several fruits, vegetables and ornamental species. The facts that RAPDs can be utilised with no prior knowledge of the DNA sequence of the species of interest and that primers are readily available commercially have no doubt been of significance. RAPD profiles can also be readily generated from dry seeds (e.g. Jianhua et al. 1996). There is no doubt that RAPDs can be very discriminating. Lee et al. (1996b), for example, could readily discriminate over 95% of a collection of 50 oilseed rape cultivars with one primer, and a combination of two or more primers achieved complete discrimination. However, as more work utilising RAPDs has been published, it has become apparent that the technique suffers somewhat from a lack of robustness, manifested as difficulties in reproducing equivalent data in different laboratories (see, for example, Fowler and Kijas 1994; Lee et al. 1996b; Morell et al. 1995). The varying intensities of the amplified bands (see Figure 3.7) can also cause problems in interpretation. Although these issues can be addressed (Lee et al. 1996b) and standardisation has been suggested (Lowe et al. 1996), it seems unlikely that RAPDs will be the profiling method of choice for most cultivar identification applications.

'Second Generation' Profiling Techniques

Although there is no doubt that RFLPs and RAPDs can be used to distinguish between cultivars, these DNA profiling techniques are now recognised as having limitations that might restrict their more general use. Consequently, there is increasing interest in the development and utilisation of so-called second generation methods, which retain the discriminating power

of RFLPs and RAPDs but which are less difficult technically, more robust, more amenable to automation, transportable between laboratories and species, employ publicly available materials and do not involve the use of radioactivity. Of these, other techniques using PCR are of particular interest. The analysis of simple sequence repeats (microsatellites) may prove to be especially useful for cultivar identification applications. Microsatellites are tandemly repeated DNA sequences, usually with a repeat unit of two to four base pairs (e.g. GA, CTT and GATA). The polymorphism found in microsatellites is due to variations in the number of the basic repeat units. In many crop species, multiple alleles have been shown to exist for many microsatellites, arising from these differences in copy number (Morgante and Olivieri 1993). These alleles can be separated by agarose or polyacrylamide gel electrophoresis (Figure 3.8).

The PCR-based analysis of microsatellites is known as the sequence-tagged site approach. In order to develop a sequence-tagged site microsatellite (STMS) system for cultivar identification, information about the sequence of the DNA flanking the microsatellite is needed. This information can sometimes be obtained from DNA sequence databases such as EMBL and GenBank, because of the existence of serendipitously sequenced microsatellites arising from other research. Otherwise, the sequences have to be determined experimentally. The tremendous potential offered by STMS for plant breeding and cultivar-related work in general has fuelled an upsurge in research activity in this area. As a result there are now microsatellites available in several species, including wheat, barley, maize, oilseed rape and other brassicas, soyabean, sugar beet, sweet potato, grapes, tomatoes, yams, citrus fruits and some ornamentals (see Morell et al. 1995; Smith et al. 1997).

The fact that microsatellites are often multi-allelic makes them very attractive for cultivar identification. A further significant advantage of STMS markers is that they can be multiplexed, i.e. a number of markers (loci) can be evaluated in a single PCR. In addition to improving efficiency,

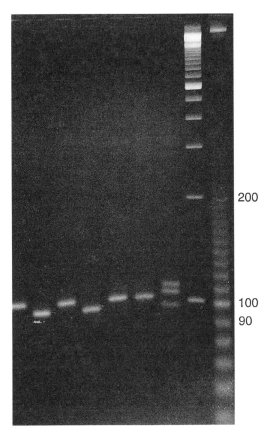

200

100
90

FIGURE 3.8 A simple example of STMS analysis in seven maize inbred lines. DNA was extracted from maize leaves and used for PCR using the microsatellite primer pair D01299 and 300. The products were analysed on an agarose gel and stained with ethidium bromide. Different lines have different alleles (bands of different size) of this particular microsatellite and hence can be distinguished. Higher levels of discrimination between alleles (and hence cultivars) can be obtained by using other electrophoresis and detection systems such as denaturing acrylamide gels and silver staining (photograph kindly supplied by Dr Stephen Smith, Pioneer Hi-bred International, USA)

multiplexing is very suitable for automation, which both enhances the data-gathering process and improves cost-effectiveness. Automated and more rapid methods of STMS analysis, based on the fluorescent labelling of microsatellites and separation of multiplexed products using DNA sequencers or capillary electrophoresis, are now becoming more widespread (e.g. Frazier et al.

1996; Wang et al. 1996) and are being increasingly used for analysis of crop species. An excellent example is provided by the work of Thomas and co-workers on grapevines (Thomas et al. 1994; Botta et al. 1995), who have clearly demonstrated the potential of such an approach for database production, automated profiling and computer-based cultivar identification.

Another second generation technique that promises to be extremely useful for cultivar identification is the analysis of amplified fragment length polymorphisms (AFLP) (Vos et al. 1995). AFLP analysis involves the selective amplification of DNA restriction fragments by PCR. Profiles are produced with no prior sequence information using a limited set of primers. The number and nature of fragments amplified are altered by the choice of primer. The technique has the advantage of sampling many loci simultaneously (Figure 3.9), although this can complicate data collection and it is probable that computerised systems will be needed to derive the maximum benefit from the technology. The same fluorescent labelling and automated analysis techniques can be used for AFLP as for STMS. In addition, AFLP analysis is more robust than other random priming techniques such as RAPDs, since stringent conditions are used in the PCR.

Although protected by a patent and only available for use under licence, AFLP technology is now being sold in kit form, and hence can be utilised more generally. Relatively few direct reports of AFLP analysis for cultivar identification so far exist, although the indications from related work on diversity and phylogeny in wheat, maize, beans, lentils, tomatoes and other species (Vos et al. 1995; Sharma et al. 1996; Tohme et al. 1996; Donini et al. 1997) are very promising.

Whilst these are rather early days for a proper evaluation of the use of DNA profiling for cultivar identification, enough is already clearly known to indicate the tremendous potential of these approaches. Substantial effort is now being made in many laboratories, particularly with regard to the second generation techniques (STMS and AFLP). What is now required is a systematic study of the various technologies

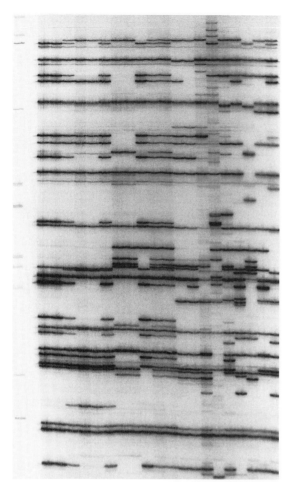

FIGURE 3.9 AFLP analysis in 20 inbred lines of maize. DNA was extracted from maize leaves and subjected to the AFLP analysis procedure (Vos et al. 1995). In this case the primers were radioactively labelled and the fragments separated on a denaturing polyacrimalide sequencing gel followed by autoradiography. This procedure reveals a number of polymorphisms between the lines which could thus be used to distinguish between them (photograph kindly supplied by Dr Stephen Smith, Pioneer Hi-bred International, USA)

available, to compare the degrees of discrimination achievable within much wider collections of cultivars, to obtain data on the question of intra-cultivar polymorphisms (cf. biotypes in protein electrophoresis; see Lee et al. 1996b) and to produce simple yet robust analytical procedures.

Practical Applications of the New Methods

Perhaps unsurprisingly, the most widely applied of the foregoing new methods in seed production and the seed trade generally has been gel electrophoresis of proteins. All of the different kinds of question posed in the introduction to this section regarding cultivar identity have been addressed by the application of electrophoresis. The issues of identity and purity are important commercially, for instance in industries that require good quality cultivars for processing (flour milling, bread baking and malting, oilseed processing). Increasingly, companies in these industries operate quality control procedures based on electrophoresis. A particularly important use of electrophoresis in the seed industry is in the measurement of hybrid purity (Cooke 1988). The use of a suitable electrophoresis technique allows the purity of seed stocks to be determined rapidly and without recourse to field growing-out procedures. Protein electrophoresis has also been used for the documentation of genetic resources and for cultivar registration in certain crops (Cooke 1995b).

It is reasonable to suppose that the other new methods will find similar uses. Already, the use of HPLC for hybrid purity determination in wheat has been investigated (McCarthy et al. 1990) and the image analysis-based examination of F1 hybrid populations of brassicas reported (Fitzgerald et al. 1997). Simple biochemical methods of flavonol analysis have been used to distinguish related species of *Lotus*, which is useful for seed production in these species (Kade et al. 1997). The use of various DNA profiling methods to assess hybrid purity has been reported in brassicas, tomatoes, cucumbers, peppers and water-melons (Cooke 1995b, c) and further developments in other crops, perhaps especially with regard to the use of STMS markers, can be anticipated. The identification of malting barley cultivars by DNA analysis has recently been reported (Tsuchiya et al. 1995) and the uses of PCR-based analyses to address problems encountered in seed production discussed (Poulsen et al. 1996). Hence there seems to be little doubt that the future will bring many more practical applications of DNA profiling technologies.

Future Prospects for the Applications of Molecular Methods

Since the advent of molecular technologies there have been considerable developments, both of the techniques themselves and of their application to seed and cultivar work. Their power and utility for a multiplicity of uses are becoming widely recognised by a diverse group of end-users. For example, the European Commission has been funding a pan-European generic project, involving 35 research groups, examining molecular screening tools for use in measuring biodiversity in a range of organisms, including horticultural and agricultural crops. This research effort has recently been extended and output from this consortium has acted as a stimulus for the formation of the Biotechnology for Biodiversity Platform (BBP). This has the objective of providing an extended audience for the results and experience from the generic project to make sure that maximum benefit is obtained from the research. The BBP has established informal links with both UPOV (the International Union for the Protection of New Varieties of Plants) and ISTA, both of which clearly have considerable interests in new techniques for cultivar registration and identification.

UPOV has the role of facilitating the uniform and efficient protection of new cultivars in its member states. As a consequence of this, UPOV maintains a continuing interest in the scientific and technical developments which underpin its mandate and has established a Working Group on Biochemical and Molecular Techniques and DNA Profiling in Particular (BMT) to examine progress in this general field. This group is considering the applicability of DNA profiling for use in cultivar registration procedures but must also be aware of the potential consequences of the adoption of such techniques for plant breeders, both in terms of the implications for the breeding process and of the quality of the protection offered within PBR

schemes. PBR is a form of intellectual property protection which is mediated in the UPOV system by DUS testing of a newly bred cultivar. As mentioned previously, in many species the amount of variability available in morphological characteristics is limited and distinctness of new cultivars can be difficult to establish. In addition, the expression of these characteristics interacts with environmental conditions leading to a lack of consistency across testing sites and years. Both of these problems could be minimised by the use of DNA profiling methods.

However, UPOV also has to be conscious that there could be disadvantages from the use of DNA profiling techniques to establish DUS. If the discriminatory power of the testing procedures were to be excessively increased, it could lead to an erosion of what is known as the 'minimum distance' between cultivars, thereby devaluing the quality of intellectual property protection. In addition, new cultivars could be required to demonstrate sufficient levels of uniformity for a particular profile, which could present problems for breeders. There are ways in which these disadvantages can be handled (Lee et al. 1996c) and discussions within the BMT are underway to establish how widely these and other approaches can be realistically applied. It is possible that an integrated approach (Mudzana et al. 1995) employing morphological characters, perhaps measured by IA, in conjunction with biochemical and molecular techniques represents a sensible way forward.

A related issue in which DNA profiling is expected to play a significant role is that of the determination of essential derivation. Both UPOV and plant breeders have an interest in this concept although UPOV considers breeders to have the primary responsibility. An essentially derived cultivar has been defined as one which is: (i) predominantly derived from the initial cultivar or from a cultivar that is itself predominantly derived from the initial cultivar, while retaining the expression of the essential characteristics that result from the genotype or combination of genotypes; (ii) clearly distinguishable from the initial cultivar; (iii) except for the differences which result from the act of derivation, conforms to the cultivar in the expression of the essential characteristics that result from the genotypes or combination of genotypes of the initial cultivar. DNA profiling techniques offer perhaps the only means by which the problem of establishing essential derivation can be addressed and hence will inevitably be the methods of choice. However, there remains much discussion as to which of the profiling technologies will be the most appropriate and how essential derivation will be defined in molecular terms.

There are other applications beyond registration and its associated issues where cultivar identification is important and consequently where new identification technologies can have an impact. These include, for example, seed production, seed certification, seed trading, crop production, grain and produce trading and processing, and consumer protection through authentication and quality management systems. ISTA is concerned with many of these and has a Variety Committee which has added investigations of DNA profiling technologies to its deliberations concerning new methods for the verification of cultivar identity.

It is difficult to accept that the widespread interest in profiling techniques will not result in their ultimate application in both cultivar registration, essential derivation and cultivar identification. Molecular markers of various types are already in use by many plant breeders and are the subject of considerable research activity. Even so, there remain difficulties in their wider adoption. Apart from the problems identified by UPOV, other disincentives exist. Technology is developing very rapidly and there is no clear consensus as to which particular DNA profiling method is most appropriate for a given purpose. In addition, the capital investment required to establish a technology and then to achieve sufficient sample throughput for routine use can be rather high. Data analysis methods are not yet established or agreed internationally. Nevertheless, the pace of development is rapid and these difficulties will be overcome. Also, as already mentioned, methods of analysis that do not require the use of gels are becoming available, and both capital and unit costs will decline. As technology develops and access to it increases, the pressure for the use of new techniques in many areas of seed and cultivar technology and testing will become increasingly difficult to resist.

VIABILITY AND PHYSICAL PURITY

In so far as the individual seed is concerned, acceptable quality is the ability to germinate rapidly once sown in a seed bed and to produce a seedling which is capable of emerging rapidly through the surface of the seed bed and establishing a healthy plant which will then contribute effectively to yield. This ideal is not met directly by commercial seed quality testing practices, however. First, seeds are tested, purchased and sold by the lot (i.e. as populations of seeds), not as individuals, the seed lot being a homogeneous population which has been treated identically throughout its history (i.e. sown from a single parent stock, plural in the case of hybrids only) in a single field or location, harvested at the same time and subsequently dried and stored in a single common environment. Second, descriptions of seed lot quality are based on samples drawn at random from the lot. Third, the seed trade is international and so must rely on tests which provide results that are both repeatable (i.e. in time at one site) and reproducible (i.e. at other sites, including those in different countries). Since seed bed environments themselves are neither repeatable nor reproducible, emergence ability from seed beds cannot be tested directly in commercial seed testing. For this reason emergence ability is generally considered under two principal separate but related topics: viability and vigour.

Viability is simply the proportion of seeds in the lot which is alive. In seed testing the principal method of assessing viability is to determine whether or not the seed is able to germinate. There are two caveats to this. First, the definition of germination in seed testing is more stringent than simply germination *sensu strictu*; in seed testing germination, described as 'normal germination', is the emergence from the seed and subsequent development of those essential structures which indicate whether or not the resultant seedling would be able to develop further into a satisfactory plant under favourable conditions in soil (ISTA 1996a).

The second caveat concerns seed dormancy, specifically the need to avoid confounding seed dormancy with viability. Being a business, commercial seed testing requires results quickly and seed lot dormancy can be problematic. For this reason, various procedures are recommended to break dormancy in germination tests. The alternative is to estimate viability by means other than germination tests. This is most common when testing forest tree seeds in which species the excised embryo test and the topographical tetrazolium test (Gordon 1992) are often employed.

Vigour is a more difficult term to describe briefly, simply because there are many different ways of defining and assessing seed vigour. One definition that was agreed internationally is that 'seed vigour is the sum total of those properties of the seed which determine the potential level of activity and performance of the seed lot during germination and seedling emergence' (Perry 1978).

Ultimately we are concerned with field emergence ability, and the relevance of the concept of seed vigour to the grower (i.e. the ultimate purchaser of seeds) is the extent to which problems occur in crop establishment and the extent of any discrepancies between field emergence (percentage of seeds sown from which seedlings emerge from the seed bed) and laboratory germination. For example, Figure 3.10 presents a graphical illustration of seed vigour which Heydecker (1977) suggested provides an agronomist's definition of seed vigour, whereby the lower the relative field emergence (percentage of *viable* seeds sown from which seedlings emerge from the seed bed), the lower the seed vigour. This is only part of the story, however, because it considers only one seed bed. An important aspect of seed vigour is the fact that in poor (or low) vigour seed lots, emergence may vary greatly among different sowings. This aspect of seed vigour can be represented

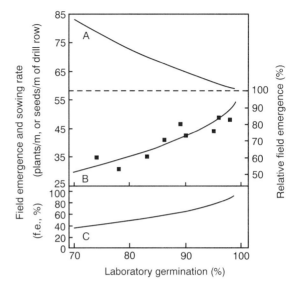

FIGURE 3.10   Field emergence data (plants per metre of drill row) of nine barley seed lots at one sowing (squares) reported by Ellis and Hansan (1974). The sowing rates (curve A) were adjusted to ensure that the same number of germinable seeds were sown per metre. The horizontal broken line indicates 100% relative field emergence and the variation in vertical distance between this line and the symbols below has been suggested as an agronomist's definition of seed vigour by Heydecker (1977). Curve B is the product of sowing rate and a model of field emergence in which it is assumed that the difference in probits between the field emergence and the laboratory germination of each lot is 0.88 (from Ellis and Roberts 1981. Reproduced by permission of ISTA)

graphically by plotting the seed lot × seed bed environment interaction (Figure 3.11). In this diagram it is clear that in good seed bed environments all seed lots (with viability above the legal minimum germination level for sale) show acceptable emergence. However, the sensitivity (or stability) of emergence to seed bed environment varies among seed lots such that in very poor seed bed environments differences among seed lots are considerable: good quality (or high vigour) seed lots provide acceptable emergence but poor quality (or low vigour) seed lots do not.

Vigour remains a contentious subject. Whereas viability is comparatively simple to define (being the proportion of live seeds in a seed lot), vigour

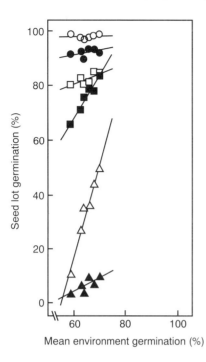

FIGURE 3.11   Relations between the germination of six different seed lots of barley in six different environments with the mean of all seed lots in each environment. The difference in the slopes of the fitted regression lines shows the seed lot × seedbed environment interaction. This interaction is considered by many to be a graphical definition of seed vigour (from Ellis and Roberts 1981. Reproduced by permission of ISTA)

has tended to be open to several different interpretations. For example, to some, seed vigour comprises several different criteria, including ability to germinate (Perry 1978), but others have argued that ability to germinate is widely recognised as a component of seed quality and so differences in vigour among seed lots should not simply reflect differences in germination (Matthews 1980). In addition, some (but by no means all) researchers have found correlations between normal germination and field emergence (e.g. MacKay and Tonkin 1965; Matthews 1980). At present, therefore, although many vigour tests exist (e.g. Hampton and TeKrony 1995), seed vigour is not included in statutory seed testing procedures.

The four principal components of statutory seed testing are seed purity, seed health, seed germination, and (in certain species) seed moisture content. All seed producers generally have to meet strict contractual obligations on these four components as to the standard of the seed lots they produce: it is common for these contractual standards to be rather more onerous than those enshrined in legislation.

Seed purity implies that all the seeds present are of the cultivar named and that no other matter, whether plant (e.g. stems) or inert (e.g. stones), is present in the seed lot. Strict generation control of the seed multiplication process and also field inspections of the seed crop are required to ensure cultivar authenticity (see the section of 'Generation Control' in this chapter). The remaining aspects of seed purity are assessed during the seed lot certification procedures. Examples of the UK requirements for freedom from seed impurities are provided in Table 3.9 for various pasture species.

Field inspection will provide some indication of the likelihood of seed health problems, but for certification purposes seed lot health is assessed post-harvest. The general requirement for seed lot certification is that the seeds shall be of a satisfactory state of health in so far as seed-borne diseases and organisms affecting the seeds are concerned, with specific requirements provided for certain species (Table 3.10).

The standard for seed germination in seed lot certification varies considerably (Table 3.11). These minimum germination levels for seed offered for sale are in effect a pragmatic compromise between the ideal of 100% normal germination and the relative difficulties of providing enough seed of adequate quality to satisfy growers' demands for planting material. In practice, market requirements (especially those of large-scale purchasers of seed such as vegetable seedling producers using modules) are much more rigorous than these minimum germination levels for sale.

The inclusion of a standard for seed moisture content is species-dependent and also varies among countries. For example, in the UK there is a requirement that the moisture content in cereal seed lots shall not exceed 17%, while in France the requirement for oilseed rape is that moisture content shall not exceed 8%.

Table 3.9 Permitted impurities allowed in seed lots of pasture species for certification in the UK

Kind	Weight of sample for determination of foreign seeds by number grams	Maximum permitted content of seed impurities (by number or as a percentage of weight)																	Other Requirements
		Avena fatua A. ludoviciana and *A. sterilis* (Wild Oat) or *Cuscuta* spp. (Dodder)		*Rumex* spp. (Docks and Sorrels) excluding *R. acetosella* (Sheep's sorrel) and *R. maritimus* (Golden dock)			*Alopecurus myosuroides* (Blackgrass)			*Agropyron repens* (couch)			*Melilotus* spp. (Sweetclovers)		Maximum content of any one other plant species			
		Basic Seed	Certified Seed Minimum Standard and Higher Voluntary Standard	Basic Seed	Certified Seed Minimum Standard	Higher Voluntary Standard	Basic Seed	Certified Seed Minimum Standard	Higher Voluntary Standard	Basic Seed	Certified Seed Minimum Standard	Higher Voluntary Standard	Basic Seed	Certified Seed Minimum Standard and Higher Voluntary Standard	Basic Seed	Certified Seed Minimum Standard	Higher Voluntary Standard	
		No.	No.	No.	No.	No.	No.	%	No.	No.	%	No.	No.	%	No.	%	%	
(1)	(2)	(3)	(4)	(5)	(6)	(7)	(8)	(9)	(10)	(11)	(12)	(13)	(14)	(15)	(16)	(17)	(18)	(19)
Perennial ryegrass	60	0	0a	2	5	5	5	0.3	10	5	0.5	10	—	—	20b	1.0	0.5e, g	d
Italian ryegrass	60	0	0a	2	5	5	5	0.3	10	5	0.5	10	—	—	20b	1.0	0.5e, g	
Hybrid ryegrass	60	0	0a	2	5	5	5	0.3	10	5	0.5	10	—	—	20b	1.0	0.5e, g	
Cocksfoot	30	0	0a	2	5	5	5	0.3	10	5	0.3	10	—	—	20b	1.0	0.5	
Timothy	10	0	0a	2	5	4	1	0.3	10	1	0.3	10	—	—	20	1.0	0.5e	
Meadow fescue	50	0	0a	2	5	5	5	0.3	10	5	0.5	10	—	—	20b	1.0	0.5f, g	
Smooth-stalked Meadowgrass	5	0	0a	1	2	2	1	0.3	3	1	0.3	3	—	—	20	1.01	0.5m	h
Red fescue	30	0	0a	2	5	5	5	0.3	10	5	0.5	10	—	—	20b	1.0	0.5g	
Festulolium	60	0	0a	2	5	—	5	0.3	—	5	0.5	10	—	—	20b	1.0	—	
Tall fescue	50	0	0a	2	5	5	5	0.3	10	5	0.5	10	—	—	20b	1.0	0.5f, g	
Red clover	50	0	0	5	10	10	—	—	10	—	—	10	0j	0.3	20	1.0	0.5	
White clover	20	0	0	5	10	10	—	—	10	—	—	10	0h	0.3	20	1.0	0.5	
Sainfoin Seed/fruit	400/600	0	0k	2	5	5	—	—k	10	—	—k	10	0	0.3k	20	1.0k	0.5	
Lucerne	50	0	0	3	10	10	—	—	10	—	—	10	0j	0.3	20	1.0	0.5	

a   For Certified Seed minimum standard, 1 seed of dodder in a sample of the size specified in column (2) shall not be regarded as an impurity if a second sample of the same weight is free from dodder.

b   The presence of a maximum of 80 seeds of *Poa* spp. in a sample of the size specified in column (2) shall not be regarded as an impurity.

c   There shall be no more than 0.4% by weight of annual meadowgrass.

d   In Basic Seed awned ryegrass seeds shall not exceed 5 seeds in a sample of the size specified in column (2) of seeds of a variety known not to produce seeds with awns. In Certified Seed awned ryegrass seeds shall not exceed 1% by weight in seeds of a variety known not to produce seeds with awns.

e   There shall be no more than 0.3% by weight of *Agrostis* spp.

f   There shall be no more than 0.3% by weight of ryegrasses.

g   There shall be no more than 0.3% by weight of rough-stalked meadowgrass.

h   For Certified Seed, Higher Voluntary Standard of red fescue, there shall be no more than a total of 4 seeds of ryegrass, cocksfoot and meadow fescue in a sample of the size specified in column 2.

j   One seed of *Melilotus* spp. in a sample of the size specified in column (2) shall not be regarded as an impurity if a second sample of twice that size is free of *Melilotus* spp.

k   Including Commercial Seed.

l   In Certified Seed a maximum of 0.8% by weight of seeds of other meadowgrasses shall not be regarded as an impurity.

m   In Certified Seed a maximum of 0.4% by weight of seeds of other meadowgrasses shall not be regarded as an impurity.

Table 3.10 Standards for seed health in selected vegetable crops for certification in the UK

Kind	Disease	Category	No. of seeds to be examined	Standard
Brassicas	*Phoma lingam* (Canker)	Basic Seed	1,000	*Nil infection
Red beet or beetroot	*Phoma betae* (Blackleg)	Basic Seed	200	*Nil infection
Celery	*Septoria apiicola* (Leaf blight)	Basic Seed Certified Seed	400	*Nil infection
Peas	*Phoma apiilcola* (Root rot) *Asocochyta pisi* *Mycosphaerella pinodes,* *Ascochyta pinodella* (Leaf and pod spot)	Basic Seed	200	*Nil infection
		Certified Seed	200	†Not more than 20 seeds
Lettuce	Lettuce mosiac virus	Basic seed Certified Seed	5,000	Nil infection
French beans	(i) *Colletotrichum lindemuthianum* (Anthracnose)	Basic Seed	600	Nil infection
	(ii) *Pseudomonas phaseolicola* (Halo blight)	Basic Seed	5,000	Nil infection
Broad beans	*Ascochyta fabae* (Leaf and pot spot)	Basic Seed	600	Nil infection

*   Where infected seeds are found an effective treatment approved by the Minister must be applied before the seeds can be officially certified.
†   Where not more than 20 infected seeds are found in Basic Seed, or more than 20 infected seeds are found in Certified Seed an effective treatment approved by the Minister must be applied before the seeds can be officially certified.

Table 3.11 Minimum germination levels (normal germination in laboratory tests) for the certification in the UK of seed lots of contrasting species

Crop	Minimum germination (% by number of pure seeds or pure pellets)	Crop	Minimum germination (% by number of pure seeds or pure pellets)
Oilseeds		Pea	80
Brassica species	85	Radish	70
Sunflower	85	Red Cabbage	75
Flax	92	Runner bean	80
Linseed	85	Savoy cabbage	75
Soya bean	80	Spinach	75
White Mustard	85	Sprouting broccoli	75
Vegetable seeds		Tomato	75
Asparagus	70	Turnip	80
Beet	70 (Clusters)	Pasture Seeds (a)	
Broad bean	80	Perennial ryegrass	80
Brussels sprouts	75	Italian ryegrass	75
Cabbage	75	Hybrid ryegrass	75
Calabrese	75	Cocksfoot	80
Carrot	65	Timothy	80
Cauliflower	70	Meadow fescue	80
Celeriac	70	Smooth-stalked meadowgrass	75
Celery	70	Red fescue	75
Chicory	65	Festulolium	75
Chinese cabbage	75	Tall fescue	80
Cucumber	80	Red clover (b)	80
Curly kale	75	White clover (b)	80
Endive	65	Sainfoin (b)	75
French bean	75	Lucerne (b)	80
Gherkin	80	Cereal Seeds	
Gourd	80	Barley	85
Kohlrabi	75	Durum wheat	85
Leek	65	Maize	90
Lettuce	75	Naked oat	75
Marrow	75	Oat	85
Melon	75	Rye	85
Onion	70	Spelt wheat	85
Parsley	65	Wheat	85

(a) All fresh and healthy seeds which do not germinate after pre-treatment to remove dormancy are assumed to be dormant but viable and thus considered as seeds which have germinated.

(b) Hard seeds are considered as seeds capable of germination.

REFERENCES

Ainsworth, C.C. and Sharp, P.J. (1989). The potential role of DNA probes in plant variety identification. *Plant Varieties and Seeds*, **2**, 27–34.

Allard, R.W. (1960). *Principles of Plant Breeding*. John Wiley, New York.

Anon (1979). *Revised General Introduction to the Guidelines for the Conduct of Tests for Distinctness, Homogeneity and Stability of New Varieties of Plants*. Document TG/01/2, International Union for the Protection of New Varieties of Plants, Geneva, 2–9.

Anon (1980). *International Code for the Nomenclature of Cultivated Plants*. The International Bureau for Plant Taxonomy and Nomenclature, Utrecht, 12–13.

Anon (1991). *International Convention for the Protection of New Varieties of Plants*. International Union for the Protection of New Varieties of Plants, Geneva, 6, 12–14.

AOSCA (1971). *AOSCA Certification Handbook*. Publication No. 23, Association of Official Seed Certifying Agencies, USA, 172 pp.

Asker, S. (1980). Gametophytic apomixis: elements and genetic regulation. *Hereditas*, **93**, 277–293.

Barker, D.A., Vuori, T.A. and Myers, D.G. (1992). The use of slice and aspect ratio parameters for the discrimination of Australian wheat varieties. *Plant Varieties and Seeds*, **5**, 47–52.

Becker, H.C. (1989). Breeding synthetic varieties in partially allogamous crops. *Vortrage fur Pflanzenzuctung*, **16**, 81–90.

Bickelmann, U. (1993). On fatuoids and hybrids of *Avena* spp. and their significance for seed production. 1. Flowering behaviour of oats and origin of fatuoids and hybrids. *Plant Varieties and Seeds*, **6**, 21–26.

Bietz, J.A. and Huebner, F.R. (1995). Variety identification by HPLC. In *Identification of Food Grain Varieties* (Ed. C.W. Wrigley). AACC, USA, 73–90.

Biolley, J.P. and Jay, M. (1993). Anthocyanins in modern roses: chemical and colorimetric features in relation to the colour range. *Journal of Experimental Botany*, **44**, 1725–1734.

Bornman, Ch.H. (1993). Micropropagation and somatic embryogenesis. In *Plant Breeding: Principles and Prospects* (Eds M.D. Hayward, N.O. Bosemark and I. Romagosa). Chapman and Hall, London, 246–260.

Botta, R., Scott, N.S., Eynard, I. and Thomas, M.R. (1995). Evaluation of microsatellite sequence-tagged site markers for characterizing *Vitis vinifera* cultivars. *Vitis*, **34**, 99–102.

Caetano-Anolles, G. (1996). Fingerprinting nucleic acids with arbitrary oligonucleotide primers. *Agro-Food Industry Hi-Tech*, Jan/Feb, 26–35.

Choo, T.M., Reinbergs, E. and Kasha, K.J. (1985). Use of haploids in breeding barley. *Plant Breeding Reviews*, **3**, 219–252.

Chtioui, Y., Bertrand, D., Dattee, Y. and Devaux, M.-F. (1996). Identification of seeds by colour imaging: comparison of discriminant analysis and artificial neural network. *Journal of the Science of Food and Agriculture*, **71**, 433–441.

Cooke, R.J. (1988). Electrophoresis in plant testing and breeding. *Advances in Electrophoresis*, **2**, 171–261.

Cooke, R.J. (Ed.) (1992). *Handbook of Variety Testing – Electrophoresis Testing*. ISTA, Zurich, 1–42.

Cooke, R.J. (1995a). Introduction: the reasons for variety identification. In *Identification of Food Grain Varieties*, (Ed. C.W. Wrigley). AACC, USA, 1–18.

Cooke, R.J. (1995b). Variety identification: modern techniques and applications. In *Seed Quality: Basic Mechanisms and Agricultural Implications* (Ed. A.S. Basra). Food Products Press, New York, 279–318.

Cooke, R.J. (1995c). Varietal identification of crop plants. In *New Diagnostics in Crop Sciences* (Eds J. Skerritt and R. Appels). CAB International, Wallingford, 33–63.

Cooke, R.J. (1995d). Gel electrophoresis for the identification of plant varieties. *Journal of Chromatography*, **698**, 281–299.

Cooke, R.J. (1995e). Distribution and diversity of hordein alleles in barley varieties. *Seed Science and Technology*, **23**, 865–871.

Cooke, R.J. (1995f). Allelic variability at the *Glu-1* loci in wheat varieties. *Plant Varieties and Seeds*, **8**, 97–106.

Dark, S.O.S. (1971). Experiments on the cross-pollination of sugar beet in the field. *Journal of the National Institute of Agricultural Botany*, **12**, 242–266.

de Nettancourt, D. (1993). Self- and cross-incompatibility systems. In *Plant Breeding: Principles and Prospects* (Eds M.D. Hayward, N.O. Bosemark and I. Romagosa). Chapman and Hall, London, 203–212.

Donini, P., Law, R.J., Stephenson, P. and Koebner, R.M.D. (1997). Analysis of genetic diversity in past and present varieties of *Triticum aestivum* using AFLP and microsatellites. *Plant and Animal Genome V*, San Diego.

Draper, S.R. and Keefe, P.D. (1989). Machine vision for the characterisation and identification of cultivars. *Plant Varieties and Seeds*, **2**, 53–62.

Driscoll, C.J. (1986). Nuclear male sterility systems in seed production of hybrid varieties. *CRC Critical Reviews in Plant Sciences*, **3**, 227–256.

Ellis, J.R.S. and Hanson, A.D. (1974). Tests for cereal yield heterosis based on germinating seeds: a warning. *Euphytica*, **23**, 71–77.

Ellis, R.H. and Roberts, E.H. (1981). The quantification of ageing and survival in orthodox seeds. *Seed Science and Technology*, **9**, 373–409.

Evola, S. (1996). Transgenic European corn borer resistant maize. *Vorträge für Pflanzenzüchtung*, **33**, 22–33.

Fitzgerald, D.M., Barry, D., Dawson, P.R. and Cassells, A.C. (1997). The application of image analysis in determining sib proportion and aberrant characterization in F1 hybrid *Brassica* populations. *Seed Science and Technology*, **25**, 503–510.

Frazier, R.R.E., Millican, E.S., Watson, S.K., Oldroyd, N.J., Sparkes, R.L., Taylor, K.M., Panchal, S.P., Bark, L., Kimpton, C.P. and Gill, P.D. (1996). Validation of the Applied Biosystems PRISM™ 377 automated sequencer for forensic short tandem repeat analysis. *Electrophoresis*, **17**, 1550–1552.

Fowler, J.C.S. and Kijas, J.M.K. (1994). Current molecular methods for plant genome identification. *Australasian Biotechnology*, **4**, 153–157.

Giles, R.J. (1989). The frequency of natural cross-fertilisation in sequential sowings of winter barley. *Euphytica*, **43**, 125–134.

Gilliland, T.J. (1989). Electrophoresis of sexually and vegetatively propagated cultivars of allogamous species. *Plant Varieties and Seeds*, **2**, 15–26.

Gordon, A.G. (1992). Seed testing. In *Seed Manual for Forest Trees* (Ed. A.G. Gordon). Forestry Commission Bulletin 83, HMSO, London, 105–115.

Gorg, R., Schachtschabel, U., Ritter, E., Salamini, F. and Gebhardt, C. (1992). Discrimination among 136 tetraploid potato varieties by fingerprints using highly polymorphic DNA markers. *Crop Science*, **32**, 815–819.

Greneche, M., Lallemand, J. and Migaud, O. (1991). Comparison of different enzyme loci as a means of distinguishing ryegrass varieties by electrophoresis. *Seed Science and Technology*, **19**, 147–158.

Hampton, J.G. and TeKrony, D.M. (Eds) (1995). *Handbook of Vigour Test Methods*. International Seed Testing Association, Zurich.

Hanson, M.R. and Conde, M.F. (1985). Functioning and variation of cytoplasmic genomes: lessons from cytoplasmic-nuclear interactions affecting male fertility in plants. *International Review of Cytology*, **94**, 213–267.

Hanson, P. (1973). The production of pure stocks of self-pollinating cereal varieties. *Annals of Applied Biology*, **73**, 111–117.

Hérouart, D., Bowler, C., Willekens, H., Van Camp, W., Slooten, L., Van Montau, M. and Inzé, D. (1994). Genetic engineering of oxidative stress resistance in higher plants. In *The Production and Uses of Genetically Transformed Plants* (Eds M.W. Bevan, B.D. Harrison and C.J. Leaver). Chapman and Hall, London, 47–52.

Heydecker, W. (1977). Stress and seed germination: an agronomic view. In *The Physiology and Biochemistry of Seed Dormancy and Germination*. (Ed. A.A. Khan). North-Holland Publishing Company, Amsterdam, 237–282.

Heyne, E.G. and Smith, G.S. (1967). Wheat breeding. In *Wheat and Wheat Improvement* (Eds K.S. Quisenberry and L.P. Reitz). American Society of Agronomy, Madison, 269–306.

House, L.R. (1985). *A Guide to Sorghum Breeding*, 2nd edition. ICRISAT, Hyderbad.

Huaman, Z. and Stegemann, H. (1989). Use of electrophoretic analyses to verify morphologically identical clones in a potato collection. *Plant Varieties and Seeds*, **2**, 155–161.

ISTA (1996a). International Rules for Seed Testing: Rules. *Seed Science and Technology*, **24** (Supplement), 1–86.

ISTA (1996b). International Rules for Seed Testing: Annexes. *Seed Science and Technology*, **24** (Supplement), 87–335.

Jianhua, Z., McDonald, M.B. and Sweeney, P.M. (1996). Soybean cultivar identification using RAPD. *Seed Science and Technology*, **24**, 589–592.

Kade, M., Wagner, M.L., Strittmatter, C.D., Ricco, R.A. and Gurni, A.A. (1997). Identification of *Lotus tenuis* and *Lotus corniculatus* seeds by their flavonols: testing procedure. *Seed Science and Technology*, **25**, 585–587.

Kaul, M.L.H. (1988). *Male Sterility in Higher Plants*. Monographs on Theoretical and Applied Genetics, Vol 10, Springer-Verlag, Berlin.

Keefe, P.D. (1992). A dedicated wheat grain image analyser. *Plant Varieties and Seeds*, **5**, 27–33.

Kelly, A.F. (1988). *Seed Production of Agricultural Crops*. Longman Scientific and Technical, Harlow, 16–26.

Kelly, A.F. and Boyd, M.M. (1966). The stability of cultivars of grasses and clovers when grown for seed in differing environments. *Proceedings of the X International Grassland Congress*, Finnish Grassland Association, Helsinki, 777–782.

Konishi, T. (1995). Isozyme variation and analysis in agriculturally important plants. In *Seed Quality: Basic Mechanisms and Agricultural Implications* (Ed. A.S. Basra). Food Products Press, New York, 303–318.

Lee, D., Reeves, J.C. and Cooke, R.J. (1996a). DNA profiling and plant variety registration: 2. Restriction fragment length polymorphisms in varieties of oilseed rape. *Plant Varieties and Seeds*, **9**, 181–190.

Lee, D., Reeves, J.C. and Cooke, R.J. (1996b). DNA profiling and plant variety registration: 1. The use of random amplified DNA polymorphisms to discriminate between varieties of oilseed rape. *Electrophoresis*, **17**, 261–265.

Lee, D., Reeves, J.C. and Cooke, R.J. (1996c). DNA

profiling for varietal identification in crop plants. *Proceedings of BCPC Symposium No. 65: Diagnostics in Crop Production*, 235–240.

Levings, C.S.III and Dewey, R.E. (1988). Molecular studies of cytoplasmic male sterility in maize. *Philosophical Transactions of the Royal Society, London B*, **319**, 177–185.

Lookhart, G.L. and Wrigley, C.W. (1995). Variety identification by electrophoretic analysis. In *Identification of Food Grain Varieties* (Ed. C.W. Wrigley). AACC, USA, 55–72.

Loos, B.P. and Degenaars, G.H. (1993). pH-dependent electrophoretic variants for phosphoglucoisomerase in ryegrasses (*Lolium* spp.): a research note. *Plant Varieties and Seeds*, **6**, 55–60.

Lowe, A.J., Hanotte, O. and Guarino, L. (1996). Standardization of molecular genetic techniques for the characterization of germplasm collections: the case of random amplified polymorphic DNA (RAPD). *Plant Genetic Resources Newsletter*, **107**, 50–54.

MacKay, D.B. and Tonkin, J.H.B. (1965). Studies in the laboratory germination and field emergence of sugar beet seed. *Proceedings of the International Seed Testing Association*, **30**, 661–676.

Mailer, R.J., Daun, J. and Scarth, R. (1993). Cultivar identification in *Brassica napus* L. by RP-HPLC of ethanol extracts. *Journal of the American Oil Chemists Society*, **70**, 863–866.

Mariani, C., de Beuckeleer, M., Truettner, J., Leemans, J. and Goldberg, R.B. (1990). Induction of male sterility in plants by a chimaeric ribonuclease gene. *Nature*, **347**, 737–741.

Marshall, D.R. and Brown, A.H.D. (1973). Stability of performance of mixtures and multi-lines. *Euphytica*, **22**, 405–412.

Matthews, S. (1980). Controlled deterioration: a new vigour test for crop seeds. In *Seed Production* (Ed. P.D. Hebblethwaite). Butterworth, London, 647–660.

Mayo, O. (1987). *The Theory of Plant Breeding*, 2nd edition. Oxford Science Publications.

McCarthy, P.K., Cooke, R.J., Lumley, I.D., Scanlon, B.F. and Griffin, M. (1990). Application of reversed-phase high-performance liquid chromatography for the estimation of purity in hybrid wheat. *Seed Science and Technology*, **18**, 609–620.

McRae, D.H. (1985). Advances in chemical hybridisation. *Plant Breeding Reviews*, **3**, 169–191.

Morell, M.K., Peakall, R., Appels, R., Preston, L.R. and Lloyd, H.L. (1995). DNA profiling techniques for plant variety identification. *Australian Journal of Experimental Agriculture*, **35**, 801–819

Morgante, M. and Olivieri, A.M. (1993). PCR-amplified microsatellites as markers in plant genetics. *The Plant Journal*, **3**, 175–182.

Mudzana, G., Pickett, A.A., Jarman, R.J., Cooke, R.J. and Keefe, P.D. (1995). Variety discrimination in faba beans (*Vicia faba* L.): an integrated approach. *Plant Varieties and Seeds*, **8**, 135–145.

Nielsen, G. (1985). The use of isozymes as probes to identify and label plant varieties and cultivars. *Current Topics in Biological and Medical Research*, **12**, 1–32.

Nybom, H., Rogstad, S.H. and Schaal, B.A. (1990). Genetic variation detected by use of the M13 'DNA fingerprint' probe in *Malus, Prunus* and *Rubus* (Roseaceae). *Theoretical and Applied Genetics*, **79**, 153–156.

Ockenden, D.J. and Smith B.M. (1993). Brussels sprouts *Brassica oleracea* var. *gemmifera* D.C. In *Genetic Improvement of Vegetable Crops* (Eds G. Kallo and B.O. Bergh). Pergamon Press, Oxford, 87–112.

OECD (1995a). *OECD Scheme for the Varietal Certification of Herbage and Oil Seed Moving in International Trade*. Organization for Economic Co-operation and Development, Paris.

OECD (1995b). *OECD Scheme for the Control of Vegetable Seed Moving in International Trade*. Organization for Economic Co-operation and Development, Paris, 12–14.

Payne, R.C. (Ed.) (1993). *Handbook of Variety Testing – Rapid Chemical Identification Techniques*. ISTA, Zurich, 1–52.

Perry, D.A. (1978). Report of the Vigour Test Committee 1974–1977. *Seed Science and Technology*, **6**, 159–181.

Poulsen, D.M.E., Ko, H.L., van der Meer, J.G., van de Putte, P.M. and Henry, R.J. (1996). Fast resolution of identification problems in seed production and plant breeding using molecular markers. *Australian Journal of Experimental Agriculture*, **36**, 571–576.

Priestley, R.H., Bayles, R.A. and Parry, D. (1988). Effect of mixing cereal varieties on yield and disease development. *Plant Varieties and Seeds*, **1**, 53–62.

Riley, R. and Kimber, G. (1961). Aneuploids and the cytogenetic structure of wheat varietal populations. *Heredity*, **16**, 275–290.

Sapirstein, H.D. (1995). Variety identification by digital image analysis. In *Identification of Food Grain Varieties* (Ed. C.W. Wrigley). AACC, USA, 91–130.

Sharma, S.K., Knox, M.R. and Ellis, T.H.N. (1996). AFLP analysis of the diversity and phylogeny of *Lens* and its comparison with RAPD analysis. *Theoretical and Applied Genetics*, **93**, 751–758.

Smith, J.E. and Whittington, W.J. (1991). Effects of roguing on the frequency of atypical wheat (*Triticum aestivum* L.) plants. *Plant Varieties and Seeds*, **4**, 77–86.

Smith, J.S.C. (1995). Identification of cultivated varieties by nucleotide analysis. In *Identification of*

Food Grain Varieties (Ed. C.W. Wrigley). AACC, USA, 131–150.

Smith, J.S.C. and Smith, O.S. (1991). Restriction fragment length polymorphisms can differentiate among U.S. maize hybrids. *Crop Science*, **31**, 893–899.

Smith, J.S.C. and Smith, O.S. (1992). Fingerprinting crop varieties. *Advances in Agronomy*, **47**, 85–140.

Smith, J.S.C. and Wych, R.D. (1986). The identification of female selfs in hybrid maize: a comparison using electrophoresis and morphology. *Seed Science and Technology*, **14**, 1–8.

Smith, J.S.C., Chin, E.C.L., Shu, H., Smith, O.S., Wall, S.J., Senior, M.L., Mitchell, S.E., Kresovich, S. and Ziegle, J. (1997). An evaluation of the utility of SSR loci as molecular markers in maize (*Zea mays* L.): comparisons with data from RFLPs and pedigree. *UPOV BMT Meeting, Cambridge*, March 1997.

Snape, J.W., de Buyser, J., Henry, Y. and Simpson, E. (1986). A comparison of methods of haploid production in a cross of wheat, *Triticum aestivum. Zeitschrift Pflanzenzuchtung*, **96**, 320–330.

Sovero, M. (1996). Commercialization of laurate canola. *Vorträge für Pflanzenzüchtung*, **33**, 8–21.

Staub, J.E., Serquen, F.C. and Gupta, M. (1996). Genetic markers, map construction and their application in plant breeding. *HortScience*, **31**, 729–741.

Stegemann, H., Shah, A.A. Krogerrecklenfort, E. and Hamza, M.M. (1992). Sweet potato (*Ipomoea batatas* L.): genotype identification by electrophoretic methods and properties of their proteins. *Plant Varieties and Seeds*, **5**, 83–91.

Thomas, M.R., Cain, P. and Scott, N.S. (1994). DNA typing of grapevines: a universal methodology for describing cultivars and evaluating genetic relatedness. *Plant Molecular Biology*, **25**, 939–949.

Tohme, J., Orlando Gonzalez, D., Beebe, S. and Duque, M.C. (1996). AFLP analysis of gene pools of a wild bean core collection. *Crop Science*, **36**, 1375–1384.

Tsuchiya, Y., Araki, S., Sahara, H., Takashio, M. and Koshino, S. (1995). Identification of malting barley varieties by genome analysis. *Journal of Fermentation and Bioengineering*, **79**, 429–432.

van der Burg, W.J. and van Zwol, R.A. (1991). Rapid identification techniques used in laboratories of the International Seed Testing Association: a survey. *Seed Science and Technology*, **19**, 687–700.

van de Vooren, J.G. and van der Heijden, G.W.A.M. (1993). Measuring the size of French beans with image analysis. *Plant Varieties and Seeds*, **6**, 47–53.

Vos, P., Hogers, R., Bleeker, M., Reijans, M., van de Lee, T., Hornes, M., Frijters, A., Pot, J., Peleman, J.,

Kuiper, M. and Zabeau, M. (1995). AFLP: a new technique for DNA fingerprinting. *Nucleic Acids Research*, **23**, 4407–4414.

Wang, Y., Wallin, J.M., Ju, J., Sensabaugh, G.F. and Mathies, R.A. (1996). High-resolution capillary array electrophoretic sizing of multiplexed short tandem repeat loci using energy-transfer fluorescent primers. *Electrophoresis*, **17**, 1485–1490.

Warren, D.E. (1997). Image analysis research at NIAB: chrysanthemum leaf shape. *Plant Varieties and Seeds*, **10**, 59–61.

Welsh, J. and McClelland, M. (1990). Fingerprinting genomes using PCR with arbitrary primers. *Nucleic Acids Research*, **18**, 7213–7218.

White, J. and Cooke, R.J. (1992). A standard classification system for the identification of barley varieties by electrophoresis. *Seed Science and Technology*, **20**, 663–676.

Williams, J.G.K., Kubelik, A.R., Livak, K.J., Rafalski, J.A. and Tingey, S.V. (1990). DNA polymorphisms amplified by arbitrary primers are useful as genetic markers. *Nucleic Acids Research*, **18**, 6531–6535.

Worland, A.J. and Law, C.N. (1985). Aneuploidy in semi dwarf wheat varieties. *Euphytica*, **34**, 317–327.

Wrigley, C.W. (Ed.) (1995). *Identification of Food-Grain Varieties*. AACC, USA, 1–283.

Further Reading

Anon (1971). OECD Standards, Schemes and Guides Relating to Varietal Certification of Seed. *Proceedings of the International Seed Testing Association*, **36**(3).

Basra, A.S. (Ed.) (1995). *Seed Quality: Basic Mechanisms and Agricultural Implications*. The Haworth Press, New York.

Copeland, L.O. and McDonald, M.B. (1995). *Principles of Seed Science and Technology*. Chapman and Hall, New York, Chapter 12.

Douglas, J.E. (Ed.) (1980). *Successful Seed Programs: a Planning and Management Guide*. Westview Press, Boulder, Colorado, 111–114.

Feistritzer, W.P. and Kelly, A.F. (Eds) (1978). *Improved Seed Production*. FAO Plant Production and Protection Series No. 15, Food and Agriculture Organization of the United Nations, Rome, 16–18.

Kelly, A.F. (1994). *Seed Planning and Policy for Agricultural Production*. John Wiley and Sons, Chichester.

Mathur, S.B. and Mortensen, C.N. (Eds) (1997). *Seed Health Testing in the Production of Quality Seed*. International Seed Testing Association, Zurich.

4

Principles of Seed Production

ECOLOGY OF THE CROP

To produce good quality seed it is essential that the crop is grown in a situation giving the correct conditions. Climate is obviously of first importance but latitude and altitude are also important considerations. These factors determine where different crop species can be grown to maturity. Within those areas with suitable climate, latitude and altitude for the production of seed of a particular crop species, considerations such as soil type and uniformity determine where the best sites are situated within which suitable fields are likely to be found.

Climate

Ideally, the climate of an area where seed crops are to be grown should be such that conditions are likely to allow sowing into a suitable seed bed, followed by good growing conditions (suitable temperature and rainfall) to give uninterrupted growth to maturity. Dry harvesting conditions are also required. Conditions during crop growth can be ameliorated by the use of some measures to improve on possibly detrimental climatic effects; for example, protection against cold or wind can be given in the early stages of growth by the use of

plastic sheeting or throughout growth by arranging shelter belts, and irrigation can be used to improve dry conditions. However, these measures add expense to the seed crop which gives an advantage to those areas where they are not essential.

It is not easy to describe climate with any precision but there are certain generalised types which can be classified as follows (Kelly 1994).

1. There are four main divisions: *Tropics; Subtropics with summer main rainfall; Subtropics with winter main rainfall; Temperate.*
2. Within each of these main divisions there are general temperature regimes: *Warm; Cool; Cold.* These are distinguished as follows.
   (i) Warm: 24-hour mean daily temperature more than 20°C during the growing season.
   (ii) Cool: 24-hour mean daily temperature from 5 to 20°C during the growing season.
   (iii) Cold: 24-hour mean daily temperature less than 5°C during the growing season.
   All three of these occur in the tropics and subtropics with summer rainfall, but in the subtropics with winter rainfall and in temperate areas only cool or cold conditions are generally found.

Encyclopaedia of Seed Production of World Crops.
Edited by A.F. Kelly and R.A.T. George. © 1998 John Wiley & Sons Ltd

3. A further subdivision can be made by describing these 10 classes in terms of the amount of rainfall in each: *Arid; Semi-arid; Subhumid; Humid*. In terms of rainfall these are defined as follows.
   (i)   Arid: less than 400 mm per year.
   (ii)   Semi-arid: 400–599 mm per year.
   (iii) Subhumid: 600–1200 mm per year.
   (iv)   Humid: more than 1200 mm per year.

This classification is of course very approximate. However, in very general terms these 40 classes cover the main types of climate throughout the world in those areas where crops can be grown. Within areas defined by these criteria it will be necessary to identify more closely the average weather conditions during the growing season for each locality, based on data collected over a number of years (normally at least 10). These data cannot predict with certainty what conditions will be like in any one year, but they will indicate those areas where there is reasonable chance that the conditions required for seed production of a particular crop species are likely to occur.

Another aspect of climate not covered by the criteria above but important for some seed growing activities is prevailing wind speed and direction. For species which are pollinated by wind-borne pollen this is especially important, to ensure that pollen is dispersed within the seed crop, and to ensure that the seed crop is situated with adequate isolation from the direction of the prevailing wind.

Sites in very exposed positions may experience high wind speeds as the seed crop reaches maturity and this can cause loss of seed by shedding. Wind can also exaggerate cool or cold temperatures (the so-called *chill factor*), and cause more rapid drying of the soil surface. For some high value seed crops such as vegetables it may be worthwhile to install windbreaks. George (1985) describes different types of windbreak: a permeable windbreak provides the best conditions since it does not produce turbulence like a solid break. George notes that a permeable break can reduce wind speed in a horizontal direction downwind for a distance equivalent to up to 30 times its height, although he suggests that in practice a leeward break of a height to give shelter for a distance of 10 times its height should be the aim. Breaks can be either temporary or permanent structures. The former may be either tall plants such as maize or screens made of plastic mesh or hessian. The latter are usually trees planted as shelter belts.

Shelter belts are advantageous in that they reduce transpiration and evaporation of soil water and protect seed crops from wind damage to leaves and flowers; they can also reduce the isolation distance required to prevent undesirable cross-pollination in crops such as sugar beet and pepper. Permanent windbreaks of trees can cause damage to water courses and drains by root growth, and in some circumstances may act as hosts of pests and pathogens such as aphis and virus; these points should be borne in mind when siting the break.

Latitude

Latitude is of great importance because it determines the length of day at any particular time of the year. The tropics have no seasonality and the day length varies not at all throughout the year at the equator and hardly at all at very low latitudes. In the subtropics and the higher latitudes there is increasing seasonality as the day length varies, being shorter in winter and longer in summer.

These factors have considerable influence on the ability of some seeds to germinate or of some plants to change from the vegetative to the reproductive phase. Crop species may be *day neutral* and so can be grown in the tropics or at any latitude; or they may require short or long days to effect these changes and must be grown at a latitude which gives the required contrast of day length between winter and summer. Seeds of some species will only germinate when light intensity and quality and day length are suited to their requirement (Copeland and McDonald 1995). Plants adapted to short-day conditions at lower latitudes will not flower and produce seed if grown at higher latitudes. Likewise, plants adapted to long-day conditions at higher latitudes will not flower and produce seed if grown at lower latitudes. Day-neutral species can be grown at any latitude with the correct climatic conditions. It is thus very important to ensure that the crop species is adapted to the latitude at which it is to be grown, particularly when producing seed away from the area where the cultivar to be grown was selected. Particular requirements for each species are given in Section II.

Vernalisation

The term *vernalisation* is used to describe the combined influence of day length (although strictly it is the length of the period of darkness which is implicated) and temperature on the ability of plants to change from the vegetative to the reproductive phase. Temperature is involved in two ways. First, it is necessary for some species to go through a cold period, and second, some species are stimulated to flower by warmer temperatures in addition to day length. For both these requirements it is not usually markedly lower or higher temperatures which are needed but the accumulated 'day degrees' above or below a certain base figure over a period of days. Species vary in their requirements and, where appropriate, details are given in Section II.

Some species which require a cold spell can be grown as annuals since the cold period can occur at any time and may be during or soon after germination or at an early growth stage; provided they are sown early enough in the growing season to ensure a cold period they will flower and produce seed, or they can be vernalised in a germination chamber before sowing. Seed vernalised before sowing, however, is very susceptible to damage during handling and may produce a reduced plant population if treated roughly.

For some other species the plants must have achieved a sufficiently advanced growth stage before the cold period, which must provide sufficient accumulation of low temperature. These usually have to be sown in the autumn (for example, winter wheat or autumn-sown oilseed rape). There are also species in this category which are biennial or perennial (i.e. they are sown in the spring of one year but will not start to produce seed until the following growing season after the required cold period).

Knowledge of vernalisation requirement is therefore required in deciding whether or not a particular species can be grown for seed in a particular area.

Altitude

The effects of altitude are mainly climatic. At higher altitudes temperatures are usually lower, rainfall may be higher and less predictable, and solar radiation may intensify in periods free from cloud cover. Chandler and Gregory (1976) suggest a 10% increase in radiation per 1000 m increase in height in the British Isles in the absence of cloud. These effects may influence the yield of seed crops but generally will not affect the species that can be grown, given that the weather conditions and day length are satisfactory for flowering.

FIELD FACTORS

Soil Type

Species vary in the type of soil which is best suited to their production. Some prefer dry, free-draining soils, others wetter conditions. Soils are difficult to classify because the types tend not to be separated by definite boundaries. There are, however, some criteria which can be identified with reasonable accuracy.

1. Soils may be either *mineral* or *organic*.
    (i)   Mineral soils vary in texture and may be described as *sandy* (at least 85% sand), *loam, silt* or *clay* (at least 40% clay).
    (ii) Organic soils may be either *well decomposed* or *not well decomposed*.
2. Soil may be either *acid* (pH below 6), *neutral* (pH from 6 to 7.5) or *alkaline* (pH above 7.5).

3. Some soils are *saline*. A specific conductivity
    of 4 m. hos. per second corresponds roughly
    to less than 3000 parts per million salts. Those
    with less than 4 m. hos. are considered normal
    and those with more, salty.

These broad categories give some indication of
the type of soil in a given area, but they are
generally not sufficiently precise to determine
whether or not a soil is suited to a particular crop.
In many countries, soil surveys have enabled soil
maps to be prepared. The most useful classification for locating suitable areas for seed production
is the 'land use capability classification'. According to Russell (1973), the usual method for this
kind of classification is based on a system developed in the USA which divides land into seven
classes. In the UK, Class 1 land has soils which
are deep and well drained and with a good
available water-holding capacity; these are the top
quality soils and they allow a wide range of crops
to be grown. Class 2 soils are less favoured but
generally provide conditions for almost as wide a
range of crops; some root crops may be difficult to
harvest because the soil may become waterlogged
in a wet autumn. Class 3 soils are generally only
suitable for crops which can be harvested before
the onset of wet conditions in winter. Classes 4–7
are increasingly less suited to arable crops.

Classification of soils takes into account the soil
profile to rooting depth. Texture is particularly
important; soil should be capable of retaining a
good crumb structure under heavy rainfall to
prevent capping. This is particularly important
when irrigation is to be used (see, for instance,
Bailey 1990).

Soil Fertility

Fertile soils are generally to be preferred for seed
production. They are easier to work and provide
the best environment for plant development so
that the characteristics of cultivars can be seen to
best advantage. Soils low in fertility or which
suffer from some deficiency should be avoided, as
the plants will present an atypical appearance.

However, soils exceptionally high in fertility
should also be avoided as excessive vegetative
growth can produce an atypical appearance in
plants.

Crop Rotation

Previous cropping is an important consideration
in planning a seed crop. Apart from the factors
which are taken into account in deciding upon
any rotation, the following points should be
considered when planning a seed crop.

The crop rotation should be arranged so as to
reduce to a minimum the chance that volunteer
plants of the same species will occur in the seed
crop. Some seed is very long-lived in soil; for
example, *Brassica* or some *Trifolium* species can
remain viable for 10 years or more. It is also
necessary to take account of those species which
have similar-sized seed to that of the species to be
grown; for example, barley can occur as volunteers in a following wheat crop and the seeds can
be difficult to separate after harvest.

Rotation should take account of good husbandry. Opportunity should be provided in a suitable
preceding crop to clean the field of weeds, particularly those which might be difficult to eliminate in the seed crop. This is also an important
consideration for those weeds which have seeds
difficult to remove from the crop seed or for
which selective herbicides are not available or are
difficult or expensive to use (for example, grass
weeds in a grass seed crop).

Planning previous cropping should also ensure
that the preceding crop will not cause conditions
prejudicial to the sowing of the seed crop. For
example, late harvest of a root crop may cause the
preparation of a good seed bed for a following
seed crop to be delayed; some species may act as
hosts to pathogens or pests which will be harmful
to a following seed crop.

Isolation

Seed crops have to be isolated from all other crops
(whether for seed or not) to avoid all possibility of

mixture during harvest. For self-fertilised species it is important that a seed crop is clearly separated from all other crops, especially those with seeds similar in size and shape to those of the species being grown for seed.

The strict interpretation of cross-fertilisation in the botanical sense is that pollen is transferred from one flower to another to effect fertilisation; the two flowers involved may be on the same plant or they may be on different plants. In seed growing there is a particular need to ensure that the donor flower is carried on a plant from which the pollen will create a seed that will produce a plant typical of the desired cultivar. This is illustrated in the extreme case of the artificially created hybrid cultivar where the female parent is emasculated to ensure pollination by the selected male parent. Emasculation or its equivalent may be achieved by natural means (for example, cytoplasmic male sterility as in maize, or self-incompatibility as in some brassicas); or it may be achieved by manipulation (for example, tomato hybrid cultivars are created by emasculation of the female parent and transference of pollen from the male by hand).

For cross-fertilised species there is therefore the additional need to isolate the seed crop from sources of undesirable pollen in order to prevent the production of plants which are not true to the cultivar being grown for seed. The distance required is different for different species and may also vary according to the size of the field. Pollen from outside the seed crop will generally be more effective around the margins of a field, and fields of 2 ha or less normally require twice the isolation distance needed for larger fields. Generally, the distance required when seed is being produced for further multiplication is double that required when the seed to be produced is intended only for the production of food or industrial crops. Distances are also adjusted to take account of the method of pollination: greater distances are required for species pollinated by insects than for those which are wind pollinated.

The isolation distance is normally specified to separate crops (fields) which can cross-pollinate. However, for some species it is also necessary to take account of plants growing in the vicinity of the seed crop in hedgerows, road verges or gardens (for example, some grass species or brassicas).

In some instances it is possible to reduce the isolation distance when there are features such as windbreaks which reduce the chance of wind-borne pollen or insects reaching the seed crop. It is also possible sometimes to arrange isolation in time by growing cultivars which differ markedly in period of flowering.

Isolation distances required for individual crops are specified in Section II.

FIELD PRACTICES

Sowing or Transplanting

Seed crops are established either by sowing the seed *in situ* or by transplanting, generally in accordance with the usual practice for crops grown for purposes other than seed, although there are some exceptions; for example, beetroot (*Beta vulgaris* L. subsp. *esculenta*) may be transplanted as stecklings when growing for seed. A main consideration is to ensure that no mixture with seed or transplants of other cultivars occurs. All equipment brought onto the seed production field must be scrupulously cleaned before use to ensure that there are no foreign seeds or plants which might be dropped on the field to grow into plants which would contaminate the cultivar under production.

When sowing seed it is usual to adjust the depth of sowing to take account of the needs of the species. Some species require light to enable them to germinate. Bewley and Black (1994) note that light can penetrate up to 2 mm in sand, but only to 1.1 mm in loam; shallow sowing is therefore required for some species, and where necessary this is noted for particular species in Section II.

Plant Density

Plant density (plant population per unit area) in a seed crop is usually the same as that required in crops grown for purposes other than seed, but there are some exceptions; for example, seed crops of some forage grasses are sown to produce a lower plant density and seed crops of swedes are sown to produce a higher plant population than for crops not grown for seed. Plant density is determined by the amount of seed sown per unit area (seed rate) except for those crops which are thinned after seedling emergence or are transplanted. For both direct sown and transplanted crops, the distance between the rows will also determine plant density per unit area. When seedlings are raised in a seed bed for transplanting, the amount of seed sown per unit area of seed bed can affect the size of the transplants; if the seed is sown too thickly some of the plant population of the cultivar may be suppressed by competition.

Plant density is important in a seed crop for several reasons, and suggestions for optimum seed rates are given for many of the species described in Section II. Optimum plant density for seed production of a particular species balances the ability of each plant to produce seed against the number of plants per unit area which produce seed. Plants that are isolated one from another may individually produce all the seed of which they are capable, but such plants may not achieve collectively the highest seed weight per unit area. The production per unit area is governed by the interaction of the number of plants which are able to produce seed effectively within that area and the number and size of seed produced by each plant.

Generally, crops should not be too dense as this increases the stress on each individual plant caused by competition for water and nutrients. Highly stressed plants may produce seed of low viability and vigour; stress during reproductive growth can limit the ability of seed to develop properly or conditions can cause seed to deteriorate after the seed has reached physiological maturity (Dornbos 1995). Dense crops will increase the chance of disease spread and may reduce the chance of effective pollination in cross-pollinated crops.

On the other hand, crops which are not dense enough may not yield well. However, it is often desirable in the early stages of multiplication of a new and important cultivar to achieve the highest possible multiplication rate rather than the highest possible seed yield per unit area of a crop. To do this it is necessary to adopt a plant density which is much lower than normal so as to give each plant the possibility to produce the maximum amount of seed of which it is capable. For example, in the UK a seed rate as low as 30 kg/ha with wide spacing between the rows has given satisfactory yields of wheat seed and a high multiplication rate, although yield per hectare was lower than with a normal seed rate.

There have been interesting developments from plant density studies in some vegetable crops, notably those in the Umbelliferae. In principle, the seed from a primary umbel has a larger embryo than seed from subsequent umbels. Studies have therefore been directed at determining plant populations which achieve the highest proportion of seed in the harvested seed lot from primary umbels. For example, increased plant density of seeding parsnips increased the proportion of seeds obtained from primary umbels (Gray et al. 1985).

For crops which are transplanted or those which are thinned after seedling emergence, it is possible to control the plant density in a seed crop more precisely. In large-scale seed crops, care must be taken that the plants which are retained to produce seed are not selected deliberately as this may upset the balance of plants in the cultivar population; for example, if only large plants are retained some characteristics of plants which are naturally smaller may be lost. On a small scale, however, it is possible to use the process of transplanting or thinning as a way to select plants with the characteristics of the cultivar. For example, some vegetables such as beetroot can be selected as to cultivar during the time that roots are stored over winter for transplanting as a seed crop, and bulb selection in onions is an important procedure in Basic Seed production of that crop.

In arriving at an optimum seed rate it is necessary to know the germination potential and vigour of the seed lot to be sown. Given that the seed bed provides optimum conditions for germination, the germination test (see Chapter 3) will give a good indication of eventual plant density at any given seed rate. However, if (as is usual) conditions are less than optimum, allowance must be made for some seedlings not being able to survive.

An indication of the vigour of the seed lot is desirable and this is determined by a vigour test. When the germination or vigour test reveals a potential which is lower than required, it may be possible to compensate by increasing the seed rate. However, such seed lots are often low in vigour and, unless seed bed conditions are very good, an optimum plant density may not be achieved.

An aspect which has not been fully investigated is the possible effect of plant density on the maintenance of cultivar purity. Under stress conditions, some plants in the cultivar population may be suppressed and those which survive may not reproduce all the characteristics of the cultivar. Attention was drawn earlier to the risks involved in growing cultivars in areas widely removed from that in which they were selected; changes induced in cultivars under these stressful conditions may also be induced by any stress which eliminates some plants from the cultivar population during growth, so causing 'shift' in the characteristics of the cultivar.

Fertiliser Application

The nutrient status of soil should be determined and fertiliser application adjusted accordingly. Fertiliser requirements for seed crops are generally similar to those for food or industrial crops. However, it is advisable to monitor carefully the use of nitrogen, as excessive use can cause seed crops to develop excessive top growth. This can make crop inspection difficult and may affect pollination in cross-fertilised crops by inducing early lodging. When lodging occurs before flowering, adequate pollen may not penetrate to inflorescences which are almost touching the soil below the lodged crop. For those crops where the seed is not the part of the plant used for food or other purposes (e.g. sugar beet, brassicas, radish, fodder grasses), special fertiliser treatments may

be required and are detailed for individual species in Section II.

Weed Control

Although control of weeds in seed crops is essentially similar to that in the wider context of crops grown for food or industrial purposes, there may be differences in the emphasis on control of particular weeds or in the timing of control measures. Some weeds are of greater importance in seed crops because they cause problems at harvest or because they have seeds which are similar in size, weight or shape to the crop seed in which they occur. In crop species in which the seed is not the part of the plant which is used for food or other purposes, weeds may occur which cause difficulties when the plants are grown to maturity in a seed crop; for example, *Galium aparine* (cleavers) does little harm, except in excessively large numbers, in a crop of swedes for fodder, but is difficult to handle in a swede crop at seed harvest, and the seeds of the weed are difficult to remove from the crop seed. For some crop species there are weeds which are extremely difficult and costly to remove completely from the crop seed: for example, grass weeds such as *Alopecurus myosuroides* (blackgrass) in a ryegrass seed crop. There are some weeds which are parasitic on crop plants, and some which act as alternative hosts to pathogens; these will be discussed in the section in this chapter on 'Diseases and pests and their control'.

The best way to ensure clean seed at harvest is to put the seed crop into a clean field. The less weed control needed after a crop is sown, the better. As mentioned in the section on 'Crop Rotation', the selection of appropriate preceding crops to allow for the control of weeds is an important consideration in planning a seed crop.

Once a field is selected to grow seed, there are three ways in which weeds can be controlled during the growing period: hand weeding, mechanical cultivation, and herbicide application.

In crops grown in extensive fields, hand weeding is practical when there are only a few important weeds present. Whereas in an ordinary crop hand weeding is normally not economic, in a seed crop it may be well worthwhile if the weed is particularly objectionable in the crop species being grown for seed, but occurs so infrequently that other measures would not be appropriate: for example, *Avena fatua* (wild oats) in a barley crop when there are less than 50 plants per hectare. In cereal or other crops sown in close-spaced rows it is worth leaving wider spaces at suitable intervals to allow access for hand weeding and other operations. In vegetable crops grown for seed in relatively small gardens, hand weeding is the norm either by hoeing or hand pulling.

In crops sown broadcast, mechanical cultivation after sowing is not possible. However, for those sown in rows, inter-row hoeing or other cultivation is practical if enough space is left between rows, although in many cases the practice has been replaced by the use of herbicides. In many crops, inter-row cultivation is a practicable alternative to herbicides but must be executed with care as in some instances it may damage the crop plants; for example, inter-row cultivation of grass seed crops can damage the root system if too deep.

There are numerous herbicides available to deal with most weed problems, but their efficiency varies as does their cost. Herbicides are applied mostly as sprays in water either to the soil or to the growing plants, or in pelleted form to the soil. They may be divided into the following main types:

1. non-selective treatments which will eliminate all plants growing in the treated area;
2. selective herbicides which are intended to kill weed plants in a crop without damaging the crop plants.

In each case, treatment may be applied to the soil as 'residual treatments', or to the foliage of the plants. Soil treatments are absorbed by the roots of plants. The foliar applications may operate either on contact by killing the foliage or they may be translocated within the plant to the site where the chemical is effective.

Time of application may be either pre-sowing, pre-emergence or post-emergence, depending on the chemical being used. Pre-sowing treatments will kill weeds during seed bed preparation and are usually used in a 'stale seed bed' technique. Pre-emergence treatments are applied after sowing but before the seedlings have emerged. In both cases, disturbance of the soil after treatment should be kept to a minimum so as to reduce or eliminate further weed seed germination. Timing is very important when using selective herbicides on growing crops as the stage of growth of the crop plants may determine whether or not the crop will be damaged.

There are many types of sprayer available, from the small hand-held and operated model to the large mechanical sprayer, either tractor-mounted or self-propelled. As well as overall spraying it is also possible, when the rows are sufficiently widely spaced, to direct a spray only onto the inter-row space, so avoiding the crop plants. When weeds are widely spaced and few in number, spot treatment with a hand-held spray is possible.

For seed crops it is well worthwhile to create a 'sterile strip' at least one metre wide around the crop, by eliminating all plants in the strip by cultivation. This prevents weeds from spreading into the crop from field boundaries and is also an advantage to wildlife in the area as it provides a buffer to prevent sprays applied to the crop from being too near the field boundary.

When deciding on the appropriate treatment for a seed crop, it is necessary to consider the effect on the production of seed as well as the efficacy of the treatment. For example, some hormone-type sprays used for broad-leaved weeds in cereals may cause distortion of the ear if applied at the wrong time, so making the identification of off-types in the seed crop impossible; or some selective herbicides used for treatment of grass weeds in grass seed crops may markedly reduce seed yield if applied at the wrong growth stage.

The use of herbicides therefore needs very careful consideration in the context of the particular crop, and the advice of the chemical manufacturer should always be followed.

Diseases and Pests and their Control

An important aim of plant breeding is to produce cultivars which are resistant to or tolerant of diseases or pests. The introduction of genetic resistance provides the commercial grower with the most cost-effective and environmentally friendly method of control. It is essential that the seed production process should be so organised that genetic resistance or tolerance is preserved in the cultivar under production. The crop must be established using seed entirely free from disease or pest organisms on a field which has no history of such organisms. The seed crop must also be suitably isolated so as to avoid any chance of infection from other crops.

Types of Organism and their Method of Spreading

There are four types of diseases and pests:

1. fungal diseases;
2. bacterial diseases and viruses (including those caused by mycoplasma-like and rickettsia-like organisms);
3. animal pests (including aphids, weevils, nematodes and birds);
4. parasitic weed plants.

Each type of organism can be spread either as seed-borne pathogens or by other means. In the latter case there are several possibilities: wind dispersal, rain splash, soil contamination, or transport by animals or birds. For these diseases or pests the control measures used do not differ from those used in a commercial crop. However, the damage caused to a seed crop by a disease or pest can be more severe than in a commercial crop because often a diseased crop will produce shrivelled seed which is likely to have lower germination or vigour than fully developed seed. Treatment may therefore be economic in a seed crop which would not be so in a commercial crop, but applications of insecticides to crops cross-pollinated by insects must be done with great caution.

Seed-borne pathogens are, however, more important in a seed crop. It is essential to produce seed which is healthy and will not transmit diseases or pests to the following crop, especially if the seed is to be used in an area far removed from the area of production where the disease or pest in question has not previously been identified.

Control Measures

The primary need is to ensure that the seed used to sow the seed crop is healthy and that the field in which the seed crop is to be grown is free from contamination with soil-borne pathogens. However, there are some control measures which can be taken.

Seed-borne pathogens may be carried either with or inside the seed. Those carried with the seed may be either free-living or attached to the seed coat.

Free-living organisms can sometimes be removed by seed cleaning; for example, seeds of parasitic weeds can be removed by cleaning in some cases. The majority can be treated by a variety of chemical seed dressings applied before the seed is sown. A few can be treated by fumigation; for example, stem eelworm of lucerne, onion or leek, although this is a dangerous operation and must only be carried out by highly trained personnel because the fumigant is highly toxic to humans.

Organisms attached to the seed coat are usually treated by chemical seed dressings.

Pathogens carried inside the seed are generally more difficult to deal with. Some can be treated by chemical seed dressings that contain systemic fungicides which are translocated to the infected site. Alternatively, in some cases heat can be used but requires very careful control as the temperature which kills the pathogen is generally very close to that which will kill the seed. Loose smut of barley has been successfully treated in this way, but heat treatment has now been superseded by translocated chemicals. Heat is normally applied by soaking the seed in water at controlled temperature followed by immediate cooling and drying (but some reduction in germination usually occurs).

Chemical seed dressings may be in the form of powders, slurries or liquids. For small quantities of seed the treatment may be applied by hand by placing the seed and a measured quantity of the chemical on a clean floor or in a suitable container and mixing thoroughly (for example, concrete mixers have been used when other equipment is not available). For larger quantities of seed there are several types of equipment available which meter the seed and the chemical into a mixing chamber to give a continuous flow. Chemical seed dressings are generally toxic to humans and those handling them should be provided with suitable protective clothing and face masks.

In some cases the seed treatment can be incorporated into a seed coating. However, seed coating or pelleting is an operation requiring special equipment and is usually only done by specialist companies.

Seed Inoculation of Legumes

The nitrogen-fixing *Rhizobium* associated with legume species may be inoculated onto the seed when amounts occurring naturally in the soil are not adequate. Inoculum is usually added to the seed immediately before sowing, but is not compatible with some chemical seed treatments. When inoculation is intended, therefore, it is necessary to select with care any other seed treatment which may be applied.

Current Developments in Farming Methods

There are currently two developments which may affect seed growing: organic farming or gardening and precision farming.

For organic farming it is necessary to avoid the use of all inorganic fertilisers and not to use herbicides, fungicides or pesticides. The organic farmer or gardener is thus required to use organic manures or naturally occurring fertilisers such as rock phosphate or rock potash. For nitrogen it is

necessary to rely on residual nitrogen from preceding crops or to use farmyard manure. Both pose difficulties for seed growing if not carefully planned. Preceding crops of legumes to provide nitrogen may build up soil-borne diseases or pests; following legumes with legumes too closely should therefore be avoided. Farmyard manure may contain undesirable weed or other seeds and should therefore be clamped for a season to permit the heat generated to kill them.

At the other end of the scale, precision farming is an attempt to use inputs more efficiently by mapping fields precisely to identify areas where there might be a problem; for example, areas of lower fertility or pockets of disease (Graham and Dawe 1995). It is then possible to give these areas special treatment. Fertilisers and other treatments can be applied in differing amounts to areas with different requirements. Currently, mapping has only been done in cereal and oilseed rape crops by fitting a combine harvester with continuous-flow yield-measuring equipment and also with a satellite navigation device which accurately plots its position in the field. These data are fed into a computer which produces an accurate map of yields from all parts of the field. Investigation of areas showing low yields may reveal ways in which they might be improved. This could be important to the seed grower as the proportion of undersized seed which is lost in seed cleaning might be reduced.

SELECTION PRESSURE AND ROGUING

There are two ways in which cultivar purity can be maintained by selection: *positive selection* and *negative selection*.

Positive Selection

Positive selection involves the selection of a limited number of plants typical of the cultivar, which are then isolated as a group to produce seed. Isolation may be achieved either by leaving the selected plants *in situ* and eliminating all unwanted plants within the group, or by lifting the desired plants and replanting them as a group in isolation, either outside or within insect-proof and protected structures. For bred cultivars it is usually the plant breeder who exercises this type of selection during the maintenance programme, but there are also some situations when the seed grower may use this method. For example, in Canada there is the designation Selected Seed which is applied to some self-fertilising species and is recognised in the seed legislation.

Negative Selection

In this method, a limited number of unwanted plants are removed from a seed crop. Plants which are judged to be not typical of the cultivar are removed by hand and the majority which are typical plants are left to produce seed. This operation is called *roguing* (Laverack and Turner 1995). To remove atypical plants it is necessary to walk through the crop at times when some characteristics of the cultivar can be seen. Some off-types are difficult to see in a seed crop, particularly those which are sown densely in close spaced rows or broadcast. In such crops it is necessary to leave paths at intervals to enable roguers to walk through the crop, taking care not to space the paths too wide apart, so the off-types can be observed and reached without difficulty.

In some species, off-types are difficult to spot, and may be more so at certain times of day when light is unfavourable. In such crops roguing has to be regarded as a cosmetic exercise to a crop which is already largely satisfactory.

In other species negative selection is a useful tool to ensure cultivar purity, especially when the plants in the crop are spaced reasonably far apart. However, roguing should normally be regarded as fine-tuning of an already largely satisfactory crop.

Roguing should be the responsibility of the seed grower who may require training in recognition of off-types. In some instances, particularly during the production of the earlier generations, the plant breeder or the seed production company may supply a roguing team. All rogue plants should be removed completely from the field and destroyed. Roguing can also include the removal of undesirable weeds and individual plants infected with seed-borne pests or pathogens.

Local cultivars are generally grown in a designated area where the plants are subjected to particular natural selection pressures imposed by the local growing conditions, including traditional methods of crop management. Roguing in these cultivars should be carefully controlled so as to avoid changing the balance within the cultivar population.

REFERENCES

Bailey, R. (1990). *Irrigated Crops and their Management*. Farming Press Books, Ipswich, Chapters 1 and 2.

Bewley, J.D. and Black, M. (1994). *Physiology and Development of Germination*. Plenum Press, New York, 275.

Chandler, T.J. and Gregory, S. (1976). *The Climate of the British Isles*. Longman Group Ltd, London, 273.

Copeland, L.O. and McDonald, M.B. (1995). *Principles of Seed Science and Technology*. Chapman and Hall, New York, 69–76.

Dornbos, D.L. Jr (1995). Production environment and seed quality. In *Seed Quality* (Ed. A.S. Basra) Haworth Press, New York, Chapter 4.

George, R.A.T. (1985). *Vegetable Seed Production*. Longman, Harlow, 44–47.

Graham, R. McD. and Dawe, A.F. (1995). Yield mapping for precision farming – operating principles and equipment. *Journal of the Royal Agricultural Society of England*, **156**, 35–42.

Gray, D., Stekel, J.R.A. and Ward, J.A. (1985). The effect of plant density, harvest date and method on the yield of seed and components of yield of parsnip (*Pastinaca sativa*). *Annals of Applied Biology*, **107**, 547–558.

Kelly, A.F. (1994). *Seed Planning and Policy for Agricultural Production*. John Wiley and Sons, Chichester, 138–140.

Laverack, G.K. and Turner, M.R. (1995). Roguing seed crops for genetic purity: a review. *Plant Varieties and Seeds*, **8**, 29–46.

Russel, E.W. (1973). *Soil Conditions and Plant Growth*. Longman, London, Chapter 26.

Further Reading

Fenner, M. (Ed.) (1992). *Seeds, The Ecology of Regeneration in Plant Communities*. CAB International, Wallingford.

Jeffs, K.A. (Ed.) (1986). *Seed Treatment*. The British Crop Protection Council, Thornton Heath.

Kelly, A.F. (1988). *Seed Production of Agricultural Crops*. Longman, Harlow, 6–8.

Martin, T. (Ed.) (1994). *Seed Treatment: Progress and Prospects*. British Crop Protection Monograph No. 57.

Webster, C.C. and Wilson, P.N. (1980). *Agriculture in the Tropics*. Longman, Harlow, Chapter 1.

5

Harvesting, Processing and Storage

WHEN TO HARVEST: SEED DEVELOPMENT AND RIPENING

The main objective of harvesting a seed crop is to secure the maximum amount of high quality seed with the minimum amount of any other undesirable materials (the descriptions of the factors involving seed quality are given in Chapter 3).

The time to harvest a seed crop in relation to its stage of maturity is often a difficult management decision. For maximum yield of quality seed it is necessary to ensure that the bulk of seed is sufficiently mature, particularly in those species of indeterminate habit which tend to ripen over an extended period. If the decision to harvest is too premature, the subsequent yield of good seed will be depressed, mainly due to the presence of immature seed material. However, if the decision to harvest is delayed there is likely to be a loss resulting from seeds dropping (usually referred to as 'shedding'). Seed which has been shed is not normally recoverable, although suction or vacuum harvesting equipment is used for some of the forage legumes, for example *Macroptilium atropurpureum* (DC) Urban.

The moisture content of the individual seeds is an important consideration in deciding when to harvest. The optimum of safe moisture content varies with the crop species. For example, for wheat it is 20% or less, while for tetraploid ryegrass it is 40% or less. Moisture contents at harvest of less than 10% for wheat or 35% for ryegrass are likely to present problems with shedding or other losses. In practice, the decision for most crops whose seeds dry on the straw is based on the moisture content determined from a sample rubbed out in the field. The hardness is assessed by pressing individual seeds with a thumb-nail. Seed should usually have a doughy texture before harvest, but should not be so hard that the thumb-nail does not mark it. Seed which is too hard will be more susceptible to mechanical damage during subsequent operations.

The decision as to when to harvest will depend on the prevailing weather conditions and also on the harvesting method to be used. Time of day can also be used to advantage; although the seed material will become drier as the temperature increases, crops most prone to shattering can be harvested with the dew still on them to minimise losses.

The natural drying process of standing seed crops can be accelerated in some species by the use of a desiccant such as diquat. However, this increases the production cost and there is a possibility of reducing potential germination and

Encyclopaedia of Seed Production of World Crops.
Edited by A.F. Kelly and R.A.T. George. © 1998 John Wiley & Sons Ltd

of causing seed discolouration if not applied correctly. Desiccants are used successfully in the seed production of *Beta* spp. and some of the forage legumes.

There has been some interest in the use of water-soluble plastic resins such as polyvinyl acetate sprayed onto standing seed crops to prevent shedding (Williams 1977) but this has not been developed to any appreciable extent.

It is essential that all machines, appliances and containers used for harvesting and subsequent operations are clean and free of seeds and other debris likely to contaminate the potential seed crop.

METHODS OF HARVESTING

The seed for harvesting is contained either in the heads of relatively dry plant material, for example cereals and brassicas in which the seed drying process has already commenced, or within fruits such as tomatoes and cucurbits from which the wet seed has to be extracted before it is dried.

The simplest system is to gather the seed-bearing material by hand, using tools such as knives, sickles and secateurs, or in some cases shaking the heads into a bag. The tool used usually depends on the height above ground at which the seed-bearing material is to be cut. Hand harvesting is frequently used for the horticultural crops whose fruits mature and ripen over an extended period. It is also widely used for any seed crop produced in areas where hand labour is readily available and cheap, and allows for the removal of only ripe seed at each harvest, with repeated visits as more seeds ripen. Small areas of seed crops produced by plant breeders during cultivar development and maintenance are usually hand harvested; this is especially useful when a species has a sequential seed development and ripening pattern on individual plants. Seed heads from several plants can be tied together in bundles for further drying and threshing later.

It is generally accepted that while hand harvesting ensures the gathering of the maximum amount of seed material per unit area, it is a relatively slow process for larger areas where the use of mechanisation will secure the potential seed harvest more quickly, and this is important when the duration of satisfactory harvest time is limited by inclement weather.

Mechanical Methods for Harvesting Dry Seed

There are three alternatives for the mechanical harvesting of dry seed crops: (i) cutting the plant material but leaving the seed on the straw to be threshed later; (ii) using a combine harvesting system which cuts and threshes out the seed in one operation; (iii) brush ('stripper') harvesters which remove ripe seed and leave the stems in the field.

Cutting

There are two main systems associated with cutting seed crops prior to threshing. One is for the standing crop to be cut and left in windrows and the other is for the cut material to be automatically tied into sheaves or bundles. In both cases, further drying of the seed and plant material can take plant *in situ*.

When the cut crop is in sheaves it is usual practice for them to be propped up in groups (known as stooks). When done on a small scale it is possible to move the sheaves to a protected area for further drying or place the bundles on raised racks in the field.

Threshing

Threshing can be achieved by hand or machine. Hand threshing is done for very small seed lots and is still used in many countries for large amounts of cut material where hand labour is readily available; in these cases the seed is removed from the bulk of the dry mother plant material by flailing, beating or rolling.

Machines designed for threshing have a revolving drum with either pegs or spikes attached; some models have aspiration which assists the

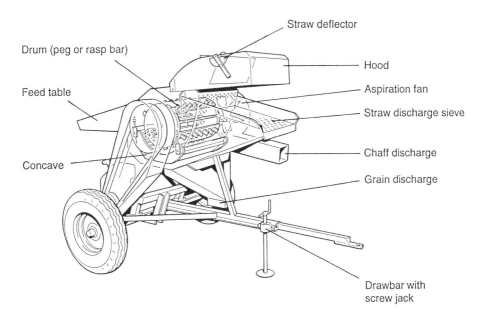

FIGURE 5.1   Mobile thresher (figure kindly supplied by Alvan Blanch Ltd, UK)

overall separation of seed from the lighter plant materials. A diagram of a mobile thresher with aspiration is shown in Figure 5.1.

Combining

Combine harvesters are designed as self-propelled machines to cut and thresh in a single operation while passing over the standing crop. In principle, combines are suitable for relatively large production areas but, while capable of effecting a single-pass operation, they require a high level of investment in machinery with trained staff to operate and maintain them. Small combines have been developed for harvesting cultivar trials and can also be used for small-scale seed production of suitable crop species (for example, during development by a plant breeder).

The seed crop should be sufficiently dry for efficient threshing when using the combine harvester's cutting and threshing capabilities simultaneously. This combined operation enables the producer to secure the seed crop from the field and move it to the next stage.

Seed crops which ripen over a relatively short time-scale are especially suited for combining, but the technique is also used for lodged seed crops where the seed may have shed but is still held within the laid plants.

Combine harvesting machines can also be used as stationary threshers or for picking up and threshing seed-bearing material which has been previously cut and left in windrows. In the latter case the combine is fitted with a pick-up header or pick-up reel.

Alternative harvesters have been designed for making more than one pass over a field; they use a brushing or shaking action to remove only the ripe seed at each pass, leaving immature seed on the plants to ripen further.

Harvesting Seed from Fleshy Fruit

Fruits are either removed from the parent plant by hand or by machine. Mobile seed extractors are used for large-scale production, which pick up the fruit by front elevators and crush them. The seed is separated from the debris by passing the mixture through a revolving cylindrical screen. Extra water is added for some crops, such as squashes, when the fruits are relatively dry. Stationary crushers use the same separation principle.

SEED HANDLING AFTER HARVEST

Extraction of Seed from Fleshy Fruits

Some of the vegetable crop species produce seed in a fruit from which the seed has to be extracted. In some crops the seed is extracted from the fruit immediately after it is harvested (e.g. tomatoes and cucumbers), while in others the fruit is partially dried first (e.g. small-fruited peppers).

The general principle is to chop the fruit and mechanically separate the seeds from the debris. The machines used are generally referred to as 'fleshy fruit seed extractors' or 'wet vegetable seed separators'. The wet seeds are dried before further processing. Small quantities can be extracted by hand. It is necessary to remove the gelatinous coating which surrounds tomato and some cucurbit seeds by either fermentation or a chemical process prior to drying. The techniques used for individual seed crops are described in Section II.

Processing

The term 'processing' covers a wide range of operations which enhance the physical properties of a seed lot. When seed has been separated from the mother plant, some plant debris and other impurities still remain mixed with it and further processing is necessary. The cleaning or 'upgrading' of a seed lot is done in stages according to crop species and the condition of the individual seed lot. In most climates it is also necessary to ensure that the seed is further dried in order to retain its potential germination. Only when the threshed seed has been produced in arid conditions is immediate further drying unnecessary. A series of machines and operations for sequential seed processing is usually called a 'processing line'; Figure 5.2 shows this diagramatically.

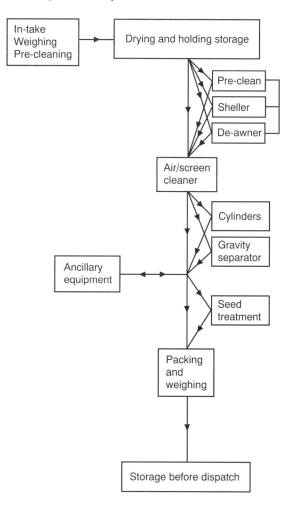

FIGURE 5.2   Example of a seed processing line

Seed Moisture Content and Pre-cleaning

It is important that the seeds' moisture content be reduced to a safe level immediately following threshing or extraction from fruit.

The impurities in the seed lot, such as pieces of leaf, straw and weed seeds, are likely to have a relatively high moisture content and will contribute to the overall moisture content.

The duration between seed threshing or extraction and its first cleaning should be planned within hours rather than days, especially when the material has a high moisture content, otherwise the seed will be damaged by the material's natural process of heating up. Material should not be left in dense heaps or closed containers.

Initial Post-harvest Processing

The first processing operation is usually to separate all of the crop seeds from plant debris and the non-seed materials in order to prepare for the initial post-harvest drying process. The main objective is to reduce the bulk so that the drying process is more efficient, especially when it is to be dried by warm air. This operation is often referred to as 'pre-cleaning' or 'scalping'. The combination of pre-cleaning and the initial post-harvest drying is called 'conditioning'.

There are many methods of pre-cleaning; the one used will depend on climate and scale of operation. Those available include hand sieving, winnowing (including mechanised winnowing) and pre-cleaning screen machines with a range of throughput rates (with or without aspiration).

The methods or systems for drying seed at the conditioning stage are also diverse. They include sun drying, ventilated buildings, in-sack driers, structures and buildings with purpose-built floors of fine mesh through which warm air is forced, ventilated bins with air introduced through either a perforated floor or a central perforated duct which passes the exhausted air through perforations in the bin walls. A system for drying large quantities is the continuous-flow drier through which the seed material moves either vertically or horizontally. A wide range of drying machines and systems has been described (ISTA 1963).

The necessary drying facilities must be available on seed farms. Farmers producing commercial quantities of seed in remote areas or in fragmented seed production schemes should also have access to appropriate driers.

The upper moisture content to be obtained at the completion of conditioning is considered to be 3% above the level required for storage when all processing has been completed. The suggested maximum warm air temperatures for drying seed are given in Table 5.1.

TABLE 5.1 Suggested maximum warm air temperatures for drying seed (source: Kelly 1988)

Crop	Initial moisture content (%)	Maximum temperature of air reaching seed (°C)
Wheat, barley, oats, rye	Over 24	44
	24 and below	49
Brassicas and clovers	18–20	27
	10–17	38
Peas	Over 24	38
	24 and below	43
Chaffy grasses (temperate), e.g. *Lolium* spp., *Dactylus glomerata*	45	38
	40	49
Tropical grasses, e.g. *Chloris gayana*, *Bracharia* spp.	Over 18	32
	10–18	37
	Below 10	43
Soyabean	20	40
Maize (usually dried on the cob)	25–40	35
	Below 25	40

Seed Processing Equipment for Specific Purposes

There is a wide range of seed processing equipment; Delhove and Philpott (1983) listed and described 78 types and they also listed some 325 suppliers covering 18 countries.

The following items of equipment are considered to be the most commonly used during seed processing for the purposes given; these are in addition to the types of air and air/screen machines and fleshy fruit seed extractors mentioned above. Figure 5.3 shows the principles involved in an air/screen cleaner.

Indented Cylinder

This machine can be used to separate seeds of different species which may have similar girths but different lengths; it can also be used for grading seed of some species.

Each indented cylinder is made of steel and has precision-made, equal-sized indents or 'pockets'

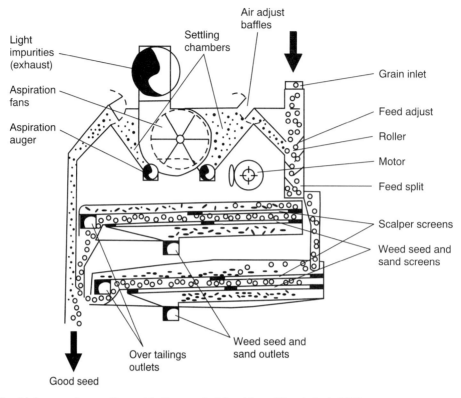

FIGURE 5.3   Air/screen cleaner (figure kindly supplied by Alvan Blanch Ltd, UK)

on its inner surface. Cylinders are available with different sizes of indent and some have detachable covers for use with different species. The cylinder, which is at an angle, revolves; seed is fed in from the higher end and the longer seeds move towards the lower end while the shorter seeds each lodge in an indent. As the cylinder revolves, the seeds (or other pieces of material) drop out as the indent approaches the higher position. There is a trough running through the cylinder which catches the seeds from the indent as they drop, and they pass out of the cylinder separately from the longer seeds, or other materials, which have not entered the indents.

Disc Separator

The disc separator works on the same principle as an indented cylinder machine except that there are a series of discs with indents which revolve on a

shaft within a hollow cylinder. The seeds are fed into the cylinder. Smaller seeds enter the pockets and are carried upwards until they drop into catching trays, while the longer seeds travel down the cylinder and are collected separately from the seed and other materials which did not enter indents on the revolving discs.

Gravity Separator

Seeds which are the same size and shape, but of differing weights, can be separated by a gravity separator; for example, light-weight seeds of low potential germination can be removed from a seed lot. The machine consists of a perforated deck covered with cloth or wire mesh which can be oscillated and set at a slope. Air is blown from below the deck and passes through the permeable surface. The principle of separation is that the lighter materials float on the air cushion and move down the gradient while the denser seeds remain on the deck and pass up the slope as a result of the oscillations. The separated fractions of seeds are collected at different points at the sides of the deck. Adjustments can be made by interchanging the table surface, the angles of tilt sideways and/or lengthways, oscillation rate, and rate of air passing through the permeable deck. The types of deck-covering used relate to the separation requirements. Options for positions of guide plates or dividers provide a range of density fractions.

There are two types of gravity separators, usually referred to as triangular or rectangular, according to the table's basic shape, although their general principles are the same.

De-awner

The de-awner is used for the removal of appendages, such as awns, from the seeds of a range of crop species including carrot, celery and dill. It is also used for de-bearding barley seeds and the machine may therefore be referred to as a de-bearder. It is also used to improve the potential flow characters during drilling of some species' seeds by trimming off excess awns or loose glumes during seed processing.

The machine consists of rotating beater arms, or blades, on a steel cylinder which rotates within a tilted cylindrical cage. The blades rotate and, as the seeds are thrown against the interior of the cage, appendages are clipped off by the perforations. The speed of the blade cylinder is adjustable. There is an inlet for unprocessed seed at the top of the higher end of the outer cylinder and an outlet for cleaned seed on the underside of the lower end.

Spiral Separator

A spiral separator, sometimes referred to as a 'spiral gravity separator', relies on gravity for its function. It is capable of separating seed of the same size but different densities or seeds of different shapes or degrees of 'roundness'; for example, separating whole legume seeds from those which are broken or split.

The apparatus consists of a vertical spiral trough which is narrow at the top and widens towards the base. As the seed materials travel by gravitation down the trough to the basal outlet, the more rounded or denser seeds travel to the outer edge of the spiral trough while the flatter, less dense or broken seeds and materials remain closer to the centre of the spiral. Splitter chutes segregate the two main fractions which are collected in separate outlets, or spouts, at the base. Some spiral separators have more than two divisions formed by increasing the number of splitter chutes near the base of the spiral.

Roll Mill

This item of seed processing equipment is sometimes referred to as the 'velvet roll mill' or 'dodder mill'. It works on the principle of differential seed coat characters, i.e. rough or smooth. It may also be used for separating immature, damaged or wrinkled seed material from a smooth-seeded species. A common application is the separation of dodder seed from a clover seed lot. The roll mill is only used when indicated following processing of the seed lot by other types of cleaners.

The roll mill principle is based on two parallel and inclined velvet-covered rollers or cylinders. The covered rollers revolve outwards from each other. The seed material is fed at the higher end of the pair of rollers. Smooth seeds gravitate down the trough formed by the two rollers and are collected at the bottom. The contaminating seeds and other materials with rough surfaces adhere to a roller's velvet surface and are deflected into a discharge spout. The speed of the rollers and feed rate can be adjusted according to the specific requirements of the seed lot being processed.

Magnetic Separator

This machine is used as a final cleaning process in order to accomplish separation of seeds of specific contaminants, such as *Galium* spp. ('cleavers') which have a rough seed coat, from seeds of smooth-coated crops, e.g. *Brassica* spp. The principle of operation is that iron powder is introduced into the seed lot which has been moistened with

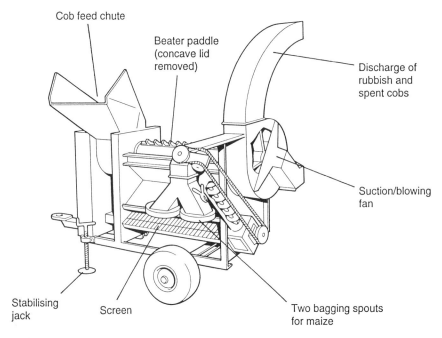

FIGURE 5.4   Maize sheller (figure kindly supplied by Alvan Blanch Ltd, UK)

either water or high viscosity oil; oil is used in preference to water for crop seeds such as flax, which absorbs water following the formation of a slime on the seed coat when wetted by water. The iron particles adhere to rough-coated seeds and any other irregular-surfaced material, including cracked and damaged crop seeds.

The seed lot is passed over a magnetised roller from a feeder unit via a mixing unit. The seeds and other materials with iron particles on them adhere to the magnetic roller and are brushed off into spouts for the contaminant and/or reject materials. The smooth crop seeds are collected separately as they fall from the roller.

Inclined Draper

The inclined draper is a finishing machine which is used to separate seeds with different potential rolling or sliding physical characters. It mainly functions on the principle that different species' seeds in a mixture have different specific gravities; it also operates, to some extent, on differing surface textures of seed or other contaminants.

This piece of equipment consists of a tilted flat belt covered with velvet, canvas or plastic. Seed is fed onto the upward moving belt. Smooth or round seeds move to the lower end where they are collected separately from the rough-surfaced or non-spherical materials at the top end.

The seed input rate from the hopper, the belt's angle of inclination and speed can each be adjusted according to the physical properties of the seed lot to be processed.

Maize Sheller

When maize ears have been dried they are shelled either by hand or machine. Figure 5.4 shows a typical machine designed for maize shelling. With extra concaves the same machine can be used for sorghum or millet.

Picker Belt

This is a simple machine, consisting of a moving belt on which the seed materials are conveyed to the end where they are collected. Workers stand, or sit, on either side of the belt and hand pick materials to be discarded; rejected materials are placed in suitable containers beside them. It is usually used to remove contaminants such as stones or larger pieces of plant debris significantly larger than the seed to be retained, or any other specified off-type material. The efficiency of this machine is highly dependent on the diligence of operators.

Electronic Colour Separator

The electronic colour separator, also sometimes referred to as a 'colour sorter', is used for the final processing operation of relatively large-seeded crop species to remove discoloured seeds from an otherwise normal seed lot. Examples of its use are the separation of bean seeds with halo blight from those which are uninfected, and removal of discoloured, brown-stained pea seeds from a normal green or light green seed lot.

The principle of operation is that a regulated stream of seeds passes a photoelectric cell; individual 'off-colour' seeds activate an eliminating mechanism such as a jet of air, which removes the relevant seed from the main seed lot. Instruments can incorporate monochromatic and bichromatic sensors so that different colours and varying intensities of the same colour can be detected and removed.

Electrostatic Separator

In the electrostatic separator, seeds are segregated according to their natural charge or their ability to disperse when placed in an induced charge.

Seeds are charged and passed through an electric field and those with an insulating coat hold their charge. The charged seeds are diverted from the main stream of seeds, which do not retain their charge.

While the principle is widely accepted, the practical application of this method for seed processing remains relatively little used.

Needle Drum Separator

This machine has a series of needles housed on the inside of a drum. Seed is passed through as the drum revolves. Those seeds which have holes in

them adhere to the needles and are removed as the drum passes a rotary brush. This type of separator is used to remove weevil-damaged seeds from relatively large-seeded species, such as beans and peas.

Machines for Treatment against Pests and Diseases

It is essential that machines and equipment used to apply seed treatments or dressings to seeds be both efficient and safe. The important considerations include: accurate application of the relevant material to each seed in the lot being treated; efficient adherence of the chemical; safety of operators; and avoiding environmental pollution. The machines should be installed with appropriate dust extraction facilities.

There is a range of methods used for the application of pesticides or fungicides to seed. A comprehensive review has been made by Jeffs and Tuppen (1986). The simplest models are hand operated and are very appropriate for use where there is a low technical back-up. There are also types of treaters which are attached to drill boxes. Other seed treaters which are generally more sophisticated fall into the following groups: *auger mixers* which generally apply powders, although some models also cope with slurries; *revolving drum*, usually for the application of liquids, but some models are also designed to apply slurries; *spinning disc* for the application of liquids; *rotostats* which are capable of applying powders, liquids or emulsions. There are also other types of seed treaters, based on modified or additional principles to the above, which have been designed and developed by individual pesticide manufacturing companies. Figure 5.5 shows a liquid seed treater designed primarily for on-farm use, but which can also be incorporated within fixed or mobile seed processing plants.

Seed Inoculation

There are situations where it is necessary to inoculate seed of leguminous crop species with *Rhizobia* spp. in order to ensure satisfactory

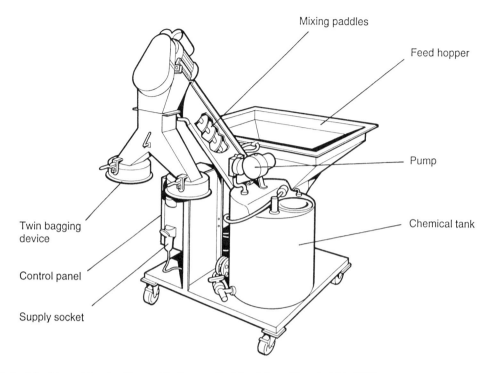

FIGURE 5.5   Liquid seed treater (figure kindly supplied by Alvan Blanch Ltd, UK)

nitrogen fixation. This is an important consideration when introducing a leguminous species into
an area where it has not been previously grown.
Many of the chemical seed treatments are incompatible with inoculation since they are toxic to the
Rhizobia spp., and care must therefore be taken in
the selection of suitable treatments. Stovold and
Evans (1980) have described the effects of fungicide
seed dressings on the nodulation of *Pisum sativum*
and *Glycine max*. The effects on *Rhizobia* spp. of
fungicides applied to legume seed have been
discussed by Staphorst and Strijdom (1976).
Inoculum is best added to the seed just before
sowing and is usually done on the farm. The
technique is well developed in many countries and
the inoculum is available commercially from
specialist companies. Finely ground peat is usually
used as a carrier for the inoculum. The peat carrier
is usually dampened with water to form a slurry
which will adhere to the seed. Other materials, such
as molasses, may be added to the mix to cause it
to stick to the seed. An alternative method is to
sprinkle the seed with water and then to add the
dry peat powder plus the inoculum, although this
is not generally as effective. Some growers add the
dry powder in the seed drill at the time of sowing,
but this is not as effective as the slurry method.
Methods have also been developed for soil
application.

Pelleting

Seeds of some crops are pelleted in order to
modify the seed's shape or to increase its size. The
pellets are of a uniform size and shape within each
crop species, which facilitates mechanical handling
for direct sowing in the field (for example, sugar
beet) or in soil blocks (for example, lettuce).
During the pelleting process the seeds are coated
with an adhesive and then coated with clay and/or
fine fibre. In some cases, crop protection chemicals
and/or nutrients are incorporated in the pelleting
materials.

Coating

When a seed is coated it retains its general shape. The coating material can contain specific additives such as crop protection chemicals, nutrients and nitrifying bacteria.

Treatments to Improve Seedling Establishment

Various treatments have been developed which are designed to improve the rate and uniformity of germination and the early establishment of seedlings. The results have been variable, even between different seed lots of a single cultivar. Three treatments have been suggested.

1. *Prehydration*. The seeds are soaked in water for a strictly controlled period which allows the initial processes of germination to start; the seeds are then either sown immediately or dried for storage and sowing later.
2. *Osmoconditioning*. The seeds are treated with low water potential osmotica allowing pregerminative metabolic processes to start, but are prevented from germinating by controlling the treatment time.
3. *Matric priming*. The seeds are treated with low water potential solids or gels; this has the same potential effect as osmoconditioning.

Best results have usually been achieved when the seed is sown soon after treatment. A system for which special equipment has been developed, known as 'fluid drilling', involves sowing seeds which have been suspended in a protective carrier gel.

These treatments, sometimes collectively known as 'priming', are costly and have been used only on high value seeds.

PACKAGING

Seed is normally packaged when the final processing operation has been completed. The size of the package for an individual crop species will depend on whether it is destined for distribution to the ultimate user or is to be distributed to reputable organisations which will repack it in smaller quantities. Some cereal seed is delivered to large-scale farmers and wholesalers in bulk or bagged lots of one tonne, in which case the seed is handled either mechanically or by front-end loaders.

The final quantity of seed in a given packet size usually relates, as far as is commercially possible, to the requirements of recipient farmers and growers in the various marketing areas. However, individual packets must not exceed a weight which can be safely handled. Within these limits, packages of an individual crop's seed can match farmers' plot or field sizes, or for some of the high value vegetable seeds may relate to the number of plants required per unit area.

The type of packet or container used should protect the seed from contamination, mechanical damage and storage pests including rodents.

A wide range of materials is used for packaging seed. These include jute, woven polypropylene, multiwalled paper, aluminium foil, plastic and tinplate. Combinations of two of these basic materials are frequently used: for example, a jute or woven polypropylene outer layer over a plastic inner container; multiwalled paper with aluminium lamination; aluminium foil with polyethylene. The choice of materials used may depend on the seed industry's level of development in an area, available materials, the value of individual seed lots, and whether vapour-proof packaging is required.

Although most modern packaging materials will repel moisture to some extent, it is possible to use a vapour-proof packaging system. This depends on the use of vapour-proof materials, packing seed which has a moisture content not exceeding 8%, and immediately sealing the package so that it is airtight. When the seed has been stored in a controlled atmosphere, the packaging process is done in rooms with controlled humidity and airlock entrances. The use of vapour-proof packaging systems enhances the storage life of the contained seed lot and ensures that the individual seed lot's moisture-proof environment is intact right up to the time that the package or container is opened.

Basic information relating to the contained seed should be clearly stated on the outside of the packet or on a firmly attached label. This includes crop species, cultivar, supplier and any chemical seed treatments applied. This information must comply with any prevailing legal requirements, including taking into account OECD (Organisation for Economic Co-operation and Development) regulations where appropriate. Reputable commercial seed companies generally provide their logo as part of their policy to maintain a mutual trust between farmer and supplier, as do seed organisations in the public sector. An additional advantage is that farmers can easily identify seed from suppliers whom they have found to be reliable.

SEED STORAGE

The main objective of seed storage is to retain seed stocks securely and to ensure that, as far as is biologically possible, the stored seed retains its potential germination and vigour.

It is essential that seed which is to be stored is mechanically pure because plant and seed debris will be prone to attack by micro-organisms which produce both heat and moisture. Fungal growth will commence on stored cereal seed when moisture content exceeds 14% and on groundnut above 12.5%. It is generally accepted that for vegetable seed the critical moisture content is 12%. These differences appear to depend on crop species and interactions with individual species of storage pathogen; the topic has been reviewed by Justice and Bass (1978).

Insect pests of stored seed can be a major cause of deterioration of seed quality during storage. The actual physical damage to a crop species will depend on the pest's mode of action; for example, some feed on the seed's endosperm, others on the embryo. In addition to this type of direct effect on the seed's potential vigour or germination, the respiration of insect populations will significantly increase the moisture level in the atmosphere of the stored material and in turn lead to increased activity of storage fungi and even trigger germination of the stored seed.

The main environmental factors which affect the storage of seed are temperature, moisture and oxygen pressure; in practice, only temperature and moisture content are taken into account. Other factors include the pre-storage history of the seed lot (including any detrimental effects of harvesting and processing) and the crop species being stored. Most geographical areas of the world experience seasonal and daily fluctuating temperatures; in many locations there are also periods of high relative humidity which can cause the moisture content of the stored seed to rise. As it is the seed's moisture content coupled with storage temperature which generally determine the seed's storage life, the two rules of thumb described by Harrington and Douglas (1970) are frequently applied to demonstrate storage potential. These are that:

(i)   for every decrease of 1% in seed moisture content the life of the seed is doubled;

(ii) for every decrease of 5°C in storage temperature the life of the seed is doubled.

These rules of thumb apply when the seed moisture content is between 14 and 5% and within the temperature range of 50 to 0°C. Generally commercial seed stocks have to be stored from one season to the next, which is a duration of six to 12 months, depending on whether the seed is to be used locally or exported to another region. Most crop species will maintain satisfactory viability for this duration when stored with seed moisture contents of between 10 and 14%. Their moisture should ideally be reduced to below 10% when seed stocks have to be stored beyond this duration, and the seed stored in vapour-proof containers in order to avoid the seed moisture content being at equilibrium with that of the ambient atmosphere. High value seed stocks, such as Basic and Breeder Seed categories, are either stored in vapour-proof containers or controlled atmosphere rooms in which the relative humidity is maintained at optimum levels to enhance the seeds' storage life. Long-term storage of germplasm is usually achieved using combinations of vapour-proof containers and low temperature in specially designed buildings (Cromarty et al. 1982).

The International Seed Testing Association's rules for determination of seed moisture content should be followed in order to achieve both accuracy and repeatability (ISTA 1996). Results from the continuous use of methods using other instruments should be observed with caution, especially when changing between seed lots of different species and particularly if routine recalibration is not done according to manufacturer's instructions.

There are no recognised recalcitrant seed crops (for example *Citrus* spp. and *Hevea* spp.) within the agricultural and horticultural species covered in the scope of this work. The reader is referred to Roberts (1972) for a comprehensive review of storage environment and the control of viability.

SECURITY OF STORED SEED

Buildings designed or adapted for use as seed stores should be secure, relatively fire-proof, insect-, rodent- and bird-proof, and on a site free from flooding. In areas where high temperatures prevail, the exterior surfaces should minimise the absorption of solar radiation. The interior surfaces, including the floor, should be of smooth finish and constructed so that they exclude moisture arising from the exterior.

Seed store discipline should be enforced in order to minimise fire hazards, maintain cleanliness, maintain security of seed stocks, and exclude unthreshed seed materials, chemicals, fertilisers and machinery not directly associated with seed storage.

REFERENCES

Cromarty, A.S., Ellis, R.H. and Roberts, E.H. (1982). *The Design of Seed Storage Facilities for Genetic Conservation*. International Board for Plant Genetic Resources, Rome.

Delhove, G.E. and Philpott, W.L. (1983). *World List of Seed Processing Equipment* FAO, Rome.

Harrington, J.F. and Douglas, J.E. (1970). *Seed Storage and Packaging Applications for India*. National Seeds Corporation and Rockefeller Foundation, India.

ISTA (1963). Drying and storage. *Proceedings of the International Seed Testing Association*, **28**.

ISTA (1995). International Rules for Seed Testing. *Seed Science and Technology*, **24**(Supplement), 49–52, 271–273.

Jeffs, K.A. and Tuppen, R.J. (1986). Application of pesticides to seeds. In *Seed Treatment*, 2nd edition (Ed. K.A. Jeffs). The British Crop Protection Council, Boston Heath, 17–45.

Justice, O.R. and Bass, L.N. (1978). *Principles and Practices of Seed Storage*. United States Department of Agriculture, Washington DC.

Kelly, A.F. (1988). *Seed Production of Agricultural Crops*. Longman Group, Harlow.

Roberts, E.H. (1972). *Viability of Seeds*. Chapman and Hall, London.

Staphorst, J.L. and Strijdom, B.W. (1976). Effects on *Rhizobia* of fungicides applied to legume seed. *Phytophylactica*, **8**, 47–54.

Stovold, G.E. and Evans, J. (1980). Fungicide seed dressings: their effects on emergence of soybean and nodulation of pea and soybean. *Australian Journal of Experimental Agriculture and Animal Husbandry*, **20**, 497–503.

Williams, C.M.J. (1977). Glueing inflorescences increases yield in vegetable crops. In *Proceedings of Australian Seeds Research Conference, Tamworth, New South Wales*.

Further Reading

Copeland, L.O. and McDonald, M.B. (1995). *Principles of Seed Science and Technology*. Chapman and Hall, New York, Chapter 11.

Culpin, C. (1981). *Farm Machinery*. Blackwell Scientific, London.

ISTA (1973). Seed storage and drying, *Seed Science and Technology*, **28**(3).

ISTA (1987). *Handbook for Cleaning of Agricultural and Horticultural Seeds on Small-scale Machines. Part 1 and Part 2*. International Seed Testing Association, Zurich.

Pill, W.G. (1995). Low water potential and presowing germination treatments to improve seed quality. In *Seed Quality* (Ed. A.S. Basra). The Haworth Press, New York, Chapter 10.

6

Seed Security

This chapter describes the way in which plant breeders secure seed stocks for future use, and the link with genetic conservation. Various methods are detailed for the production and maintenance of pure seed stocks for different types of cultivar, with some discussion of the implication of new breeding methods for seed production. Finally, there is a review of the need to provide reserve seed stocks against major disasters.

PLANT BREEDERS AND CULTIVAR MAINTENANCE

Germplasm Protection

All plant breeding programmes need access to germplasm with genetic variability which is used to produce recombinants that are an improvement over the original parents. However, plant improvement in this way is a relatively slow process, and most progress is made by a series of small changes to the genetic base of commercial crop species by hybridisation and recurrent selection. Sources of germplasm for the breeder are various: commercially available cultivars, older cultivars no longer in use, land races and 'exotic' types, wild relatives, and parental lines selected from within the breeding programmes.

Established plant breeding programmes all maintain some kind of germplasm collection in order to provide a useful in-house resource, although this is usually a limited range of parental lines, named cultivars, and other selected lines not commercially available. The wider range of existing and potential genotypes for possible future use is normally included in purpose-run genebanks which are funded by national governments or international organisations.

The Use of Genebanks

During the 1970s, many areas of the world saw significant changes in agricultural practices, particularly in Third World countries, as growers changed from using traditional techniques to more intensive methods. This involved a change to using seed of new hybrid cultivars, along with fertilisers, pesticides and increased mechanisation, in order to obtain higher yields to satisfy the demands of a rapidly growing population and world markets in general. It resulted in the rapid decline in use of many traditional cultivars, and great efforts were made to collect them from many of the affected areas before they were lost completely. Much of this work was co-ordinated

Encyclopaedia of Seed Production of World Crops.
Edited by A.F. Kelly and R.A.T. George. © 1998 John Wiley & Sons Ltd

by the International Plant Genetic Resources Institute (IPGRI) based in Rome, and resulted in large numbers of additional cultivars coming into the genebanks that had already been established, and the creation of new ones in some countries. The initial emphasis was on creating effective systems for long-term storage, but since then efforts have been directed towards a more detailed evaluation of selected parts of the collections, and the production of informative catalogues.

Genebanks will hopefully provide a valuable future resource, although breeders in general tend to use material that has been generated from within their own programmes, as this is best adapted to the local environment. Generally speaking, 'exotic' types are used to introduce specific character traits, for example a novel source of disease resistance, and after the initial cross with an adapted genotype, subsequent back-crossing is usually necessary in order to select out the undesirable characters.

Genebanks in the UK

In the UK the major genebanks are at the John Innes Centre (JIC) in Norwich, and at Horticultural Research International (HRI), Wellesbourne, near Warwick. The JIC has an extensive collection of wheat, oats and barley, including over 50 000 accessions of the wild barley *Hordeum spontaneum* Koch, and is known internationally for its large collection of *Pisum* types. Regular requests for seed and information are made to the genebank from breeders and research organisations and, in turn, breeders are asked to deposit small quantities of seed entering the UK National List system for peas and cereals.

The vegetable genebank at HRI Wellesbourne was started in the early 1980s after concerns that the new requirements for national list testing of vegetables would lead to a huge loss of many old 'varieties', both in the UK and elsewhere. Initially funded jointly by Oxfam and the UK government, and now wholly funded by the UK government, it contains a range of species, including *Allium, Brassica* and *Daucus*, as well as a wide range of lettuces and tomatoes from all over the world.

Seed is stored at a temperature of 1°C and a humidity of less than 10%, and should remain viable for at least 25 years. These conditions help to reduce the pressure for regeneration of seed stocks, and renewal takes place only when they are depleted by requests, or the germination falls to a critical level. These requests for material are increasing both in the UK and internationally, from breeders and other researchers in the public and private sectors, and seed companies are regularly asked to donate small samples of commercial cultivars.

Genebanks Worldwide

IPGRI keeps a directory of some 87 institutions in 43 countries around the world, which have important germplasm collections. Major crop species include maize, soyabean, cotton, rice, sorghum, wheat, tomatoes and *Brassica* species, several of which are held at more than one location. In general, funding is provided internationally from a variety of sources such as the World Bank, the European Commission and national governments. Networking between them is now well established; for example, there are 30 member countries of the European Co-operative Programme for Crop Genetic Resources (ECP/GR). The aim is to ensure long-term conservation, and to encourage increased use of plant genetic resources in Europe. The programme is entirely financed by the participating countries, and co-ordinated by IPGRI.

Security for the Future

Genebanks have an important role to play in the future as the providers of the best known source of genetic diversity in economically important crop species. This is especially true now as advances in biotechnology make it possible for many crop species to be screened using restriction fragment length polymorphism (RFLP) analysis in order to establish the level of genetic variability. This technique, combined with the greater use of databases (for example, putting information onto CD-ROM), should help to increase our knowledge of the material, and make the information more accessible. The increasing evidence of climatic change in many areas of the world, in addition to population increases, loss of soil fertility from erosion, desertification, salinisation, and urban encroachment, continue to put pressure on plant breeders everywhere to produce improved cultivars. Many crops also have potential for non-food uses in a world where renewable resources will become increasingly important. There is no doubt that material held in the genebanks will be a valuable resource in trying to meet the demands of the future.

Pure Stock Production and Maintenance

Plant breeding programmes consist of many stages, from the initial hybridisation to final marketing. The initial hybrid is made between two chosen parents, and the seed from each generation is grown to produce plants which are selected for a range of attributes, until homozygosity is reached, for the self-pollinating crops. By this process of elimination, small numbers of advanced lines are used to start pure stock production, providing the link between the final stages of selection and trialling, and the start of commercial seed production. In the UK and other European community countries, legislation requires breeders to submit new cultivars for official tests before they can be entered onto the national list and sold legally. In most of the developed countries there is provision for plant breeders' rights which requires official testing before a right is granted (see Chapter 2). This involves testing for distinctness, uniformity and stability (DUS) where entry onto the national list and plant breeders' rights are needed, and value for cultivation and use (VCU) trials where there is only a need for entry onto the national list, at the authorised stations. The breeder is required to submit the prescribed amounts of seed for testing, which has to meet specified levels of purity and germination. Pure stock production aims to prepare plant material from advanced lines which will achieve these standards.

Crop Husbandry

Good crop husbandry plays an important part in pure seed production. It is essential to use a rotation which minimises the risk of 'volunteer' plants appearing in the growing crop. For example, a cereal crop would require what is technically known as a 'double-break'; this requires the two preceding crops to be non-cereal, preferably a cultivated root crop, and in the case of oilseed rape (*Brassica napus* L. var *oleifera* and *B.rapa* L.) a minimum of five years after any *Brassica* seed crop and if at all practicable, much longer. Both chemical and mechanical means of weed control are often necessary throughout the growing period to avoid any weed seeds getting into the harvested sample. At all stages, machinery must be carefully cleaned and inspected to avoid the risk of contamination, particularly at sowing and harvest. Care must also be taken to avoid mechanical damage to the seed at harvest and during the cleaning process, to avoid the risk of reduced germination. Seed cleaning will also help to remove any other contaminating 'weed' seeds and broken or undersized seed present in the sample. Isolation from other crops of the same species is essential to prevent cross-pollination in out-crossing species. It is also desirable with the self-pollinating species because, even here, a low level of out-crossing can occur and will significantly affect seed purity in small populations.

Self-pollinating Crops: Wheat, Barley and Oats

The small-grained cereals such as wheat (*Triticum aestivum* L. emend Fiori et Paol.), barley (*Hordeum vulgare* L.) and oats (*Avena sativa* L.) are self-fertile, and pure seed is most easily prepared by using a continuation of the pedigree breeding method. This also normally applies to peas (*Pisum sativum* L.) and dwarf beans (*Phaseolus vulgaris* L.). Described below is the established procedure as used at Plant Breeding International (PBI), starting at the F6 or F7 generation when selected lines are generally homozygous (see Figure 6.1). Breeding lines at this stage are the product of a rigorous selection procedure, and it would normally take three or

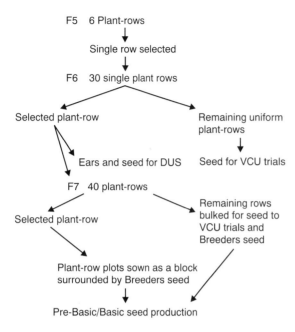

FIGURE 6.1 Pure stock production scheme for wheat, barley and oats

four years from this stage to proceed to national listing and marketing. Multiplication normally begins with 20 or 30 single plants harvested from a single uniform progeny row in the F6 generation.

Preparation for sowing starts with a detailed examination of grain and ear characters, often with the aid of a low-power binocular microscope, in order to draw up a morphological description. Every plant is examined in this way, and only those of a uniform type, conforming to the description, are used for the initial multiplication. Any off-type plants are either discarded or recycled through the breeding programme as reselections. Seed is sown using a precision drill to produce a plant spacing of between 7 and 10 cm. In this way it is easy to identify single plants and remove the off-types, if necessary, as well as giving a very high multiplication rate, often in excess of 250:1. A disadvantage of this system is that it gives rise to later-ripening tillers, and can delay harvest. Agronomic characters are recorded throughout the growing season, and any variants are removed, either as single off-type plants or as

plant-row families. If the differences can be recorded prior to flowering, it can be an advantage to remove them when recorded, as this reduces chances of out-pollination within the stock. Major characters recorded include leaf colour, time to ear emergence, glaucosity, pigmentation and plant height. The relatively small plots at this stage (less than 0.1 ha) are prone to out-pollination, particularly in winter barley if there is a prolonged cold period during flowering. For this reason it has been found that grouping stocks by character and growing these in isolation from one another is particularly effective, especially for prominent characters such as pigment and aleurone colour.

Entry requirements for national list testing usually involve both seed and ear samples for DUS and a seed bulk for VCU. Multiplication by the breeder at national list entry stage continues via the pedigree method, whereby the seed and ears harvested from a single plant the previous season are grown as separate plots and rows. These are generally referred to as progeny-rows when using seed from the whole plant, or ear-row progenies for seed from a single ear. When grown using this method, any variation due to residual genetic segregation or out-crossing can be recorded, and off-types removed from the production before harvest. In the first year of national list testing, between 30 and 60 plant progenies derived from the same material as submitted to DUS tests are grown together in a plot of about 0.2 ha. The procedures followed are the same as in the previous season, with a view to confirming the description, and removing any off-types if they occur. Sheaves and ears are harvested from a single progeny-row to provide the basis for the pedigree plots in the following season. Seed harvested from uniform progeny-rows is bulked together at harvest to provide the initial Breeder Seed stock. During this second year of tests (NL2), the selected progeny-row is threshed and seed is divided into approximately 100 small plots and drilled as a block, surrounded by the Breeder Seed crop (0.5 to 1.0 ha).

Procedures are repeated as before during the growing season. It is after completion of the second year that candidates satisfying the testing authorities can be entered onto the national list. Ongoing maintenance for the production of Pre-Basic, Basic and Certified Seed continues to follow the same basic principles of husbandry as described above. Purity is maintained by regular inspection and manual roguing aimed at removing off-types and persistent weeds such as wild oats (*Avena fatua*).

Out-pollinating Crops

Out-pollinating species require the presence of insects, wind, or a combination of both, for effective pollen transfer to produce fertile seed, and the key to production of pure seed is the use of isolation. This is achieved by the use either of minimum distances from other crops with which the seed crop is likely to out-pollinate in the field, or of physical barriers such as glasshouses or polytunnels and protection of individual flowers by bagging. For pollination in confined spaces, either bees or blowflies can be used, and the choice seems to be a matter of individual preference. However, there are some significant differences, notably that blowflies appear to be more active than bees at lower temperatures, are not subject to cultivar preferences (i.e. are more random) and are more difficult to confine. Examples of several different crops are described below.

Oilseed rape The initial stages of pure seed production are generally carried out in polytunnels or glasshouses where single plants are selected at F6, and bagged at flowering using a pollen-proof material. These are tied over the flower heads, and then shaken regularly during the flowering period. At the next generation, 400 plants are grown in an insect-proof polytunnel, and blowflies or bees are introduced for pollination. The crop is rogued carefully and any off-types are removed. Elite plants are selected and bagged, seed from bagged plants is harvested separately, and the remainder, which would normally produce around 50 kg, used to supply DUS tests, VCU trials and further yield trials. Seed from an elite plant is grown again under glass,

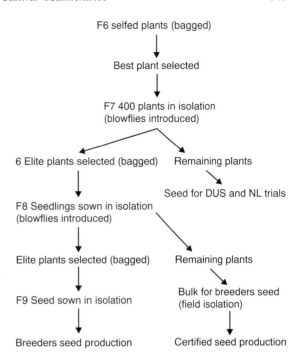

FIGURE 6.2 Scheme for oilseed rape production

rogued as before, elite plants bagged and harvested separately, and the bulk seed used to provide seed for NL2, for other trials, and to begin initial Breeder Seed production in the field. The field multiplication would usually take 1–2 ha, sown in wide rows to permit further roguing, with the aim of producing 4–8 t of Basic or Pre-Basic Seed. A minimum isolation of 500 m would be maintained from other flowering brassica crops, and seed from bagged plants in the centre of the plot used to continue the cycle of elite seed production (see Figure 6.2).

F1 Hybrids

Nowadays, nearly all commercial cultivars of maize (*Zea mais* L.), most brassicas, carrots (*Daucus carota* L.) and onions (*Allium cepa* L.) are produced by using F1 hybrids. The phenomenon of heterosis, or hybrid vigour, gives the F1 hybrid the advantage over inbred lines in that it generally produces higher marketable yields by way of better quality, earliness and uniformity.

Seed production costs are invariably higher because an F1 can only be produced from the original parental lines, which, in general, are owned by the breeder and not commercially available. Growers must therefore return to the breeder for a supply of fresh seed before each growing season. This gives absolute security of seed supply to the breeder, or his agent, because growing saved seed will produce a variable crop with little or no resemblance to the previous generation, and may be unmarketable.

Hybrids are produced when two different cultivars are crossed by transferring pollen from the male or donor to the female, whilst preventing self-pollination. This can be done by exploiting several different techniques such as genetic male steriles, self-incompatibility, chemically induced male sterility using chemical hybridising agents (CHAs), and hand emasculation. Simple hybrids are the product of cross-pollination between two inbred lines; more complex are double-cross hybrids (the product of crossing two hybrid parents) commonly used in maize, and triple-cross hybrids which have been used in producing some cultivars of kale (*Brassica oleracea* L. convar. acephala).

Maize   Maize is wind pollinated, and seed production crops need a minimum isolation of 500 m from any other maize crop, with a surround of at least three rows of pollen parent, sown 80 to 90 cm apart. Within this pollen barrier, females and males are sown in the ratio of 6:2 in clearly identified rows, timed to match the flowering dates. When cytoplasmic male sterility (CMS) has not been introduced into the female line, de-tasselling of the female parent is carried out before the anthers dehisce; this can be done by hand, as in much of India and Africa, or by using mechanisation or CHAs applied at critical developmental stages which render the pollen inactive, common in most large-scale production in the USA. This operation needs to be repeated often, to be certain that all the male inflorescences from side-tillers of the female parent have been removed. Seed from the male line is generally harvested prior to the main harvest from the female, and requires detailed instruction to harvesters as to the position of the rows. Again, this operation can be done manually, by hand picking the whole cobs, or mechanically. When hand harvested, the dried cobs can be examined for type, grain colour and diseases, and unwanted cobs discarded.

Tomatoes   Tomatoes (*Lycopersicon lycopersicon*) are naturally self-fertile, but inbred cultivars are rarely used for commercial fruit production, as most growers prefer to use F1 hybrids. All seed crops have to be hand pollinated, and the majority are grown in the field in areas of India and China. Some production is also carried out in California, USA, and under the protection of glass or polytunnels in northern Europe. Female and male plants are grown in rows supported on frames, in a ratio of 5:1; the male lines are planted to match the flowering dates of the females to ensure an adequate supply of pollen. Emasculation of the female flowers is done at the late bud stage, often in the afternoon, and pollination is carried out the following morning. This operation needs skill in manipulating the flowers and in timing the pollinations.

Brussels sprouts   The condition of self-incompatibility can be exploited for F1 production, and is due to the presence of what is known as the S-allele in some species, which acts to prevent self-fertilisation with the same genetic type. For example, in Brussels sprouts (*Brassica oleracea* L. var. *gemmifera*) this condition is widely used to produce F1 seed. In northern Europe production starts with sowing seed from the two inbred lines in nursery beds during the spring, and from there seedling plants are selected on a range of desired characteristics and planted out in polytunnels for elite seed production. The following spring, blow-flies are introduced at flowering to ensure pollination. It is possible in this situation to harvest seed from both parents, provided that a strong S-allele is present in each, and this will provide F1 seed from either parent. This can be verified by harvesting seed separately from each parent, and progeny-testing a portion by growing the seeds to assess uniformity on the seedlings. A high level of uniformity indicates strong self-incompatibility in both parents.

The inbred lines are maintained by bud pollination, which is done by hand and has the effect of overcoming the effect of the S-allele.

Male Steriles

Onions, Carrots and Parsnips Cytoplasmic male sterility (CMS) is commonly used in seed production for the biennial bulb and root crops such as onions, carrots and parsnips (*Pastinaca sativa* L.). This requires the breeder to maintain three inbred lines: the donor, male sterile and the restorer line, which is crossed with the male sterile to produce viable seed of the male sterile line.

For the production of high quality seed, the roots or bulbs of each inbred line are harvested at the end of the first growing season and selected for uniformity, trueness to type and other quality characters, before being planted for the flowering cycle in isolated pollination blocks. For the onion crop, it is common practice to plant the male steriles (MS) in a 2:1 ratio with the pollinator by placing four rows of the MS for every one row of the pollen parent on each side. The pollen donor is often planted between one and two weeks earlier than the female parent so that ample pollen is available when the MS plants flower. Because they are all insect pollinated, it is essential to provide sufficient isolation from other similar crops or species. In the UK and northern Europe, bees or blowflies are introduced into insect proof glasshouses or polythene tunnels at flowering time to effect pollination. However, there is some evidence to suggest that bees can be preferential when working some cultivars, which can lead to variable seed set, whereas blowflies are generally more random. These enclosed environments also have the effect of enhancing flowering and hastening seed maturity, and generally ensure a good quality seed harvest because sprouting can be avoided, producing high germination rates.

Composites and Synthetics

True hybrid cultivars, with their high yield potential and uniformity, are of increasing importance in some crops. However, where high production/seed costs are a constraint, then the breeding of synthetic and composite cultivars can offer an alternative with equivalent potential performance but less uniformity.

Synthetics and composites in cross-pollinated crops take advantage of heterosis due to natural combining ability, and rely on random pollination for seed yield (unlike F1 hybrids which have a strict control of pollination to produce seed, e.g. by using male sterility).

Composite and synthetic populations rely on insects or wind for pollination, and cultivar maintenance requires strict isolation. Broadly speaking, synthetics and composites are much the same thing, a synthetic being a set of inbred lines put together on the basis of tests for their combining ability, whereas a composite may comprise a set of lines with complementary characters which produce an overall advantage. In winter beans (*Vicia faba* L.), for example, composite populations are formed by bringing together up to five different inbred lines, originating from crosses made in the glasshouse, followed by subsequent field selection. Combinations are made so that, generally, characters such as height and maturity are broadly uniform, but there may be variation in less important features such as flower colour (Figure 6.3).

Pure stocks of the selected lines are grown together in pollination cages, and hives of bees are used to maximise the intercrossing. The resulting composite is multiplied in isolation for subsequent generations. After two or three generations (syn 1–3) cultivars will enter trials, and the best of them will proceed to national list trials. For DUS purposes, the breeder is obliged to maintain the constituent lines. Multiplication and maintenance of small populations can be done in field cages or soil-based glasshouses where, again, small colonies of bees can be introduced for pollination. Use of the glasshouse is more effective for *Ascochyta* control in the winter bean crop, as there are strict

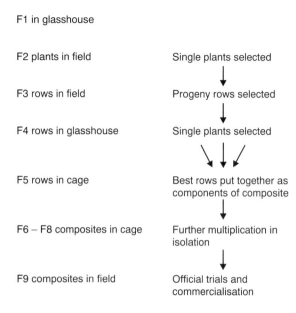

F1 in glasshouse

F2 plants in field   Single plants selected

F3 rows in field   Progeny rows selected

F4 rows in glasshouse   Single plants selected

F5 rows in cage   Best rows put together as components of composite

F6 – F8 composites in cage   Further multiplication in isolation

F9 composites in field   Official trials and commercialisation

FIGURE 6.3   Scheme for composite production in *Vicia faba*

limits on the disease levels on seed as required by the certifying authorities. Larger-scale multiplications in the field must rely on physical isolation to maintain purity. For example, beans with coloured flowers need a minimum of 400 m, and white-flowered varieties 1500 m, from another crop of the same species.

Intellectual Property, New Breeding Methods and their Implications for Seed Security

Over the past 60 years or so, genetic improvement, largely by the efforts of publicly funded plant breeders, has led to quite dramatic yield increases in many of the world's major cultivated crops, and techniques such as those outlined above have ensured a reliable supply of good seed to growers. During the 1960s national plant varieties rights legislation was introduced in a number of developed countries based on the International Union for the Protection of New Varieties of Plants (UPOV) Convention (1961), the objective being to act as an incentive to breeders to continue investing in the long-term process of crop genetic improvement. The legislation introduced tests for distinctness, uniformity and stability (DUS), which established a unique description of the cultivar using a recognised list of phenotypic characters for each crop species. Successful registration enabled the breeder or his agent to collect a royalty on sales of certified seed, but did not prevent breeders from using protected cultivars as a source of parental material in their own programmes.

Over the past decade, research in the field of biotechnology has led to a far greater understanding of genetics at the molecular level, and new techniques have been developed which have introduced new possibilities. The use of molecular markers enables rapid finger-printing of a genotype directly, and plant transformation has transferred single genes across the species barrier. Techniques such as these will enable the plant breeder to produce new cultivars (for example, with resistances to pests and diseases, herbicide tolerance, and quality traits, etc.) which would have been considered impossible by traditional methods. Along with these developments, methods which reduce the time taken from the initial hybridisation to commercial release are now more widely used, such as single-seed descent (SSD), the production of doubled haploids, and microspore culture.

However, the high cost of these technological advances and the privatisation of crop breeding has been the driving force behind the need to introduce patent-based forms of protection. The need for change is highlighted by the distinct possibility that the introduction of a novel genotype would not be exhibited as a phenotypic character in any current DUS test. This means that it would be possible for unscrupulous breeders rapidly to incorporate any new genetic trait in a UPOV-protected cultivar from a competitor into their own breeding lines, without acknowledging the originator and the development costs. This has led to a redrafting of the UPOV Convention in 1991, and has paved the way for new legislation to cover two main areas of cultivar protection. First was the right to claim royalties on farm-saved seed, and second, the right of a breeder to sue other organisations or individuals if they attempt to exploit new cultivars shown to be 'essentially derived' from that breeders own UPOV-protected material (see Chapter 2 for a detailed account of UPOV.)

As new technology becomes more important in making future breeding progress, it is likely also to play a part in the registration of cultivars and any subsequent royalty payments to plant breeders. Some kind of genetic finger-printing may well be used by breeders, so as to police the use by customers and competitors of their protected genotypes, and by the authorities responsible for DUS testing. Most breeders, particularly the plant breeding companies, would favour the rapid development of the former for obvious reasons of self-protection, but are more wary of moving away from a phenotype-based DUS test which would inevitably involve increased costs and uncertainty. However, it may be possible to maintain the present DUS test by insisting that any novel gene of economic importance should be linked with an identifiable phenotypic character.

Finally, the long-term security of the provision of reliable seed of improved cultivars to growers is further complicated by the signing of the Convention of Biological Biodiversity, at Rio de Janeiro in 1992. This gave the sovereign right to nation states to control the movement across their borders of all biodiversity within them, if they wished. Whether this will have any restrictive effect on the availability of genetic resources in food crops remains to be seen.

RESERVE SEED STOCKS AGAINST MAJOR DISASTERS

Farming systems can be seriously affected by crop failures resulting from disasters. These can be broadly attributed either to natural causes (including drought and flooding) or to human intervention (civil strife and war). Whatever the primary cause, or causes, the direct results can often include an abrupt break in seed supply at local, national or even regional level. The seed supply crisis is frequently a part of the large effect of a disaster on crop failures directly linked to food supply. This is distinct from a very local 'seed crop failure' which is a term used in the seed industry to indicate a deficient harvest of a given cultivar resulting from very local environmental or management problems during the seed crop's production. Seed shortages which result from disasters are usually much more evident in the staple crops such as cereals and grain legumes. The break in supply includes planting materials in addition to seed.

In developed countries the seed industry is normally sufficiently robust to deal with seed shortages which may arise from time to time, but in the developing countries serious shortages may occur which require contingency plans.

When the disaster area is relatively small, shortfalls in seed production may be countered by the local social network or even the affected farmers' own seed reserves.

The type of emergency seed supply will depend on whether or not there were previous arrangements for maintaining seed stocks specifically for contingencies. Thus, in addition to farmers' own reserves, the alternatives are seed aid schemes and seed security schemes.

Seed Aid Schemes

A seed aid scheme, or programme, can be defined as an emergency seed supply initiated as the result of an *ad hoc* process following a seed shortage caused by a disaster. This type of programme is usually linked with emergency food supplies and, because they are often in the form of grain for consumption, there should be a clear distinction between emergency supplies of food grain and seed intended for sowing. This is especially essential when the seed has been treated with crop protection chemicals. Misuse of seed is more likely when there are insufficient quantities of food grain distributed to subsistence farmers following a disaster. The material intended for sowing should match farmers' requirements regarding delivery time, quantity and cultivar, and ideally should be certified. However, it must be appreciated that aid agencies may not always be able to procure and deliver the same cultivar normally sown by the recipient farmers, especially when attempting to provide urgently needed seed.

Seed Security Schemes

The phase 'seed security for food security' is currently being used to indicate the need to set up schemes which would have in their objectives the requirement to ensure that there is at least one season's seed supply in store to provide for contingencies arising from disasters. It would appear that by far the majority of countries with vulnerable farming systems are still giving priority to developing their national seed industries and associated marketing programmes. However, there are indicators that a few national governments have started to address the problem of emergency seed supplies. A survey conducted by the Technical Centre for Agricultural and Rural Co-operation on national seed systems (Delhove 1991) found that in Ethiopia, where there are recurrent droughts, the Ethiopian Seed Corporation (ESC) has a strategic seed reserve project in which local land races are collected, multiplied, processed and stored for redistribution to the area farmers in drought years; however, there is no seed security programme for the normal supply of Pre-Basic, Basic or Certified Seed when there are shortages. In the same survey, Mauritius reported that a buffer stock of seeds for multiplication is kept in a cold store although there is no regular carryover of commercial seeds; Zambia replied that strategic plans were made for about 25% carryover as security against natural disasters.

Non-government organisations (NGOs) are also involved in seed security schemes, usually at local community level. Louwaars and Marrewijk (1996) reported that NGOs had set up local seed banks in Mali and The Sudan. The problems arising from these types of schemes are discussed by Cromwell et al. (1993).

It has been argued that in many cases following disasters, farmers are able to obtain sufficient seed from their own or neighbours' stocks and it would therefore be more economic to provide them with cheaper food grain for consumption, which would enable them to use their seed for planting (Anon 1985). While this may be true in some cases, the need for a more formal approach to seed security continues to receive attention.

An international lead is required to establish a system or systems of seed security. These could be centred on regions and make use of any existing networks or international groups such as the Southern African Development Community (SADC).

The Food and Agriculture Organisation of the United Nations (FAO) may act as a catalyst and play a key role in advising on the setting up of seed security schemes. Its Office for Special Relief Operations (OSRO) has considerable experience in dealing with seed shortages caused by different types of disasters.

Any proposals for seed security schemes should take into account the following points:

- the best use should be made of locally available seed materials which are preferred by farmers;
- existing national facilities should be involved, including seed storage potential for each seed category;
- potential multiplication requirements and sites should be identified;
- links should be established between vulnerable areas and developed seed industries in other parts of the world, including facilities for out-of-season production;
- the Quality Declared Seed Scheme should be introduced for ensuring minimum seed quality control during post-emergency recovery periods (see Chapter 2);
- farmers affected by disasters do not usually have sufficient finances for the purchase of seed;
- international organisations and agencies may have to assist with initial financing of proposed schemes;
- information from multicentre trials should be used to identify substitute cultivars following shortages;
- replacement and emergency seed supplies must reach the most vulnerable farming communities.

REFERENCES

Anon (1985). *Proceedings of the Seminar on Seed Production, Yaounde, Cameroon, Vol. 2*. Technical Centre for Agricultural and Rural Co-operation and The International Agricultural Centre, Wageningen.

Cromwell, E., Wiggens, S. and Wentzel, S. (1993). *Sowing Beyond the State*. Overseas Development Institute, London.

Delhove, G.E. (1991). *Seed Programmes and Projects in Developing Countries*. Technical Centre for Agricultural and Rural Co-operation ACP-EEC, Wageningen.

Louwaars, N.P and Marrewijk, G.A.M. (1996). *Seed Supply Systems in Countries*. Technical Centre for Agricultural and Rural Cooperation ACP-EU, Wageningen.

Further Reading

Brown, A.H.D. (Ed.) (1989). *The Use of Plant Genetic Resources*. Cambridge University Press, Cambridge.

Bond, D.A. et al. (1991). Breeding for improved quality in field beans. *Aspects of Applied Biology*, **27**, 31–35.

George, R.A.T. (1985). *Vegetable Seed Production*. Longman, Harlow.

IPGRI (1995). *Annual Report*. International Plant Genetic Resources Institute, Rome.

Riggs, T.J. (1988). Breeding F1 hybrid varieties of vegetables. *Journal of Horticultural Science*, **63**, 369–382.

Stefferud, A. (Ed.) (1961). *Seeds: The Yearbook of Agriculture*. USDA, Washington.

SECTION II

Seed Growing of Particular Species

ALLIACEAE: VEGETABLE CROPS

Allium cepa L. Common Name: Onion

Onions are grown in most areas of the world, either for their dry bulb production or as young, green plants which are used as salads. Onion is a biennial and the formation of the bulb is dependent on day length. The photoperiodic requirement for bulbing was described by Jones and Mann (1963) and the physiology of flowering reviewed by Peters (1990). Onion is cross-pollinated by insects.

Classification of Cultivars

Mature bulb
  Shape: ranging from flat, globe to cylindrical with intermediates
  Skin colour: white, yellow, brown, red or green
  Flesh colour
  Pungency
  Dry matter content
Foliage
  Colour
  Pose

Environmental Requirements for Seed Production

Onion seed is normally produced in rain-fed areas or where sufficient irrigation is available during first- and second-season plant growth and development. However, it is important that relatively dry conditions prevail during bulb and seed ripening.

The crop grows best on soils with a pH between 6.0 and 6.8. NPK ratios of 1:2:2 are generally applied, although some seed growers produce the crop with a higher nitrogen level, which is usually applied as top-dressings after plant establishment.

Previous Cropping and Isolation

Satisfactory rotations are necessary to minimise the incidence of soil-borne diseases such as basal rot and pink root rot, and also nematode pests.

Individual countries will usually have stipulated minimum rotation requirements according to seed class.

The usual minimum isolation distance is 1000 m, although the minimum distances required by some authorities are usually greater for cultivars in different bulb colour groups, and zoning of these groups may be enforced.

Seed Crop Establishment

There are two basic methods of producing onion seed, namely 'seed-to-seed' and 'bulb-to-seed'.

In the seed-to-seed method the crop is sown in the late summer or early autumn and is either transplanted, or grown *in situ* through to flowering and subsequent seed production. The sowing rate is 4 to 5 kg/ha with rows 70 to 100 cm apart.

For the bulb-to-seed method the crop is sown sufficiently early to produce cured bulbs by the end of the first year; these are usually referred to as 'mother bulbs'. Seeds are either sown on the flat in rows 40 to 55 cm apart or in beds with 90 to 100 cm between centres, with rows in the beds 30 to 40 cm apart. At the end of the first growing season the mother bulbs are lifted and sorted in order to make selections on bulb characters. The selected bulbs are subsequently replanted. The mother bulbs are either stored before replanting or are replanted almost immediately, depending on the climate. In some onion seed production schemes, the mother bulbs are produced in a different area from where they are grown on for seed production; this is especially the case in the hilly areas of countries such as Pakistan and Nepal, where all the farmers in one village may be participating in a seed production enterprise. The bulb-to-seed method provides a higher level of selection while the seed-to-seed method is less labour intensive.

In the more advanced and established seed production areas, where there is satisfactory control of the successive seed categories, Basic Seed of open-pollinated cultivars is produced by the bulb-to-seed method and the final seed category is produced by seed-to-seed.

When F1 hybrid seed is to be produced, the ratio of rows of female to male plants is usually 4:1, although the maintaining breeder should advise on this and also on ways of ensuring synchronisation of flowering. The breeding mechanism for hybrid onion is described in Chapter 6.

The agronomy of tropical onion production has been reviewed by Currah and Proctor (1990) following a comprehensive survey of 46 countries, which included seed production.

Seed Crop Management

Irrigation is usually necessary in most onion seed producing areas. Systems which apply water at ground level should be used in preference to over-head irrigation in order to reduce the incidence of foliar pathogens. The importance of adequate available water throughout anthesis has been clearly demonstrated by Millar et al. (1971). It is usual practice to reduce or cease irrigation during the ripening of mother bulbs, and also when the seed is ripening.

Allium species are poor competitors with most weed species and efficient weed control should be exercised, especially during young plant establishment. There will usually be a high requirement for hand weeding where appropriate herbicides cannot be used.

Seed Crop Roguing

Seed-to-seed crops There are two roguing stages, the first during the autumn of the first season when plants with off-type foliage, off-type bulb or stem colour should be removed, in addition to any plants showing signs of bolting. The second is just before flowering in the second year when plants with off-type foliage, off-type bulb or incorrect stem colour for the cultivar are removed.

Bulb-to-seed crops There are four possible roguing stages.

(i) Before bulb maturity, when plants with off-type foliage, off-type bulb or stem colour and plants which are late maturing or bolting are removed.
(ii) When sorting the lifted mother bulbs, when the shape, colour and relative size of each mother bulb is checked; any showing signs of bolting, thick necks ('bull necks'), doubling or splitting, damage or disease are discarded.
(iii) At planting of mother bulbs, when bulbs with the same characters as described under (ii), and in addition any early sprouting bulbs, are discarded.
(iv) Immediately prior to the start of flowering, when any plants showing atypical umbels or symptoms of nematode infection are removed.

It is important in the production of hybrid seed to observe roguing requirements stipulated by the maintenance breeder; this includes checking within the female parent rows for any pollen-bearing plants.

Harvesting

Onion umbels have traditionally been hand harvested when approximately 5% of the capsules on individual heads have commenced shedding seeds. Ripe seed heads shatter very readily, and experience, combined with a knowledge of the local climate, is very helpful in coming to a decision on the timing of the operation. The ripening umbels are individually cut from the mother plant with approximately 15 cm of seed stalk attached. Selective umbels harvesting can be done as individual heads start to shed, or alternatively a once-over harvest is made. However, the harvesting of individual heads as they ripen is likely to achieve the maximum potential seed yield provided that every care is taken to minimise loss from shedding.

After cutting, the severed umbels are further sun dried on sheets. The depth of cut material should not exceed approximately 20–30 cm; if deeper than this, the material should be turned daily to avoid damage from overheating and fungal activities. In areas where rainfall or heavy dew is likely, the drying of the cut umbels is done under cover. This can be efficiently achieved in drying bins; according to Brewster (1994), the draught temperature should not exceed 32°C until the moisture content is less than 19%, 38°C until less than 10%, and 43°C when less than 10%.

Mechanical harvesting is now used in some areas where seed is produced in large fields and the cost of hand harvesting would be prohibitive; however, if not carefully organised it can lead to significant seed loss. Combine harvesters with the threshing cylinder removed are usually used, and the cut material is subsequently further dried before threshing. Globerson et al. (1981) examined flowering and seed maturation in relation to mechanical harvesting and found the best time for mechanical harvesting was when the seed had a dry matter content of 60–70% and that seeds which were dried while still in capsules attached to the stalks germinated better than those dried in their capsules after separation from the umbels.

Threshing and Cleaning

The harvested and dried material is ready for threshing as soon as seeds can be separated from their capsules by rubbing in the hand. Dry onion seeds are relatively brittle and threshing should be done as soon as this stage is detected.

The method of threshing will depend on the scale of operation. Whichever process is used during threshing and subsequent cleaning, frequent checks should be made to ensure that the seed is not mechanically damaged. Hand flailing is used on small-scale production, otherwise threshing machines with carefully adjusted concaves are used.

One of the main contaminants in seed lots of *Allium* species is the remains of inflorescences, especially the flower pedicels. It is therefore important to ensure that the process of seed extraction minimises the separation of pedicels from the stalks.

Air/screen cleaners are used for the initial cleaning after threshing. Further cleaning can be achieved by indent cylinders or a gravity separator. When significant amounts of light-weight debris still remain, additional upgrading is achieved either by floating it off by putting the seed lot in water, or by magnetic separation. When magnetic separation is used, it is preferable to treat only the light fraction from the previous process. Following either flotation or magnetic separation, the seed lot is again dried to a moisture content not exceeding 12%.

Drying and Storage

The potential storage life of seed of *Allium* species is significantly reduced by a poor storage environment. Ellis and Roberts (1981) have quantified seed ageing and survival at a range of temperatures and moisture contents. As onion seed is regarded as a relatively valuable commodity, it is usually dried down to 6% moisture content and stored in vapour-proof packets or containers.

Allium porrum L. Syn. *Allium ampeloprasum* L. var. *porrum*. Common Name: Leek

Leeks are an important crop in Europe, some of their popularity being due to their winter hardiness. This species is grown in Asia and the Middle East where the young plants are cut and used as flavourings and salads.

Classification of Cultivars

Leek cultivars have been classified into eight types by Brittain (1988), namely Bulgaarse Reuzen, Danish types, Franse Zomer selections, Swiss Giant selections, Blauwgroene Herfst, Autumn Mammoth selections, Giant Winter and Winter-reuzen.

The following characters are used in cultivar descriptions:

Maturity period, winter hardiness, resistance to early bolting

Height of shaft or column, relative thickness
of shaft
Degree of bulbing
Leaf: colour, shape and length/width proportions, prominence of leaf keel.

Environmental Requirements for Seed Production

Leeks produce a satisfactory crop in soils with a
pH between 6.6 and 6.8. The market crop
responds well to bulky organic manures and, if
used, their nutrient values should be taken into
account when calculating additional fertilisers. An
NPK ratio of 2:2:1 is applied as a base dressing.

Previous Cropping and Isolation

These are as described for onions (*Allium cepa*).
There are no special zoning arrangements.

Seed Crop Establishment

There are two systems in use, namely 'seed-to-
seed' and 'root-to-seed', which are analogous to
the two systems described for onion (*Allium cepa*).
The root-to-seed method is used for Basic Seed
production.

Seed-to-seed   Seed is drilled at the rate of 2–3 kg/
ha in rows 30 cm apart during the early summer
to achieve a final distance of 10 cm between
plants. Alternatively, a denser sowing can be
made and the young plants ('stecklings') lifted,
trimmed and replanted.

Root-to-seed   Seed is sown in early spring in seed
beds at the rate of 2–3 kg/ha. The young plants
are lifted after approximately 10 weeks, graded
and trimmed, and those which are pencil thickness
are replanted 10 cm deep, 8–10 cm apart, in rows
70 cm apart.

The demand for high quality seed for precision
sowing in northern Europe has led to the development of seed production in polythene tunnels; the
results of trials using this technique have been
described by Gray et al. (1992).

Seed Crop Management

This is as described for onion (*Allium cepa*).

Seed Crop Roguing

Seed-to-seed crops   There are two roguing stages:
the first is during the initial season when the
plants are established, when they are checked for
leaf characters; and the second is at the start of
flowering when leaf characters are again checked
and in addition umbel colour is checked.

Root-to-seed crops   There are three stages. The
first is when transplanting, when leaf characters
and the presence or absence of stem pigmentation
are checked. The second stage is in the late
summer, after the transplants are established,
when the characters are checked as for the first
stage and in addition individual plant vigour is
confirmed according to the type. The third stage is
at the start of flowering in the second year, when
all characters are checked as in the first stage, and
in addition umbel colour is checked.

Harvesting

Leek umbels are harvested by hand as described
for onion (*Allium cepa*). However, as leek umbels
are slower to ripen than those of onion, their
harvesting may continue into declining weather
conditions. The umbels are spread on plastic
sheets to dry under cover in a greenhouse or
similar structure. Gray et al. (1992) demonstrated
that when the seed heads develop and dry at
temperatures of 20°C and above, germination is
improved.

Threshing, Cleaning, Drying and Storage

These are as described for onion (*Allium cepa*).

Pathogens

The main seed-borne pathogens of *Allium* species
are listed below.

Pathogen	Common name
Alternaria porri (Ell.) Ciferri	Purple blotch
Botrytis allii Munn	Damping-off, grey mould, neck rot
Botrytis byssoidea Walker	Seedling damping-off, neck rot
Cladosporium allii-cepae (Ranojevic) M.B. Ellis, syn. *Heterosporium allii-cepae* Ranojevic	
Colletotrichum circinans (Berk.) Vogl., syns *C. dematium* (Pers. ex Fr.) Grove *F. circinans* (Berk.) Arx	Smudge, damping-off
Fusarium spp.	
Peronospora destructor (Berk.) Casp.	Downy mildew
Pleospora herbarum (Pers. ex Fr.) Rabenh., syn. *Stemphylium botryosum* Wallr.	Black stalk rot, leaf mould
Puccinia allii Rud., syn. *P. porri* Wint.	Rust
Sclerotium cepevorum Berk.	White rot
Urocystis cepulae Frost	Smut
Virus	Onion yellow dwarf virus
Ditylenchus dipsaci (Kuhn) Filipjev	Bloat, eelworm rot

References

Brewster, (1994). *Onions and Other Vegetable Alliums.* CAB International.

Brittain, M. (1988). Leeks – the long and the short of it. *The Grower*, **110**(3), 20–23.

Currah, L. and Proctor, F.J. (1990). *Onions in Tropical Regions.* Natural Resources Institute Bulletin No. 35. NRI, London.

Ellis, R.H. and Roberts, E.H. (1981). The quantification of ageing and survival in orthodox seeds. *Seed Science and Technology*, **9**, 373–409.

Globerson, D., Sharir, A. and Eliasi, R. (1981). The nature of flowering and seed maturation of onions as a basis for mechanical harvesting of the seeds. *Acta Horticulturae*, **111**, 99–114.

Gray, D., Steckel, J.R.A. and Hands, L.J. (1992). Leek (*Allium porrum*) seed development and germination. *Seed Science Research*, **2**, 89–95.

Jones, H.A. and Mann, L.K. (1963). *Onions and Their Allies.* Leonard Hill, London; Interscience Publishers, New York.

Millar, A.A., Gardner, W.R. and Goltz, S.M. (1971). Internal water status and water transport in seed onion plants, *Agronomy Journal*, **63**, 779–784.

Peters, R. (1990). Seed production in onions and some other Allium species. In *Onions and Allied Crops, Vol. 1* (Eds H.D. Rabinowitch and J.L. Brewster). CRC Press, Boca Raton, Florida, 161–176.

Further Reading

UPOV (1976). *Guidelines for the Conduct of Tests for Distinctness, Homogeneity and Stability: Onion.* TG/46/3, International Union for the Protection of New Varieties of Plants, Geneva.

UPOV (1982). *Guidelines for the Conduct of Tests for Distinctness, Homogeneity and Stability: Leek.* TG/85/3, International Union for the Protection of New Varieties of Plants, Geneva.

AMARANTHACEAE: AGRICULTURAL CROPS

Amaranthus species

There are three species of *Amaranthus* which are grown for grain; however, one of these, *A. cruentus* L. (African spinach), is more popularly grown as a vegetable and is described under that heading. The other two grain species are:

> *A. hypochondriacus* L. Common name: Grain amaranth
>
> *A. caudatus* L. Common name: Grain amaranth

These are broad-leaved, erect annuals growing up to 3 m tall with simple or branched indeterminate inflorescences. According to Williams and Brenner (1995), *A. hypochondriacus* has bracts equal to or slightly longer than the style branches; the spikes are stiff. *A. caudatus* has inflorescence arms which are indeterminate and the tapering central inflorescence branch exceeds the side branches.

No descriptors for cultivars have been published by international organisations.

Environmental Requirements for Seed Production

The grain amaranths are natives of Central and South America. *A. hypochondriacus* can be grown under a wide range of conditions from sea level up to high elevations. *A. caudatus* is favoured at high elevations.

The species do best on well drained soils but require a well distributed rainfall. Warm weather is required during the growing period with soil temperatures between 16 and 18°C. *A. caudatus* requires a day length of less than 8 h to flower.

Previous Cropping and Isolation

For seed production the land should be free from all amaranth plants, including the weed amaranth, *A. spinosa*.

Although generally self-fertilised, up to 30% cross-fertilisation may occur by wind-distributed pollen. Adequate isolation is therefore desirable.

Indian seed certification standards require isolation distances of 400 m for Foundation and 200 m for Certified Seed production (Tunwar and Singh 1988).

Seed Crop Establishment

The seeds are very small and consequently difficult to sow sufficiently thinly; seed can be mixed with sand or other material and sown to give 360 000 plants/ha in rows 45–60 cm apart with 7.5 cm between plants. In wetter areas transplanting is also possible (Williams and Brenner 1995). High plant density produces plants with single stems which are suitable for mechanical harvesting.

Seed Crop Management

Amaranths lodge easily and therefore fertiliser should be used sparingly.

Seed Crop Inspection and Roguing

Plants which produce black seeds are undesirable since seed lots contaminated with them are not suitable for use in food preparations. Such plants should therefore be removed from seed crops.

Harvesting

When harvesting by hand, individual spikes may be cut and removed for drying on two or three occasions. Dense crops are suitable for harvesting by combine, particularly those of cultivars which have been selected for more uniform ripening. Combines should be sealed to prevent loss of the small seeds through gaps, for example in elevator casings.

If the crop can be left until a first frost, the plants will dry within about 10 days, making for easier harvest. However, seed shedding can be severe and the crop must be watched carefully and harvested immediately if shedding is beginning.

Drying, Storage and Cleaning

If the crop is harvested before the rainy season, it should not require further drying. However, seed harvested with a moisture content above 10% should be dried immediately.

The seed is vulnerable to storage fungi. Stores must be kept scrupulously clean and fumigants used if necessary.

An air/screen cleaner will usually be all that is required. However, because the seed is very small and runs freely, equipment should be examined carefully and any large gaps sealed to prevent undue seed loss.

References

Tunwar, N.S. and Singh, S.V. (1988). *Indian Seed Certification Standards*. The Central Seed Certification Board, New Delhi, 273–274.

Williams, J.T. and Brenner, D. (1995). Grain amaranth (*Amaranthus* spp). In *Cereals and Pseudocereals* (Ed. J.T. Williams). Chapman and Hall, New York, 129–186.

AMARANTHACEAE: VEGETABLE CROPS

Scientific name	Common name
Amaranthus cruentus (L.) Sauer.	African spinach
Amaranthus tricolor L.	African spinach
Amaranthus dubius C. Martius ex Thell.	African spinach
Celosia argentea L.	Lagos spinach, cock's comb

These two genera are considered to be important food crops in Africa and Asia and form essential nutritional elements to supplement the main staples, especially for subsistence farmers and their dependents. Both genera tolerate temperatures of up to 30°C. Because of the close similarity of their seed production methods, they are dealt with together here.

The genera are found as common weeds and some local seed stocks may contain a range of types. However, cultivars are available in most areas.

Classification of Cultivars

Genus and species
Uses: suitability for specific day length and seasons
General morphology and plant habit, apical dominance and degree of branching
  Leaf: number from sowing to flowering, shape and pigmentation
  Inflorescence: time to start of anthesis, colour, form

Environmental Requirements for Seed Production

A wide range of soil pH values is tolerated although lime-induced chlorosis is likely to occur in soil with a pH above 7.0. The optimum NPK ratio is 1:1:2 applied as a base-dressing during preparation (Grubben 1976).

Previous Cropping and Isolation

Amaranthus spp. are mainly wind pollinated but *Celosia* spp. are insect pollinated. In areas where the wild or escape species are common, seed production should follow a series of suitable break crops. A minimum isolation distance of 1000 m should be observed; however, greater distances may be stipulated by some authorities.

Seed Crop Establishment

The seed crop is usually produced from transplants. *Amaranthus* seed is sown in beds at the rate of 2 g/m^2 and *Celosia* sown at the rate of 3 g/m^2. These sowing rates produce approximately 1000 seedlings/m^2. The young plants are transplanted when large enough to handle, which is approximately three weeks from sowing, in rows 60–80 cm apart, and 40–50 cm within the rows.

Seed Crop Management

The growing points of those cultivars which have a relatively small or weak apical inflorescence are pinched out after about four weeks in order to encourage the production of secondary shoots.

The crop should be kept free of weeds, and it is especially important to control any wild species of the same genus as the seed crop.

Seed Crop Roguing

There are three roguing stages. The first is at planting out when the transplants are checked for trueness to type and leaf morphology (including leaf colour and pigmentation); any early bolting plants are discarded at this stage. The second stage is just before flowering when the plants are checked for height and general plant morphology including the amount of branching. The third stage is when flowering starts; at this stage the plants are again checked for general plant morphology, inflorescence and flower colour.

Harvesting

There is a general yellowing of the plant foliage as the seed becomes mature. Cultivars with an apical inflorescence are cut by hand, usually once only.

The cultivars which have several axillary inflorescences are hand harvested several times as the successive inflorescences mature. The cut material is retained on clean tarpaulins and further sun dried until threshed.

Threshing and Cleaning

Relatively small seed lots can be threshed by hand, otherwise small threshers are used. The threshed seed is cleaned by air/screen machines.

Drying and Storage

The seed is dried down to 10% moisture content for storage.

Reference

Grubben, G.J.H. (1976). *The Cultivation of the Amaranth as a Tropical Leaf Vegetable*. Royal Tropical Institute, Amsterdam.

BORAGINACEAE: AGRICULTURAL CROP

Borage officinalis L. Common name: Borage

Borage is a common plant in temperate areas, but prefers drier situations. It has recently been developed for oil production, which has a high gamma linolenic acid content, and is used by pharmaceutical companies. It can only be grown as a contract crop. Improved cultivars are available.

Borage is cross-fertilised by insects, mainly bees. Isolation is therefore required for seed production.

The seed crop is sown in the spring in rows 15–50 cm apart at a seed rate of 22 kg/ha. Once established, the crop is very aggressive and provided weeds are controlled in the early stages, it will effectively suppress them.

Because of the tough nature of the plant it is difficult to walk through the crop; leaving tramlines will help crop inspection, but roguing is not generally practicable.

Borage is an indeterminate plant and so ripening extends over a long period. The seed sheds freely and harvest should begin when the first seeds have dropped. The best harvest method is to swath and pick-up after about a week.

Seed usually has a moisture content of about 20% when threshed and must be dried to 10% for storage. The seeds are small and equipment should be sealed to avoid loss.

CANNABIDACEAE: AGRICULTURAL CROP

Cannabis sativa L. Common name: Hemp

Hemp can be used for three purposes: fibre, oil, or for production of a drug. The latter is a narcotic and because of this hemp growing is banned in many countries. However, the fibre is strong, soft and of good length.

The crop originated in Asia but is well adapted to warm, humid temperate regions where most of the fibre is grown. For production of the narcotic, hotter areas are preferred; production is under licence in some countries, notably India. For seed production, a well drained fertile soil is preferred. Rainfall of about 700 mm is necessary and there must be no frost during early growth.

Cultivars for fibre are very tall and can reach 8 m in height; those for narcotic production are shorter and produce more branches.

The plant is dioecious, but some attempts have been made to select monoecious cultivars. Cross-pollination is by wind. For seed production, adequate isolation is required. Some attempts have been made to create hybrid cultivars (Simmonds 1976).

Hemp is sown in the spring at seed rates of between 60 and 100 kg/ha. For seed production the plants are spaced further apart than for fibre production, as this reduces the height of the plants. The male plants die soon after anthesis.

Small areas can be harvested by hand and the plants allowed to dry before threshing. Mechanical harvesting of larger areas is possible.

Reference

Simmonds, N.W. (1976). Hemp. In *Evolution of Crop Plants* (Ed. N.W. Simmonds). Longman, Harlow, 203–204.

CHENOPODIACEAE: AGRICULTURAL CROPS

***Beta vulgaris* L. subsp. *vulgaris*. Common Names:
Sugar beet, Fodder beet, Mangel, Mangold,
Mangelwurzel**

Subspecies *vulgaris* contains the crops used in
agriculture and there are other subspecies containing those used as vegetables. This subspecies contains a range of plant types which serve different
purposes. The main plant parts used are the roots,
although the leaves are also used as fodder. Sugar
beet, as the name implies, is used for the production of sugar and has been bred to provide
cultivars with high sugar content of good quality
in the roots. Fodder beet is bred to provide
cultivars with roots of high dry matter content;
these are based on crosses between sugar beet and
mangel. The mangel, or mangelwurzel, is the
original form from which the others have been
derived; generally, its roots have lower dry matter
content, but are larger and grow with more of the
crown exposed. Mangels are now less popular,
although they may still be used where the roots
are to be hand lifted, as they are easier to pull
than the deeper-seated sugar and fodder beet.

Different breeding systems have been used to
produce cultivars of sugar and fodder beet. The
species is cross-pollinated by wind and is normally
diploid. Cultivars have been created with different
ploidy levels as follows:

$2n$	Diploid without male sterility
$2n \times 2n$	Male sterile diploid with a male diploid
$2n \times 4n$	Triploid with a male tetraploid
$4n \times 2n$	Triploid with a male diploid
$4n$	Tetraploid without male sterility
$4n \times 4n$	Male sterile tetraploid with a male tetraploid
$4n + 2n$	Polyploid without male sterility
$2n \times (2n + 4n)$	Polyploid with a male sterile female and a male polyploid
$2n \times (2n \times 4n)$	Polyploid with male sterile female and a male triploid

The most common form is the $4n$ x $2n$ – the
triploid produced from a male sterile tetraploid
and a male diploid. For cultivars produced
without a male sterile parent, the whole crop
can be harvested for seed as one lot. However,
where male sterility is used it is generally
necessary to establish and harvest the parents
separately using an appropriate proportion of
each: usually three rows of female to one of male,
or four to one. Because of the complex breeding
systems used, it is usual for plant breeders to
supervise closely the production of seed of their
cultivars and seed growers are required to follow
their directions.

The 'seed' of beet is normally a cluster of seeds
each enclosed in the ovary; each cluster thus
contains up to seven seeds. This makes it difficult
to establish a crop as it is necessary to have plants
evenly spaced in the row; singling plants when
more than one emerge from each seed cluster is
difficult. Two methods have therefore been
developed to obtain single seeds: mechanical and
genetic.

Multigerm seed can be reduced to single seeds
by rubbing or chopping the clusters to produce
broken clusters, with each piece containing a
single seed. This mechanical treatment does not
produce absolutely single-seeded pieces and will
reduce germination. Seed so treated has been
termed 'precision seed'.

The highest germination is produced from
genetic monogerms which are bred from plants
that have only a single flower at each node.
Nearly all modern cultivars are now monogerms.
The term 'monogerm' has been reserved for such
cultivars.

Distinguishing between Cultivars

No cultivar descriptors are published by the international organisations. Originally mangel cultivars were classified by root shape and root skin
colour. However, these characteristics are no
longer suitable for modern sugar beet and fodder
beet cultivars. Ploidy is an essential first

characteristic for classification, followed by multi-germ or monogerm. Subsequently, top size and habit of growth and extent of the crown above ground are useful characteristics.

Environmental Requirements for Seed Production

Beet is biennial and, while not particularly frost hardy, requires a cold period to induce flowering. The crop is thus adapted to temperate regions with well distributed rainfall. Areas with early cold in the autumn or late spring frosts should be avoided. Cold in autumn may cause bolting in a new sown crop. In a ripening seed crop, cold in autumn may cause the seed to be vernalised while still on the plant and such seed will produce undesirable bolters in the following fodder or sugar crop. Late spring frosts may delay growth and subsequently harvest.

Fertile soils are required with a good structure; very light and very heavy soils should be avoided. Apart from the usual nutrients (NPK), beet responds well to salt. One of the plants from which subspecies *vulgaris* has evolved is subspecies *maritima* (L.) Thell., which is native on sea shores and is also an objectionable weed.

Previous Cropping and Isolation

Beet seed survives in the soil for long periods. Bolters in fodder or sugar crops may shed seed which can produce weed beet in subsequent crops. There should therefore be an interval of at least five years between seed crops and two years free from all beet crops before a seed crop is planted.

During the interval between seed crops, all weed beet must be destroyed. Weed beets are annual and may arise from wild beet or crosses between wild and cultivated beet or from bolters in a crop. They are able to mature seed which is easily shed and will therefore multiply in a field unless controlled.

Other weeds which are objectionable in a seed crop are: *Atriplex patula* (common orache), *Chenopodium album* (fat hen), *Galium aparine* (cleavers), *Polygonum* spp. (bindweed), *Stellaria media* (common chickweed). These should be controlled in previous crops.

Beet is mainly cross-pollinated by wind, although some insect pollination does occur. Adequate isolation is therefore essential. Greater isolation distances are specified between different types of beet, i.e. between sugar beet and fodder beet and between sugar beet or fodder beet and vegetable beets. OECD (1995) specifies the following distances for sugar beet.

Crop type	Distance (m)
1. No isolation is necessary between seed crops using the same pollinator	
2. All seed crops to produce Basic Seed from any pollen source of the genus *Beta*	1000
3. All seed crops to produce Certified Seed of sugar beet	
– from any pollen source of the genus *Beta* not included below	1000
– the intended pollinator or one of the pollinators being diploid, from tetraploid sugar beet pollen sources	600
– the intended pollinator being exclusively tetraploid, from tetraploid sugar beet pollen sources	600
– from sugar beet pollen sources the ploidy of which is unknown	600
– the intended pollinator or one of the pollinators being diploid, from diploid sugar beet pollen sources	300
– the intended pollinator being exclusively tetraploid, from tetraploid sugar beet pollen sources	300
– between two seed production fields in which male sterility is not used	300

The specifications for fodder beet are the same, except that in section 3 wherever the words 'sugar beet' are used the words 'fodder beet' are substituted.

In some areas there are zoning schemes which restrict seed growing to particular cultivars in specified zones. Such schemes may be voluntary agreements between seed growers and plant breeding companies, or they may have legal status.

Seed Crop Establishment

In the early stages of selection it is possible to establish small areas in the same manner as for a root crop. The roots can then be lifted, selected and replanted to produce seed.

For somewhat larger areas, seed may be sown in a seed bed to produce young plants or stecklings. After sorting to remove any off-types, the stecklings are replanted to produce seed.

For commercial seed production, when areas are larger, the seed is sown and the plants are allowed to mature *in situ*. Seed crops can be established in spring under a cover crop, usually a cereal, but the cover must not be so dense as to prevent development of the beet plants. Alternatively, crops may be sown later in the year on bare ground. In this case, timing of sowing must be adjusted to allow plants to establish, but not so early that they develop to the stage where an early cold spell in the autumn may induce flowering. Any early bolters in the year of sowing should be removed.

The aim should be to achieve a plant population which consists of plants with strong main shoots and restricted branching. Seed crops of about 300 000 plants/ha can be obtained by sowing in rows 19 cm apart with 25 cm between plants in the row. Under a cover crop it is best to sow the rows closer together, allowing more space between plants in the row, as this provides the best possibility to achieve the optimum plant population.

Crops can be sown with a precision drill using pelleted seed.

Seed Crop Management

Weeds are controlled by chemicals, although weed beet can only be eliminated by roguing. Cover crops may restrict the range of herbicides which can be used until they are removed. Inter-row cultivation may be required if the soil becomes over-compacted by rain or the use of machinery, but the young plants are easily damaged if cultivation is too close.

The diseases and pests which attack root crops will also occur in a seed crop and must be controlled. It is particularly important to control the aphid vectors of virus yellows as this virus can overwinter on a seed crop and provide a source of infection to any newly sown crops in the following spring.

All cultivars require vernalisation, with a cold spell followed by lengthening days, to induce flowering, but they differ in their exact requirements. The plant breeder will normally place seed crops in areas where the requirements of the particular cultivar to be grown are likely to be met.

It is also necessary to avoid areas where there is a possibility of an early cold spell while the seed is ripening on the plant, because this may cause the ripening seed to be vernalised. When such seed is used for a root crop there may be a high proportion of bolters, which is highly undesirable.

Fertiliser requirements for a seed crop are similar to those for a root crop. Seed crops benefit from a late application of nitrogen at the rate of 200–250 kg/ha. Application too early or at too high a rate may cause excessive top growth, with side shoots which mature seed later than the main shoot; this makes harvesting more difficult without increasing the yield of good quality seed. There may also be lodging.

Seed Crop Inspection and Roguing

Modern cultivars cannot be identified in the field except in very broad terms. Size of top and amount of the crown of the root above ground are two characters which define groups of cultivars, but the latter is not reliable in closely spaced crops. There can be no guarantee of cultivar identity from a single inspection. Generally, careful record keeping and supervision of crop establishment are the main measures to ensure that a seed crop is of the cultivar it purports to be.

Seed crops may provide a source of disease inoculum in spring. In particular, virus yellows may overwinter in plants of a seed crop and spread to root crops in the area. Therefore seed crop inspection during the winter to identify seriously infected crops is required by some certification schemes, so that control of aphids in the spring can be assured, or the crop destroyed if necessary.

Inspection over winter will also detect any contamination with red beet which show anthocyanin pigmentation; obvious rogues of this type can be removed at this time.

To check on isolation, a later inspection is required at flowering time. In areas where weed beet occurs it will be particularly important to check surrounding crops and waste areas to ensure that any plants have been destroyed.

Harvesting

Seed does not ripen uniformly on the plant and sheds easily when it is ripe. Therefore careful judgement of harvest timing is essential. The plants change colour as the first ripening seed is shed and at this stage the crop can be cut. Small areas can be cut by hand, the stalks tied in bundles and stooked. Larger areas are windrowed, leaving a long stubble to keep the crop off the ground and allowing air to circulate beneath the windrow.

After cutting, the crop should be handled as little as possible as the ripening seed will easily fall to the ground and be lost. Once the crop is dry it can be threshed. For stooked crops it is best to take a combine or mobile thresher from stook to stook to avoid moving the plants more than necessary. Windrowed crops can be picked up by combine. A dry crop is not difficult to combine.

Crop desiccants may be used to hasten drying, either on the standing crop or after windrowing. They have been found not to affect germination of the harvested seed.

Drying, Storage and Seed Cleaning

Seed harvested at more than 10% moisture content should be dried immediately to 8% or below for storage to the next sowing season. Batch or continuous-flow driers are suitable and the temperature of the drying air should not exceed 38°C.

Seed cleaning and processing is a specialist operation and should only be undertaken by those with the correct equipment and experience. Monogerm seed is size graded for subsequent pelleting and use in precision drills. In OECD (1995), the standard given is that 90% of the germinated clusters shall give single seedlings and no more than 5% shall give three or more seedlings.

Multigerm seed is first rubbed to achieve single seededness. The standards given in OECD (1995) are as follows.

1. *Precision seed of sugar beet.* At least 70% of the germinated clusters shall give single seedlings and no more than 5% shall give three or more seedlings.
2. *Precision seed of fodder beet.* In seed of cultivars with more than 85% diploids, at least 58% of the germinated clusters shall give single seedlings; in other seed at least 63% of the germinated clusters shall give single seedlings; in both, no more than 5% shall give three or more seedlings.

Durrant and Mash (1990) found that the use of a gravity separator or aspirator gave some improvement in germination: a 10% rejection of seed during cleaning gave a 2.5% improvement and a 20% rejection a further 1% improvement; the aspirator was marginally better than the gravity separator. There was no improvement in vigour or synchronization of germination.

After cleaning, the seed is pelleted. This produces uniformly sized and shaped particles for use in a precision drill; pelleting is a specialist operation and can only be done by firms with the correct equipment and with experience. Even those companies which clean seed normally send it to a specialist for pelleting.

Seed is usually mixed with granules of insecticide before sowing.

Chenopodium quinoa **Willd. Common Name:
Quinoa**

Quinoa is native to the Andes and is still cultivated there to a limited extent. Recently there
has been interest in quinoa as a possible crop
for high altitudes (2240–2580 m) elsewhere (for
example, NE India and Pakistan).

There has been some attempt to select
improved cultivars. Characteristically the stems
are erect, branched or unbranched; they may be
green, yellow, red, purple or orange with stripes of
another colour which turn pale yellow or red at
maturity; the young leaves are pubescent with
white or purple hairs; the number of indentations
on the leaf lamina border varies from 0 to 27
depending on cultivar; at maturity leaf colour
changes to yellow, red or purple. Fleming and
Galway (1995) list five types but state that there is
much overlap between them:

(i)   *Valley type*: 2000-4000 m; branched, growing
      2–3 m tall; lax inflorescences; late maturing;
      very diverse.
(ii) *Altiplano type*: 4000 m; shorter and earlier
      than valley type; branched or unbranched;
      mostly pigmented; small inflorescences.
(iii) *Solar type*: 4000 m; soils above pH 8; red
      pigmented plants; black seeds.
(iv) *Sea level type*: earliest to mature; short,
      unbranched green plants; small, yellow
      translucent seeds.
(v)   *Subtropical type*: Only one plant has been
      found, in Bolivia; intense green-orange at
      maturity; small yellow-orange seeds.

Fairbanks et al. (1990) reported that electrophoresis of seed proteins could be used for
distinguishing between cultivars.

The plants may be day-neutral, but some long-
day cultivars exist. The species is adapted to harsh
conditions and will tolerate low soil moisture. It is
reasonably frost hardy before flowering, but can
be damaged thereafter.

Quinoa has mixed hermaphrodite and female
flowers on the same plant and is generally self-
pollinated; cross-pollination occurs up to 9%
(Simmonds 1984). The seed can survive for long
periods in the soil.

Crops are established in rows 20 cm apart at a
seed rate of 12–20 kg/ha. Sowing should be
shallow (1–2 cm) in moist soil. The crop responds
to nitrogen.

The seeds are hard when ripe and difficult to
mark with the thumb-nail. Crops may be hand
harvested or larger areas can be combined direct;
however, for combining the plants must be
completely dry.

References

Durrant, M.J. and Mash, S.J. (1990). The use of a
specific gravity table or an aspirator in sugar beet
seed processing. *Seed Science and Technology*, **18**,
163–170.
Fairbanks, D.J., Burgener, K.W., Robinson, L.R.,
Anderson, W.R. and Ballon, E. (1990). Electro-
phoretic characterisation of quinoa seed proteins.
Plant Breeding, **104**, 190–195.
Fleming, J.E. and Galway, N.W. (1995). Quinoa
(*Chenopodium quinoa*). In *Cereals and Pseudocereals*
(Ed. J.T. Williams). Chapman and Hall, New York,
3–84.
OECD (1995). *The OECD Scheme for the Varietal
Certification of Sugar Beet and Fodder Beet Moving
in International Trade*. Up-to-date Version of the
Seed Schemes as of 15 June 1995. Organization for
Economic Co-operation and Development, Paris,
67–90.
Simmonds, N.W. (1984). Quinoa and relatives. In
Evolution of Crop Plants (Ed. N.W. Simmonds).
Longman Group, Harlow, 29–30.

CHENOPODIACEAE: VEGETABLE CROPS

Scientific name	Common name
Beta vulgaris L. subsp. *esculenta*	Beetroot, Red beet
Beta vulgaris L. subsp. *cycla*	Spinach beet, Chard, Swiss chard
Spinacea oleracea L.	Spinach

Beetroot is grown in temperate areas for consumption as a cooked root vegetable; it is also relatively popular in the Middle East and the cooler tropics.

Spinach beet is cultivated and cooked as a green leafy vegetable in temperate areas; the green leaf has a prominent white mid-rib which is also cooked. There are red and yellow leaved forms which are generally regarded as ornamentals.

Spinach is produced for its green leaves which are cooked as a vegetable and it has become an important crop for the frozen vegetable industry. This species is sometimes referred to as 'European spinach' in Africa and Asia to distinguish it from *Amaranthus* spp. and *Celosia* spp.

Beta vulgaris L. subsp. *esculenta*. Common Name: Beetroot

Beetroot is a quantitative long-day biennial which has a low temperature requirement for the initiation of flowers. An interesting account of the physiology of flowering has been given by Chroboczek (1934). All *Beta* spp. are wind pollinated. The 'seed' of *Beta* spp. is a cluster formed from the swollen perianths to produce a multigerm fruit. However, plant breeders have produced monogerm cultivars in which the fruits bear a single seed.

Classification of Cultivars

Season of production: resistance to early bolting
'Seed' type: Multigerm or monogerm
Leaf
  Degree of anthocyanin
  Approximate number of leaves to the crown
  Shape and degree of blistering of lamina, colour, lustre
  Petiole: relative length, thickness and colour
Root
  Shape: flat, globe, oval, conical, long
  Colour: red or yellow, intensity of colour
  Interior quality

Environmental Requirements for Seed Production

Beetroot seed can be produced on soils with a pH ranging from 6.0 to 6.8. Fertilisers with an NPK ratio of 2:1:2 are applied during preparation prior to sowing. For late summer sowings to produce 'stecklings' (see below), a lower nitrogen fertiliser level is used with a ratio of 1:1:2; in this case it is usual practice to apply supplementary nitrogenous fertilisers as a top-dressing in the spring. This crop is very susceptible to boron deficiency, and if boron-deficient soils cannot be avoided then either boronated fertilisers are used for the base-dressing or, if not available, then sodium tetraborate ('borax') is applied at the rate of 1.5–2 kg/ha.

Previous Cropping and Isolation

It is usual for seed legislation to stipulate a minimum number of years between a beetroot seed crop and previous crops of *Beta vulgaris* types. Minimum isolation distances also take into account the wind pollination and cross-compatibility of beetroot, sugar beet, fodder beet, mangolds and spinach beet. This is usually accomplished by a zoning scheme. It is usual for authorities to require isolation distances of at least 1 km between the respective subspecies of *Beta vulgaris*. The distance between the same type of beetroot cultivars is usually a minimum of 500 m, with 1000 m between different types (e.g. between a globe and a cylindrical root type). Care must be taken to ensure that the prevailing minimum seed crop isolation requirements are adhered to in seed crop planning.

Seed Crop Establishment

There are two basic methods of producing beetroot seed, namely 'seed-to-seed' and 'root-to-seed'.

In the seed-to-seed method, seed is sown in late summer at the rate of 12 kg/ha and the young plants, which are usually referred to as 'stecklings', are transplanted in the following spring; 1 ha of seedlings will normally provide sufficient stecklings to plant up 4 ha. In the milder Mediterranean areas, transplanting is done in the autumn in order to avoid the dry soil conditions in the spring.

Seed is usually drilled on a four-row bed system, with up to 30 cm between the rows in a bed 110 cm wide. The sowing rate is adjusted to take the seed type into account to give an optimum density of 200 seedlings per square metre.

The optimum steckling length is 2.5 cm. Some seed producers trim the young plants, but the swollen tap root should not be cut. The stecklings are planted out at approximately 45 cm apart in rows 60 cm apart.

In the root-to-seed method, seed is sown in mid-summer at a rate of approximately 10 kg/ha in rows 45 cm apart. After the seedlings have emerged they are singled out to enable each plant to develop its root shape.

The root crop is lifted in the autumn and, following selection (see under 'Seed Crop Roguing'), are stored until replanting in the following spring. The roots are planted approximately 50 cm apart with 60 cm between the rows.

Seed Crop Management

Irrigation may be necessary if conditions are dry following planting of the roots, with further irrigation applied until the plants are well established.

Overwintered stecklings are given a top-dressing of a nitrogenous fertiliser in the spring, at the rate of 50 kg/ha. In areas with a high leaching rate, this is repeated prior to flowering.

Some beetroot seed producers reduce the height of flowering shoots on the basis that it promotes more branching and reduces potential lodging of the crop when seed is ripening. However, this should be done according to local experience and is probably more useful on exposed sites.

Seed Crop Roguing

The root-to-seed system provides a better level of roguing or selection than the seed-to-seed system. In principle, the seed-to-seed system relies on starting with Basic Seed of a high genetic quality, and is normally only used for the production of the final seed category.

Seed-to-seed crops The only roguing stage is when the stecklings are lifted from the seed beds and sorted prior to replanting. At this time plants are checked for appropriate leaf morphology and colour. Any plants which are starting to bolt prematurely are discarded.

Root-to-seed crops There are four possible roguing stages. The first is immediately before lifting the roots; at this time the plants are checked for leaf morphology and colour. Any early bolting plants are discarded. The second stage is after the lifted roots have had their tops trimmed when they are examined for trueness to type. Following storage, the roots may be examined again but if they were checked as described for the second stage then only those roots which have storage rots are discarded at this third stage. The fourth stage is when the plants have started to bolt; they are rogued to check for leaf morphology, colour and vigour. This fourth check is done prior to reducing the height of the inflorescences.

Plants showing symptoms of 'silvering' (caused by *Corynebacterium betae* Keyworth, Howell and Dowson) should be removed when observed during roguing.

Harvesting

As the seed crop ripens, the fruits at the bases of the inflorescence side shoots change colour from green to brown. Care must be taken with this assessment as immature seeds are likely to shrivel after the mature inflorescences are cut. Fruits are milky before ripening and turn mealy when ripe. The crop may either be cut with a swather or hand cut, depending on the scale of operation. The cut material is left in windrows and turned to dry further. In areas where autumn rain prevails, the cut material is dried under cover.

Threshing and Cleaning

When the material is dry it is threshed in a combine or stationary thresher. Although there is relatively little chaff, the dry inflorescences are brittle and care must be taken to run the cylinder at a low speed and to set the concave opening wide in order to minimise the amount of broken straw in the seed lot.

Seed lots of monogerm and multigerm cultivars can be upgraded by an air/screen cleaner.

Drying and Storage

Beetroot seed is relatively quick to dry compared with many other crops; the drying temperature should not exceed 42°C. After drying, the seed is usually passed through a decorticating machine to remove excess non-seed material from the individual fruit clusters.

The moisture content of beetroot seed is reduced to 7.7% for storage in vapour-proof conditions.

Beta vulgaris L. subsp. cycla. Common Names: Spinach beet, Chard, Swiss chard

Spinach beet is a biennial, but is cultivated for its edible leaves produced in its first year.

Classification of Cultivars

Season of production: resistance to early bolting and low temperatures
Leaf characters
Leaf blade – Colour: red, green or yellow; Width, degree of crinkling
Petiole – colour: red, green, yellow-green

Environmental Requirements for Seed Production

The soil pH and nutrient requirements are the same as described for beetroot (*Beta vulgaris* subsp. *esculenta*).

Previous Cropping and Isolation

Spinach beet is a subspecies of *Beta vulgaris*; the same previous cropping and isolation requirements apply as described for beetroot (*Beta vulgaris* subsp. *esculenta*).

Seed Crop Establishment

The seed crop is produced from either the 'seed-to-seed' system or selected transplants. The seed production systems are similar to those described for beetroot (*Beta vulgaris* subsp. *esculenta*). The plants are not frost hardy.

Seed is drilled in rows 90 cm apart at the rate of approximately 6 kg/ha. The young plants are subsequently singled out to a distance of approximately 40 cm apart.

Seed Crop Management

When grown on the seed-to-seed system in frost-free areas, the plants may remain in the field. When produced in areas which experience frosts, the vegetative plants are lifted in the autumn and transferred to protected structures, such as polythene tunnels. The main selection process is done at this time.

Seed Crop Roguing

Seed-to-seed method The first stage is in the autumn when the established vegetative plants are checked for trueness to type of leaf and petiole characters. Plants which have produced inflorescences are removed at this stage. The second roguing is done in the spring, before the bulk of the plant population starts to produce visible inflorescences. At this second stage the same characters are checked as described for the first stage.

Transplanting method The plants are checked before lifting for trueness to type of leaf and petiole characters. Any plants which have started to bolt prematurely are discarded at this stage. Only those plants which correspond with the cultivar's characters are replanted.

Harvesting

The mature inflorescences of this crop can be up to 3 m high and therefore have a strong tendency to lodge as the seed reaches maturity; this is especially likely for crops which have remained in the field.

The stage of ripeness and harvesting methods are similar to those for beetroot. When spinach beet seed is produced on a large field scale it is possible to use the same harvesting system as used for sugar beet seed production.

Threshing, Cleaning, Drying and Storage

These are the same as described for beetroot (*Beta vulgaris* subsp. *esculenta*).

Spinacea oleracea L. Common Name: Spinach

Spinach is a long-day annual, although market and seed crops are sometimes overwintered from autumn sowings. Populations of spinach contain male, female and hermaphrodite plants.

Classification of Cultivars

Open-pollinated or F1 hybrid
Seed type: round or spiny ('prickly')
Production season: suitability for sowing at specified times of year, resistance to early bolting
Leaf: approximate number of leaves before bolting, shape, colour, texture, petiole characters
Resistance to downy mildew (*Peronospora spinacea* Laub.)

Environmental Requirements for Seed Production

The crop grows best in soils with a pH of 6.0–6.8. The NPK ratio of 1:2:2 is applied during soil preparations.

Previous Cropping and Isolation

Fields which have produced spinach in the previous three years should not be used for seed production; some authorities may stipulate a longer break than this and local regulations should be taken into account in seed crop planning.

Spinach is wind pollinated and isolation distances should be at least 1000 m. Some control regulations may specify greater distances for cultivars with different leaf characters than for cultivars with similar leaf characters.

Seed Crop Establishment

Sowings are made in either autumn or spring. The hardiness of the cultivar and the prevailing winter conditions are considered when deciding on a programme.

Seed is sown at the approximate rate of 6 kg/ha in rows approximately 50 cm apart. The young plants are not normally thinned except for the production of Basic Seed.

When hybrid seed is to be produced, the ratio of rows of female to male plants is usually 6:2 or 14:2, although the maintenance breeder should advise on this.

Seed Crop Management

The crop should be kept as weed free as possible. Top-dressings of nitrogenous fertilisers are applied before bolting, but as the seed crop is prone to lodging care must be taken only to give supplementary nitrogen when indicated by the soil's nutrient status.

Seed Crop Roguing

The first roguing is done when the plants have reached the rosette stage. At this time, plants displaying atypical leaf characters, non-rosetting and early bolting are removed from the crop. The second stage is at the commencement of flowering when plants displaying atypical leaf characters are removed.

Plants which are found to be infected with either downy mildew or cucumis virus 1 (mosaic) should be taken out during the roguing operation.

If seed of a hybrid cultivar is being produced, advice on roguing requirements should be obtained from the maintenance breeder.

Harvesting

The seed crop is ready for harvesting when the later portion of the plant population has started to turn yellow. However, as the crop is very susceptible to shattering, care must be taken not to leave the standing crop too long before harvesting. In extensive production fields, where calm conditions prevail, the crop can be combined, otherwise the crop is cut and further dried in windrows. The cut material should be placed on tarpaulins to minimise seed loss if stacked to dry.

Threshing and Cleaning

Cut material which has completed its drying is threshed with a drum thresher and care taken to minimise the production of pieces of broken stalk in the seed lot.

The seed lot can be cleaned with an air/screen cleaner.

Drying and Storage

Spinach seed is dried down to 8.0% moisture content for vapour-proof storage.

Reference

Chroboczek, E. (1934). A study of some ecological factors influencing seed-stalk development in beets (*Beta vulgaris* L.). *Cornell University Agriculture Experimental Station Memo*, **154**, 1–84.

Further Reading

UPOV (1977). *Guidelines for the Conduct of Tests for Distinctness, Homogeneity and Stability, Spinach.* Doc. TG/55/3, International Union for the Protection of New Varieties of Plants, Geneva.

UPOV (1978). *Guidelines for the Conduct of Tests for Distinctness, Homogeneity and Stability, Beetroot.* Doc. TG/60/3, International Union for the Protection of New Varieties of Plants, Geneva.

COMPOSITAE: AGRICULTURAL CROPS

Carthamus tinctorius L. Common Name: Safflower

Safflower is grown for the oil content of the seeds which can be 35 to 45% or more of high linoleic and low linolenic acid content. The expressed meal has a high protein value. The florets can be used as a source of dye.

Distinguishing between Cultivars

The following characteristics are based on OECD (1995a) and UPOV (1990).

> Time of flowering: 50% of plants with at least one flower open
> Plant height: at flowering – short, medium, long
> Petal colour: white, yellow, light orange base, yellow with tips of lobes orange, red/orange, pink, purple.
> Leaf colour: light green, dark green, greyish
> Leaf shape: ovate, oblong, lanceolate, linear
> Number of leaf spines: few, medium, many
> Length of middle bract of the capitulum: short, medium, long
> Width of middle bract of the capitulum: narrow, medium, wide
> Seed size: small, medium, large
> Seed colour: white, cream, brown, black, grey.

Environmental Requirements for Seed Production

Safflower requires a deep, well drained soil of good fertility with neutral pH. It will tolerate salt but not waterlogging. It prefers cool temperatures during establishment, but requires warm weather and longer days when flowering and setting seed. Knowles (1989) states that temperatures above 20°C are required for satisfactory seed production.

Previous Cropping and Isolation

The seed does not remain dormant for a long period and therefore volunteer plants are not usually a problem, provided that one year free from safflower is left before sowing a seed crop.

However, in some areas wild safflower (*Carthamus oxyacanthus* M. Bieb.) may occur and it is essential that seed production fields be free of this weed.

Safflower is self-pollinating but up to 10% cross-fertilisation may occur. Cross-pollination is by insect activity. Tunwar and Singh (1988a) give distances of 400 m for Foundation Seed and 200 m for Certified Seed production in the Indian seed certification rules. For some cultivars which are less prone to out-crossing, a shorter distance may be adequate; the advice of the plant breeder should be followed. Some instances of bees improving seed yield have been reported but this appears to depend on the cultivar.

Seed Crop Establishment

Safflower is very susceptible to weed competition in the early stages of growth and a clean seed bed is essential. Seed is usually sown in rows at a seed rate of between 15 and 40 kg/ha. However, many cultivars are spiny and this makes them very difficult to walk through at later stages, so preventing roguing or inspection. It is therefore advisable to leave suitable walkways or gaps at sowing time. The primary stems usually produce the best seed so the crop should be dense enough to discourage branching.

Seed Crop Management

Seed crop management does not differ from the management for an oilseed crop. The safflower plant has a long tap root which will normally penetrate deep into the soil so that it can withstand drought conditions.

Seed Crop Inspection and Roguing

The best time for inspection is when the crop is in full flower. At this time it is possible to assess many of the characteristics described above.

Harvesting

Safflower crops are not usually prone to lodging, and seed shedding does not usually occur to any extent. Thus in most areas it is best to wait until the seed has reached a low moisture content before harvest. In the dry areas where it is grown, the seed will usually reduce to between 5 and 8% moisture. At this stage the seed is hard and can be squeezed from the heads; the bracts will have become dark in colour and the whole plant will be dry and brittle.

Bristles on the receptacle will be dislodged during harvest and can cause irritation to the skin; harvest operators should be suitably protected.

Small areas can be harvested by hand but the most usual method is by direct combine. If the seed is below 8% moisture content there should be no problem; however, if the seed is at a higher moisture content or if there is green weed growth in the crop, it may be better to windrow and to pick up by combine when seed moisture content has reduced to below 8%. Windrowing can begin when seed moisture content is 20–25%.

The seed is thin-hulled and requires careful handling during harvest to prevent mechanical damage which can reduce germination. Equipment should not be overloaded and forward speed and cylinder speed should be as slow as possible, consistent with good threshing. As the bristles are dislodged from the plant they can reduce the free flow of material through the combine.

Drying, Storage and Seed Cleaning

Seed is usually harvested below 8% moisture and so does not require drying. However, if moisture content is higher than this the seed must be dried immediately. If there is much green matter with the seed it should be removed by pre-cleaning and the bulk should then be dried. Either bin driers or continuous-flow driers can be used. The temperature of the drying air should not exceed 40°C.

For storage to the next growing season, moisture content of the seed must be below 8%.

An air/screen cleaner will normally be all that is required. Seed-borne fungi and insect pests will readily attack safflower seed and suitable seed treatment is required.

Tenecetum cinerarifolium (Trev.) Schultz Bip. Syn

Chrysanthemum cinerarifolium (Trev.) Vis.
Common Name: Pyrethrum

Although a native of Europe, pyrethrum is now grown in Africa (notably in Kenya and Tanzania) at higher altitudes, in Central and South America and, more recently, in Tasmania. It is a perennial growing to 60 cm in height and is grown for the pyrethrin content of the flowers. The crop requires an evenly distributed rainfall of between 900 and 1250 mm per year and a well drained soil. Cool weather is required to stimulate flower production; Purseglove (1984) states that 10 days at or below 16°C is needed to stimulate, while more than a week at 25°C will inhibit flower production.

Although there has been much plant breeding and cultivars with improved pyrethrin content have been produced, no descriptors of cultivars of pyrethrum have been published internationally.

The species is self-incompatible and must be pollinated by insects, mainly bees. Adequate isolation must therefore be provided. An interval free from pyrethrum should be allowed if changing from one cultivar to another.

Pyrethrum can be propagated by 'splits' (plants divided to provide several propagules) but is now more usually grown from seed, either by transplanting seedlings from a seed bed or by sowing directly in the field. Spacing between plants is about 30 cm × 1 m.

Crops are usually harvested for pyrethrin for three successive years, but this involves harvesting the flowers before the seed is ripe. Flowering is over an extended period and for seed production the seed must be allowed to mature; this will reduce the number of flowers produced. Flowers are normally picked by hand for pyrethrin production and this course may be adopted for seed production also; the crop can, however, be harvested mechanically for seed.

After harvest the seed may require drying. Before storage any immature seed should be removed on an air/screen cleaner.

Guizotia abyssinica (L.f.) Cass. Common Name: Niger

Commercial production of niger seed for oil extraction is confined to Ethiopia and India. It grows well on poorer soils and cultivation is largely by small-holders. There are no reports of cultivar descriptions. The plant is a branched herb, growing to about 1.5 m in height. Niger is cross-fertilised, mainly by bees and Indian regulations require 400 m for Foundation and 200 m for Certified Seed production (Tunwar and Singh 1988b). It is usually sown broadcast at 4–11 kg/ha. Harvesting is usually by hand, the plants being cut close to the ground and allowed to dry in the field before threshing. The seed is winnowed by hand and stores well provided it is kept dry.

Helianthus annuus L. Common Name: Sunflower

The sunflower is an important oilseed crop which originates in temperate North America but is now grown worldwide with the main centre of production in the former Soviet Union; it is also grown extensively in Argentina and the Balkans as well as North America (Heiser 1976).

Distinguishing between Cultivars

The following is based on OECD (1995b) and UPOV (1983). Cultivars may be open-pollinated, single-cross hybrids, three-way cross hybrids, double-cross hybrids or may be produced by another breeding method.

Main features used for distinguishing cultivars in the field:

Leaf shape: oblong, lanceolate, triangular, cordate, rounded
Leaf colour: light green, medium, dark green
Time of flowering: early, medium, late
Ray flower shape: elongated, ovoid, rounded
Ray flower colour: ivory, pale yellow, yellow, orange, purple, red-brown, multicoloured
Disk flower colour: yellow, red, purple
Head attitude: horizontal, inclined, vertical, half-turned down, turned down
Plant natural height: short, medium, long

Other features used for distinguishing between cultivars:

Hypocotyl anthocyanin: weak, medium, strong
Leaf size: small, medium, large
Leaf anthocyanin: absent, present
Leaf glossiness: absent, present
Leaf blistering: weak, medium, strong
Leaf serration: fine, medium, coarse
Regularity of serration: regular, irregular
Leaf cross section: concave, flat, convex
Leaf wings: absent, present
Angle of lateral veins: acute, right angle or nearly right angle, obtuse
Height of tip of leaf blade compared with insertion of petiole: low, medium, high
Angle between lower part of petiole and stem: small, medium, large
Hairiness at top of stem: weak, medium, strong
Number of leaves on main stem: few, medium, many
Number of ray flowers: few, medium, many
Anthocyanin in stigma: weak, medium, strong
Number of bracts on back of head: few, medium, many
Bract shape: elongated, rounded
Bract anthocyanin: absent, present
Natural position of closest lateral head to central head: below, above
Head size: small, medium, large
Shape of grain side of head: Concave, flat, convex, mis-shapen
Plant branching: absent, present
Type of branching: basal, at top, fully branched with central head, fully branched without central head
Seed size: small, medium, large
Seed shape: elongate, ovoid elongate, ovoid wide, rounded
Seed thickness: thin, medium, thick
Seed main colour: white, grey, brown, black
Seed mottling: absent, present

Seed stripes: absent, present

Colour of stripes: white, grey, violet-grey, brown

Position of stripes: marginal, lateral, both marginal and lateral

Laboratory Tests

Sodium dodecylsulphate electrophoresis of helianthinin polypeptides was reported by Anisimova et al. (1991) to separate a majority of cultivars tested; they state, however, that other methods such as isozyme analysis may be useful for separation of lines which are indistinguishable by electrophoresis of helianthinins. Moreau and Berville (1991) used a dot blot assay to detect maintainer plants in the seed of cytoplasmic male sterile (CMS) stocks; 20 samples were compared by molecular assay and field observations, and it was concluded that the former was as accurate as, or better than, the latter.

Environmental Requirements for Seed Production

Although originally a temperate crop, sunflowers have been selected to adapt to a wide range of conditions and are now grown from the cool tropics at the equator, to Canada in the north and to Argentina in the south. They are not essentially frost hardy but can withstand a little frost. They are not suited to very wet conditions and can grow in low rainfall areas provided that the precipitation is evenly spread throughout the growing season or that irrigation is available. Dwarf cultivars are generally better suited to dry conditions than very tall ones. Fertile soils which are deep and well drained are to be preferred.

Previous Cropping and Isolation

Sunflower seeds do not persist in the soil and volunteer plants from previous crops are not usually a problem. From this point of view an interval of one year free from sunflowers is generally sufficient. However, there are two parasitic weeds which attack sunflower: *Orobanche* spp. and *Cuscuta* spp. The former is largely confined to

eastern Europe and the former Soviet Union and is difficult to eliminate; an interval of four years free from susceptible crop plants is required. *Cuscuta* can be controlled by herbicides. Wild sunflowers (*Helianthus* spp.) occur in some areas, and fields which are infested should be avoided.

Sunflower is self-incompatible and therefore cross-pollinated. Pollination is by insects, particularly honey bees, with very little wind-borne pollen. Up to four hives per hectare are recommended. When the seed to be produced is intended for further multiplication, isolation of 400 m is usual although 600 m or more may be required for inbred lines or hybrid seed. When the seed to be produced is intended for the production of oilseed or food crops, 200 m is generally satisfactory. In the production of hybrid seed it is good practice to surround the crop by several rows of the male pollinator.

Seed Crop Establishment

Open-pollinated cultivars do not require any treatment different from that applied to crops for oilseed production. Spacing between rows varies from 55 to 100 cm and the plants are spaced from 15 to 30 cm apart in the row. Seed rate is between 4 and 10 kg/ha. Precision drills can be used when graded seed is available.

Hybrid cultivars are produced by planting alternate groups of rows of the female CMS line and the male pollinator line. Vannozzi (1987) states that these should be in proportions which may vary from 2:1 to 7:1 (female:male). He gives plant populations as varying between 45 000 and 67 000 per hectare while Langer and Hill (1991) suggest 30 000 plants per hectare in dry areas and 60 000 under irrigation. Closer spacing between plants promotes earlier flowering. For hybrid seed production it is essential to ensure that the male and female parents flower at the same time, if necessary by adjusting sowing dates. After the ray florets have developed, the sunflower heads turn towards the east; planting rows in a north–south direction allows for inspection of the heads looking west to avoid looking directly into the sun.

Seed Crop Management

The management of seed crops of open-pollinated cultivars is the same as for a crop to produce oilseed.

Crops to produce hybrid seed are also managed similarly, the main difference being that the male rows must be removed from the field prior to harvest of the female rows.

Weed control is essential, particularly of tall weeds which may compete with the sunflower plants and reduce seed yield. Wild sunflowers (*Helianthus* spp.) must be eliminated.

Irrigation is desirable during flowering; as the heads change colour to light green, irrigation will improve the plumpness of the seed.

Fungus diseases can be troublesome; the main ones are *Botrytis cinerea* Pers. ex Fr. (grey mould), *Plasmopara halstedii* (Farl.) Berl. and de T. (downy mildew), *Puccinia helianthi* Schw. (rust) and *Sclerotinia sclerotiorum* (Lib.) de Bary (stem rot). Fungicides are available for control. The sunflower moth, *Homoeosoma electellum* (Hulst), and other insects can be controlled by application of insecticides, but care should be taken to avoid times when bees are active.

As the seed ripens, birds will feed on the developing seed heads and considerable loss may occur if they are not controlled.

Seed Crop Inspection and Roguing

For open-pollinated cultivars, roguing at full flower will enable most of the important characteristics to be seen and this will be the best time for both inspection and roguing if only one visit is possible.

For the production of hybrid seed, the crop should be visited on several occasions. The first should be when the plants have six or seven leaves to check isolation and the occurrence of diseases such as downy mildew or objectionable weeds. A second inspection should be made during flowering and a third at maturity, when ripe head characteristics can be seen. It is advantageous to visit the crop when the plants are at the bud stage to check cultivar purity.

Harvesting

When harvesting hybrid seed it is essential to remove all heads of the male parent from the field before cutting the female. During this operation the rows of female plants should be checked to ensure that no male heads have become lodged in the female rows.

Seed may be harvested when moisture content falls below 20% but it is better to wait until it reaches 10%. At higher seed moisture contents the fleshy part of the head will also be high in moisture, making threshing or drying after cutting more difficult and giving greater risk of disease or other damage to seed germination in the period immediately after harvest. Therefore it is preferable to wait until the backs of the heads have changed colour, usually to yellow or brown depending on cultivar. At this stage the seed will be hard and difficult to mark with the thumb-nail and will have a moisture content of 12% or below. Shedding is not usual except when the seed becomes very dry.

Chemical desiccants may be used to hasten drying of the standing plants and, provided they are used correctly, will not damage germination.

Small areas can be cut by hand, removing only the head and a short length of stem. The cut heads are then spread thinly on a clean floor to dry before threshing. Alternatively the whole plant may be cut, the stems tied in bundles and stooked in the field to dry.

Larger areas are direct combined. Care must be taken to reduce the risk of mechanical damage, and threshing and subsequent handling must be as gentle as possible. The combine should be fitted with dividers which match the spacing between the rows. At above 8% seed moisture, a cylinder speed of 300 rpm is generally satisfactory but with drier seed this should be reduced to 200 rpm.

Drying and Storage

After threshing, the material will contain much broken stem and the fleshy part of the heads. This will usually have a higher moisture content than the seed and can cause damage to seed germination if not dried or removed immediately. Seed harvested at the higher level of around 20% moisture will deteriorate very rapidly if left for only a few hours. Prior to drying, as much as possible of the broken stems and heads should be removed on a pre-cleaner, preferably with both screens and aspiration.

Drying on a clean floor or in ventilated bins can usually be done with ambient air. In continuous-flow or batch driers the air temperature should not exceed 50°C; lower air temperature will be required if seed moisture content is above 12%.

For medium-term storage the seed should be at or below 10% moisture. For longer-term storage less than 8% is necessary.

Seed Cleaning

The most important equipment is an air/screen cleaner. This may be followed by an indent cylinder or gravity separator and size graders. The latter are required if the seed is to be sown by a precision seeder.

References

Anisimova, I.N., Gavriljuk, I.P. and Konerev, V.G. (1991). Identification of sunflower lines and varieties by helianthinin electrophoresis. *Plant Varieties and Seeds*, **4**, 133–141.

Heiser, C.B. Jr (1976). Sunflowers. In *Evolution of Crop Plants* (Ed. N.W. Simmonds). Longman, Harlow, 36–38.

Knowles, P.F. (1989). Safflower. In *Oil Crops of the World* (Eds G. Robbelin, A.K. Downey and A. Ashri). McGraw-Hill, New York, Chapter 17.

Langer, R.H.M. and Hill, G.D. (1991). *Agricultural Plants*. Cambridge University Press, Cambridge, 155–158.

Moreau, E. and Berville, A. (1991). A dot blot assay for the routine determination of maintainer contaminants in cytoplasmic male sterile seed stocks of sunflower (*Helianthus annuus* L.). *Plant Varieties and Seeds*, **4**, 31–36.

OECD (1995a). *Control Plot and Field Inspection Manual*. Organization for Economic Co-operation and Development, Paris, 98.

OECD (1995b). *Control Plot for Field Inspection Manual*. Organization for Economic Co-operation and Development, Paris, 64–66.

Purseglove, J.W. (1984). *Tropical Crops, Dicotyledons*. Longman, Harlow, 58–64.

Tunwar, N.S. and Singh, S.V. (1988a). *Indian Minimum Seed Certification Standards*. The Central Seed Certification Board, New Delhi, 113–114.

Tunwar, N.S. and Singh, S.V. (1988b). *Indian Minimum Seed Certification Standards*. The Central Seed Certification Board, New Delhi, 108–109.

UPOV (1983). *Guidelines for the Conduct of Tests for Distinctness, Homogeneity and Stability: Sunflower*. Doc. TG/81/3, International Union for the Protection of New Varieties of Plants, Geneva.

UPOV (1990). *Guidelines for the Conduct of Tests for Distinctness, Homogeneity and Stability: Safflower* (Carthamus tinctorus L.). Doc. TG/134/3, International Union for the Protection of New Varieties of Plants, Geneva.

Vannozzi, G.P. (1987). Technical and economic aspects of seed production of hybrid varieties of sunflower. In *Hybrid Seed Production of Selected Cereal, Oil and Vegetable Crops* (Eds W.P. Feistritzer and A.F. Kelly). Food and Agriculture Organisation of the United Nations, Rome, 253–280.

COMPOSITAE: VEGETABLE CROPS

Scientific name	Common name
Lactuca sativa L.	Lettuce
Cichorium endiva L.	Endive
Cichorium intybus L.	Chicory

These vegetables are important salad crops especially in the temperate regions. Chicory is also cultivated to produce roots which are forced to produce the famous Witloof 'chicons'. All three vegetables are important in the protected cropping industry for winter production, in addition to being summer field crops.

Latuca sativa L. Common Name: Lettuce

Cultivar Classification

The cultivated lettuces are divided into four types:

1. the cabbage or head lettuce: these are subdivided into crispheads and butterheads;
2. the cos or romaine lettuces;
3. the leafy or curled lettuces;
4. those with relatively fleshy stems, usually referred to as 'celtuce'.

The schemes for cultivar classification are based on the work of Rodenburg and Huyskes (1964) and Bowring (1969). Seed coat colour of mature seed is a cultivar characteristic and can be white, black or yellow (Watts 1980). The seeds of most modern cultivars are white.

Lettuce cultivar descriptions are based on the following characters:

Colour of seed: black, brown, yellow or white

Head type: cos, butterhead or crisp; hearting or leafy

Season: summer or winter field; protected cropping, temperature requirement if protected crop

Young plant characters at three to four leaf stage

Anthocyanin pigmentation

Leaf

Colour and degree of colour

Shape: round, ovate

Form: undulation and cupping

Edge: degree of indentation

Mature plant characters

Leaf: lustre, degree of blistering

Time to maturity, relative time that plants will remain in satisfactory condition for market before bolting

Resistance to specific pathogens and their races, e.g. *Bremia lactucae* Regel (downy mildew), and pests, e.g. *Pemphigus bursarius* L. (lettuce root aphid).

Environmental Requirements for Seed Production

The lettuce seed crop requires a soil pH of approximately 6.5. If there is doubt about the available calcium level it should be determined by soil analysis. Base nutrients of NPK with the ratio of 3:2:2 are normally applied during preparation.

Previous Cropping and Isolation

The proposed seed production site should not have been used for a lettuce seed crop for a minimum of three years or a market crop for two years, except where the soil or substrate has been effectively sterilised.

The lettuce inflorescence morphology favours self-pollination. Cross-pollination between compatible *Lactuca* spp. may occur and isolation from wild cross-compatible species, especially *Lactuca serriola* L., is essential. It is generally accepted that 2 m isolation between different cultivars or seed stocks is sufficient, although some authorities may require greater isolation than this.

Seed Crop Establishment

The seed crop is sown at the rate of 2 kg/ha in rows 50–60 cm apart; the seedlings are thinned to approximately 25 cm apart.

Seed Crop Management

The crop should be kept weed free, especially during the early stages of development. Supplementary nitrogen, usually as urea, can be applied as a top-dressing or foliar application, when significant leaching has occurred. However, if excessive nitrogen is applied, the heads tend to be loose which makes confirmation of the cultivar's hearting character difficult to confirm during roguing.

The solid heads of some butterhead and crisp cultivars form a mechanical barrier to seed head emergence. Where this is likely to occur, the head is removed by breaking or cutting but without damage to the internal growing point. This operation is done after the roguing stage which checks the heading characters of the cultivar.

Adequate available water increases seed yield, but irrigation applied too frequently can delay seed maturity by up to five days. Controlled irrigation can provide a drier regime prior to harvesting and thus reduce the loss of mature seed by shattering.

Lettuce mosaic virus (LMV) can be transmitted by infected seeds. It is essential that mother plants showing symptoms are removed from the crop as soon as symptoms are identified. The aphid species *Myzus persicae* is a major vector of the virus and should be controlled by routine spraying. Lettuce seed can be produced in insect-free structures, either greenhouses or structures clad with fine mesh, to reduce the incidence of LMV.

Seed Crop Roguing

Commercial category seed crops are rogued when the plants are hearting or at the marketable head stage. Crops to produce earlier categories are first rogued when the young plants have formed a rosette; they are then checked again when hearting, and finally when bolting has commenced. Any plants which display symptoms of LMV are removed during roguing.

Harvesting

Seed is harvested under dry conditions when 50% of the inflorescences show a fully developed pappus. The crop is either hand or machine cut (depending on the scale of operation) and left in windrows for further drying. Under arid conditions the crops are threshed on the same day as cutting. On a small scale, or where there is plentiful hand labour, the standing plants are left *in situ* and shaken into a suitable bag at two to three day intervals.

Threshing and Cleaning

Seed crops left in windrows can be threshed with a combine. An air/screen machine is normally used for pre-cleaning the seed lot. Any remaining debris is removed by a disc or indent cylinder.

Drying and Storage

The seed moisture content is reduced to 5.5% for storage in vapour-proof containers.

Cichorium endiva L. Common Name: Endive

The present-day cultivars have been developed for production systems similar to lettuce, although traditionally the crop was blanched when the plants were approaching marketable size. The species is treated as either an annual or biennial.

Classification of Cultivars

Season: days to maturity, suitability for protected crop production
Head characters: relative size and form
Leaf
    Colour: extent and heart
    Shape: broad or curled.

Environmental Requirements for Seed Production

Endive has the same soil and nutrient requirements as lettuce (*Lactuca sativa*).

Previous Cropping and Isolation

The same conditions apply as for lettuce (*Lactuca sativa*), but seed crops of the different types should have a minimum isolation distance of 50 m, or the minimum distance stipulated by prevailing regulations.

Seed Crop Establishment

The endive seed crop is grown as a biennial, and is produced from an autumn sowing in frost-free temperate areas. Seed is sown in rows 75 cm apart and thinned the following spring to between 25 and 50 cm apart within the rows according to the cultivar's vigour.

Seed Crop Management

The seed crop is kept weed free, especially before reaching the rosette stage. The flowering stems of endive are more vigorous than lettuce and although generally less branched, reach heights of up to 1 m before anthesis.

Seed Crop Roguing

There are two roguing stages. The first is when the young plants have formed rosettes. At this time any early bolting plants are removed and those remaining checked for vegetative characters, including leaf type and general plant form. The second stage is at normal market maturity of the cultivar when the population is again checked for early bolters and trueness to type.

Harvesting, threshing, cleaning and storage

These are the same as described for lettuce (*Lactuca sativa*).

Cichorium intybus L. Common Name: Chicory

In addition to being a specialist crop in northern Europe for which the plants are produced for forcing, this species is grown for its roots which are dried and used as an additive to coffee. *Cichorium intybus* L. is a perennial, but the seed crop is produced as a biennial.

Classification of Cultivars

The cultivars of chicory are mainly classified according to their vernalisation requirement prior to forcing (Rodenburg and Huyskes 1964). Chicory produced for forcing is usually referred to as 'Witloof chicory'. There are no records of cultivars being specifically developed for the production of roots for processing.

Seed Crop Production

The crop is grown as a biennial but is treated as described for lettuce (*Lactuca sativa*) in all other ways. The main exception is that it is generally considered to be cross-pollinated and there should be a minimum isolation distance of 1000 m between crops. Some seed companies have developed F1 cultivars; the advice of the relevant maintenance breeder should be followed when producing these.

References

Bowring, J.D.C. (1969). The identification of varieties of lettuce (*Lactuca sativa* L.). *Journal of the National Institute of Agricultural Botany*, **11**, 499–520.

Rodenburg, C.M. and Huyskes, J.A. (1964). The identification of varieties of lettuce, spinach and Witloof chicory, *Proceedings of the International Seed Testing Association*, **29**(4), 963–980.

Watts, L.E. (1980). *Flower and Vegetable Plant Breeding*. Grower Books, London.

Further Reading

UPOV (1981). *Guidelines for the Conduct of Tests for Distinctness, Homogeneity and Stability: Lettuce*. Doc. TG/13/4, International Union for the Protection of New Varieties of Plants, Geneva.

CRUCIFERAE: AGRICULTURAL CROPS

Cruciferous Oilseed Crops and Mustard (Including Fodder Rapes)

There are six species used for these purposes which, apart from minor details, are all grown for seed in a similar manner.

Scientific name	Common name
Brassica napus L. var. *oleifera.*	Swede rape
Brassica rapa L. (including *B. campestris* L.)	Turnip rape
Brassica juncea L. Czerni et Cosson.	Brown mustard
Sinapis alba L.	White mustard
Brassica nigra (L.) Koch.	Black mustard
Eruca vesicuria (L.) Cav. subsp. *sativa* (Mill.) Thell.	Taramira, Rocket salad

Under *B. napus* and *B. rapa* there are subspecies which are grown for their roots (swedes and turnips); these are discussed elsewhere. In both swede and turnip rape there are cultivars which are grown for the oil content of their seeds and those which are grown as fodder crops; the latter are grown for seed in a similar manner to the oilseed rapes.

The mustards are grown for the seeds from which condiments are produced, although *B. juncea* is also grown for oil production, mainly in Asian countries. Black mustard is of least importance and is now little grown.

Taramira is grown as an oilseed crop in India and is also used as a salad in Asia.

Distinguishing between Cultivars

For the rapes, a first classification can be made according to alternativity. The three mustards are annuals. A further subdivision for the oilseed rapes can be made on the basis of the amount of erucic acid in the oil extract; cultivars with erucic acid content of 2% or less are classified as 'low' and those above as 'high'. Additionally, the amount of glucocinolate in the extracted meal may be either low (below 35 μmol/g (micromoles/gram)) or high; a standard of below 25 μmol/g is used for Certified Seed, while in the UK seed harvested with below 18 μmol/g is considered suitable for an additional generation of multiplication to maintain the 'low' classification in the harvested seed. Cultivars which are low in both erucic acid and glucocinolate are described as 'double low'.

The following comparison of characteristics is extracted from OECD (1995). UPOV (1978) and UPOV (1988) are also relevant. There are no descriptors published for *B. nigra*. In the table '×' indicates that the characteristic is of use for the species concerned (see p. 184).

Laboratory Tests

The erucic acid content of the oil and glucocinolate content of the meal are grouping characters which do not distinguish individual cultivars. Electrophoresis of enzymes has been reported by Munges et al. (1990) to separate rapeseed cultivars. White and Law (1991) showed that over four seasons it was possible to distinguish between a limited number of rapeseed cultivars by canonical variate analysis of the fatty acid profiles of the oil.

Environmental Requirements for Seed Production

The six species are adapted to temperate conditions and are not grown in the humid tropics. The plants are tap-rooted and prefer a deep fertile soil. Preferred soils are medium to heavy, well drained but moisture retentive. The small seed is sown into a fine seed bed at a shallow depth and is therefore vulnerable to a dry period after sowing. It is important that growing conditions are such that the seed will germinate and that the plants will establish quickly and grow without check. A

Characteristics	B. napus	B. rapa	B. juncea	S. alba
Seed				
Colour		×	×	×
Reticulation on seed coat			×	×
1000 seed weight		×	×	×
Erucic acid content	×	×		
Seedlings				
Maximum width of cotyledons	×	×	×	×
Length of cotyledons		×	×	×
Ploidy number			×	
Leaves				
Blade colour	×	×	×	×
Indentation of margins	×	×	×	×
Undulation of margins		×	×	×
Development of lobes	×	×		
Number of lobes		×	×	×
Number of leaves			×	×
Size and shape				×
Length and width			×	×
Glaucosity and pubescence			×	×
Petiole length			×	×
Stems				
Length of main stem		×	×	×
Natural height of plant		×	×	×
Thickness			×	×
Number of internodes			×	×
Flowers				
Time of flowering	×	×	×	×
Petal colour	×	×	×	×
Anther dotting	×			
Length and width of petal		×	×	×
Siliqua				
Length and width			×	×
Length of stalk			×	×
Length of beak	×		×	×
Pubescence			×	×

well distributed rainfall is necessary. However, the plants will not withstand waterlogging. *B. juncea* is better adapted to dry soils than the others.

Previous Cropping and Isolation

Brassica seed sheds easily and is long-lived in the soil. For oilseed production it is essential that the quality be preserved by avoiding mixture or cross-pollination with cultivars with different erucic acid or glucocinolate content. For the mustards, similar considerations apply and the particular pungent principle of a cultivar must be preserved. Since volunteer plants occurring in a field will cause both mixture in the harvested seed and cross-pollination with other plants in the crop seed, certification schemes usually prescribe long intervals between seed crops: five years for the earlier generations and three for the final production of Certified Seed.

There are several weed species which are very similar to the cultivated forms, for example *Sinapis arvensis* (charlock) and *Raphanus raphanistrum* (wild radish), and these have seeds which are very difficult to separate from the crop seed. For taramira, the Indian seed certification standards also mention *Argemone mexicana* L. as an objectionable weed (Tunwar and Singh 1988). Previous cropping should be so arranged that these weeds can be eliminated from the field by herbicides. *B. nigra*, although sometimes grown as a crop for mustard, sheds seed very freely and can become a weed; this too must be eliminated during the previous cropping period. Some herbicides are available for both pre- and post-emergence use on a seed crop, but require very careful use; specialist advice should be sought in the area where the crop is to be grown.

The *Brassica* species will all interpollinate at least to some extent. *B. napus* and *B. juncea* are largely self-pollinated but cross-pollination can occur to varying degrees. The other species are cross-pollinated. *S. alba* and *E. vesicuria* will not cross-pollinate with each other nor with the brassicas; both are cross-pollinated. For all six species wind pollination occurs, and in *S. alba* it is said to be the main agent (Hemingway 1976). However, Crane and Walker (1984) state that insect pollination is generally the more important; it is considered beneficial to place hives of bees in or near the crop at the rate of about five per hectare.

Because of the need to preserve the oil characteristics, it is usual to adopt good isolation distances for all seed crops. For early generations 400 m is recommended, and for Certified Seed production, 200 m. In India the seed certification standards prescribe 50 m and 25 m for Foundation and Certified Seed of self-compatible cultivars, and 100 m and 50 m for self-incompatible ones (Tunwar and Singh 1988).

Seed Crop Establishment

Forage rapes are grown for seed in a similar manner to the oilseed rapes. Establishment of seed crops of oilseed and mustard does not differ from that for crops grown for oil extraction.

Winter cultivars require early sowing so that the plants become well established before winter sets in. Spring cultivars also respond to early sowing except in situations where early growth may be checked by early poor growing conditions.

Seed size varies considerably between cultivars and between seed lots of the same cultivar; growers should therefore check the 1000 seed weight.

For winter rape, NIAB (1996) suggests sowing 120 seeds/m^2 to give 80 plants/m^2 after winter. Spring-sown cultivars have smaller seeds than winter cultivars and NIAB (1996) suggests sowing 150 seeds/m^2 as they are shorter and branch less than winter cultivars.

Hybrid cultivars are now becoming available, particularly winter cultivars of *B. napus*. The first hybrid was recommended in the UK in 1996 (NIAB 1996). In the production of hybrid seed using cytoplasmic male sterility, the seed crop is established with the appropriate proportions of female:male parents; proportions of 3:1 to 7:1 have been proposed. The crop is also surrounded by several rows of the male parent. In this system the male parent plants must be eliminated before the hybrid seed is harvested. The resulting seed is also male sterile and must be mixed with seed of a pollinator for commercial use in a proportion of 80:20.

An alternative to the hybrid cultivar is the synthetic. This is composed of a mixture of lines in predetermined proportions which is multiplied for a limited number of generations.

Seed Crop Management

Except in the case of hybrid seed production, management of a seed crop does not differ from that for an oilseed crop. The crop responds well to nitrogen, but excessive use may cause lodging.

Pests during establishment, particularly during the winter, require treatment as appropriate, but the best protection is provided by a vigorously growing crop. The main pests are flea beetles and slugs; pigeons also graze the young plants and can cause considerable damage.

The main diseases are *Leptosphaeria maculans* (Desm.) Ces. and de Not (canker), *Pyrenopeziza brassicae* (light leaf spot), *Peronospora parasitica* (downy mildew), *Alternaria brassicae* (Berk.) Sacc. (alternaria), *Sclerotinia sclerotiorum* (Lib.) de bary (sclerotinia), *Plasmodiophora brassicae* Woron (club root) and *Pseudocercosporella capsellae* (Ell. and Ev.) Deighton (white leaf spot).

Club root cannot be controlled by fungicides and it is best to use clean land; however, most cultivars are able to cope with a mild infection. For the other diseases listed there are suitable fungicides available.

There are also three viruses which attack these crops: beet western yellows, turnip mosaic virus and cauliflower mosaic virus. These are transmitted by aphids and it is important to apply aphicides as appropriate.

During flowering, pollen beetle can seriously damage the seed set and must be controlled by insecticide. However, care must be taken to avoid applying insecticide at times when insect pollinators are active. Pollen beetle can be particularly damaging in hybrid seed production if allowed to restrict pollen supply.

Seed Crop Inspection and Roguing

Roguing is only practicable during the early stages in a multiplication cycle, with flower colour being the most obvious feature.

Seed crop inspection when flower colour can be seen is also desirable, but an earlier inspection when isolation can be checked is required.

Harvesting

Seed sheds freely and it is therefore important to harvest at the correct time to avoid seed loss. The period appropriate for harvest may be short and last no more than a week so the crop should be inspected frequently as it ripens. There are three possible methods for harvest: windrow and pick up later by combine; desiccate and direct combine after a short interval; direct combine.

When small areas are to be cut by hand, the crop can be tied in bundles which are allowed to dry and then threshed on the field. The cut crop should be handled as little as possible to avoid seed loss.

Windrowing is the generally preferred method. A crop is ready for windrowing when seed moisture content is below 35% but above 20%. At this stage the plants will be beginning to turn yellow and the seed will be firm, but not so hard that it cannot be marked with the thumb-nail. When seed moisture content falls to 14% the windrows can be picked up.

Desiccation is not recommended for seed production since germination of the harvested seed may be damaged. However, if the crop is excessively weedy, with much green matter, the method may be appropriate. Desiccation will normally be possible when seed moisture content has fallen below 20% and the interval between desiccation and combining should be as short as possible.

Direct combining can begin when seed moisture content has fallen below 14%. Both plants and seed will have changed colour and the seed will be hard.

Crops can become very tangled and a vertical knife can be fitted to windrowers (or combines) to separate the crop as it is cut. Pick-up attachments should be used on combines rather than lifters when picking up from the windrow.

The seed flows easily and is small. Combines and other seed handling equipment should be sealed to prevent seed loss.

Drying and Storage

The seed of rape and mustard can quickly deteriorate if stored with too high a moisture content because of the high oil content. For storage up to six months, moisture should be below 9% and for longer, below 7%. When samples are to be kept in long-term storage, moisture content below 5% is required.

Being small, the seed packs tightly and resists airflow. If drying on a ventilated floor, the depth of seed should be about 1 m. More than 1 m depth will resist airflow while less than this may cause pockets of undried seed to form between the air inlets. Similarly, ventilated bins should not be over-filled. A maximum air temperature of 5°C above ambient has been suggested for on-floor drying, and 7°C above ambient for ventilated bins.

Continuous-flow driers can also be used for rape or mustard seed but may require attention to the drier and the ancilliary equipment to prevent loss of seed through gaps. It is essential to adjust the depth of seed on flat-bed driers to ensure that the air flows freely and that bubbling is prevented.

Maximum seed temperatures during drying in continuous-flow driers were given by Ward et al. (1985) as follows:

Moisture content of seed (%)	Maximum safe temperature (°C)
10–17	66
19	60
21	54
23	49
25	43
27	38
29	32

Seed Cleaning

An air/screen cleaner will normally clean seed satisfactorily. Seed treatment with fungicides is available for some of the diseases listed above.

References

Crane, E. and Walker, P. (1984). *Pollination Directory for World Crops*. International Bee Research Association, London, 58, 97 and 118–119.

Hemingway, L.S. (1976). Mustards. In *Evolution of Crop Plants* (Ed. N.W. Simmonds). Longman, Harlow, 56–59.

Munges, H., Kohler, W. and Friedt, W. (1990). Identification of rape seed cultivars (*Brassica napus*) by starch gel electrophoresis. *Euphtica*, **45**, 179–187.

NIAB (1996). *Oilseeds Variety Handbook*. National Institute of Agricultural Botany, Cambridge, 9–40.

OECD (1995). *Control Plot and Field Inspection Manual*. Organization for Economic Co-operation and Development, Paris, 67 (*B. napus*), 70 (*B. rapa*), 115–116 (*B. juncea*), 166–167 (*S. alba*).

Tunwar, N.S. and Singh, S.V. (1988). *Indian Minimum Seed Certification Standards*. The Central Seed Certification Board, New Delhi, 103–105 (*Brassica* spp.), 110–112 (*Eruca vesicaria*).

UPOV (1978). *Guidelines for the Conduct of Tests for Distinctness, Homogeneity and Stability*, B. napus. Doc. TG/36/3, International Union for the Protection of New Varieties of Plants, Geneva.

UPOV (1988). *Guidelines for the Conduct of Tests for Distinctness, Homogeneity and Stability*, B. rapa. Doc. TG/37/7, International Union for the Protection of New Varieties of Plants, Geneva.

Ward, J.T., Basford, W.D., Hawkins, J.H. and Holliday, H.M. (1985). *Oilseed Rape*. Farming Press, Ipswich.

White, J. and Law, J.R. (1991). Differentiation between varieties of oilseed rape (*Brassica napus* L.) on the basis of the fatty acid composition of the oil. *Plant Varieties and Seeds*, **4**, 125–132.

CRUCIFERAE: VEGETABLE CROPS AND FODDER CROPS

Brassica napus L. var. *napobrassica* (L.) Rchb.
Common Name: Swede

Swedes are cultivated for stockfeed in northern Europe and also as a winter root vegetable.

Classification of Cultivars

The following classification is based on UPOV (1984).

> Leaf lobes: present or absent
> Root: root skin anthocyanin absent or present above soil level; intensity of root skin anthocyanin above soil level; flesh colour, intensity of flesh colour

Distinguishing between Cultivars

Main characters for distinguishing cultivars in the field:

> Leaf attitude: semi-erect, indeterminate, semi-drooping
> Leaf colour: light, medium or dark green
> Leaf glaucosity: absent or very weak, weak, medium, strong or very strong
> Number of major leaf lobes: absent, few, medium or many
> Size of terminal leaf lobes: small, medium or large
> Total length of longest green leaf: short, medium or long
> Leaf width at widest point: narrow, medium or broad
> Number of minor leaf lobes between major lobes: few, medium or many
> Number of minor lobes on petiole: few, medium or many
> Thickness of petiole: thin, medium or thick
> Root
> > Anthocyanin colouration of skin above soil level: absent or present

> > Colour of skin below soil level: white, yellow or reddish
> > Shape: transverse elliptic, circular, broad elliptic, obovate or oblong
> > Length: short, medium or long
> > Width: narrow, medium or broad
> > Length of neck: short, medium or long
> > Colour of neck surface between leaf scars: uniform red or purple, green, or purple mottled with green
> > Colour of flesh: white or yellow
> > Intensity of yellow colour of flesh: weak, medium or strong
> > Dry matter content: low, medium or high

Environmental Requirements for Seed Production

The optimum soil pH is 6.5 and the base NPK fertiliser dressing during preparation is 1:1:1. Swedes are susceptible to boron deficiency, boronated fertilisers should be used where this micronutrient is deficient.

Previous Cropping and Isolation

These are as described for fodder kale (*Brassica oleracea* L. var *acephala* DC).

Seed Crop Establishment

The seed crop can be produced by either the 'seed-to-seed' or the 'root-to-seed' method. The root-to-seed method is used for the production of Basic Seed.

Seed-to-seed The crop is drilled in late summer at the rate of approximately 3 kg/ha in rows 35–50 cm apart. The plants are subsequently thinned to approximately 5 cm apart within the rows.

Root-to-seed The crop is drilled and thinned as described for the seed-to-seed crop, but starting earlier in the summer to coincide with the sowing date of a commercial root crop.

Seed Crop Management

Seed-to-seed crops This is as described for fodder kale (*Brassica oleracea* L. var *acephala* DC).

Root-to-seed crops This is as described for the seed-to-seed method of fodder kale (*Brassica oleracea* L. var *acephala* DC) except that the roots are lifted in the late autumn. During the lifting operation the roots which are true to type are selected and retained. They are then stored during the winter. The selected roots are replanted in the early spring approximately 70 cm apart in rows 75 cm apart. When the flowering shoots are approximately 30 cm high, the top 10 cm of the terminal shoot is removed from each plant; this encourages the development of axillary inflorescences and the reduced plant height will minimise the risk of the seed crop lodging.

Seed Crop Roguing

There are two roguing stages. The first is in the first year, while the crop is in active growth; at this stage the vegetative characters are checked for trueness to type. The second stage is at the start of anthesis to check general morphology, flower colour and size. There is the additional stage for root-to-seed crops when the mature roots are selected before storage.

Harvesting, Drying and Cleaning

These are as described for fodder kale (*Brassica oleracea* L. var *acephala* DC). The cleaned seed is further dried to 5% moisture content for vapour-proof storage.

Brassica oleracea L. Common Name: Brassicas, Cole crops

The following varieties of this species are considered here.

Variety	Common name
Brassica oleracea L. var. *acephala* DC	Kale, Fodder kale
Brassica oleracea L. var. *capitata* L.	Cabbage, Fodder cabbage
Brassica oleracea L. var. *botrytis* L.	Cauliflower
Brassica oleracea L. var. *italica* Plenck	Sprouting broccoli
Brassica oleracea L. var. *gemmifera* Zenker	Brussels sprouts
Brassica oleracea L. var. *gongylodes* L.	Kohlrabi

These varieties of *Brassica oleracea* L. are all important fodder and vegetable crops of temperate regions; cauliflower and cabbage are also cultivated in the tropics and subtropics. Thompson (1976) has discussed the species' history, and detailed information on each of the types has been given by Nieuwhof (1969).

All of the varieties and their cultivars will readily cross-fertilise with one another.

Fodder kale

Classification of Cultivars

The fodder kales have traditionally been divided into 'marrow stem' types, which have a tall, thick, fleshy stem, and the 'thousand head' types which are generally shorter, thin-stemmed and leafier. Recent work by plant breeders has resulted in hybrids and types which do not fit this traditional classification. A general outline scheme for the classification of cultivars is as follows.

Use and frost hardiness
Leaf: length, width, anthocyanin, waxy bloom, serration of leaf margins
Petiole: length, pose, anthocyanin
Stem: shape, colour, amount of branching
Plant height

Environmental Requirements for Seed Production

The plants require a cold stimulus for flowers to be initiated; this is usually achieved by over-wintering plants which have a minimum of 15 leaves and/or leaf scars. During this period the plants have to be exposed to a minimum night temperature of 5°C or less, for approximately 15 weeks.

Previous Cropping and Isolation

A minimum of five years should be observed between seed crops of any of the *Brassica* types and at least two years exclusion of any *Brassica* species. Special attention should be given to controlling all cruciferous weeds, escape species and related volunteers in the previous crop. Weeds which are especially undesirable include *Rumex* spp., *Chenopodium album, Galium aparine, Geranium dissectum* and *Melandrium album*. Brassicas are a host to the pathogen *Plasmodiophora brassicae* Woron, commonly called 'club root' or 'finger and toe'; fields known to have been infected should not be used. *Alternaria brassiciola* (Schw.) Wilts. (dark leaf spot) and *Phoma lingam* (Tode ex Fr.) Desm. (canker) are both seed-transmitted and can be serious pathogens of *Brassica* seed crops. Therefore starting with disease-free stock seed and care with seed crop planning, to ensure isolation from both seed and ware crops of this species, are vital.

All the types of *Brassica oleracea* L. listed above cross-pollinate very readily between the varieties and their cultivars, and also with the wild cabbage (*Brassica oleracea* L.). Therefore some seed production regulations include a zoning scheme. The minimum isolation distance for fodder kale is 400 m when producing Basic Seed and 200 m for Certified Seed, although some authorities may stipulate greater distances than these.

Seed Crop Establishment

Kale is a biennial and crops intended for seed production are sown in mid-summer. Seed is sown at a rate of 1.5–3 kg/ha in rows 35–70 cm apart.

Seed Crop Management

The crop is generally left untouched over winter. Supplementary nitrogen is only given if the crop is slow to respond to increasing temperatures in the spring.

Populations of pollen beetle (*Meligtnes aeneus* F.) may build up from anthesis; if they reach about 20 per plant then an appropriate insecticide should be applied. The same spray application can be timed to control seed weevil (*Ceuthorhynchus assimilis* Payk.) if their population exceeds two per plant. Cruciferous weeds, especially *Sinapis arvensis* L. (charlock), are alternative hosts to both of these pests and care must be taken to ensure effective weed control in preceding and current crops. The natural insect populations and hive bees are vital pollinators of *Brassica* crops and every care must be taken with both the timing of the sprays and the active ingredient used.

In addition to related weeds being alternative hosts to pests and pathogens, it is important that other weeds whose seeds can contaminate the harvested seed lot are controlled by efficient crop husbandry.

In order to ensure that the maximum amount of pollination occurs, up to three bee hives per hectare are placed in close proximity to the crop.

Seed Crop Roguing

The fodder kales are normally only rogued once at flowering time. It is not possible to view each plant separately as is the case with the vegetable brassicas because of the relatively high plant population. Therefore the main criteria for identifying off-types are inflorescence characters; flower size and colour are especially useful characters for this purpose.

Harvesting

The seed crop ripens over a long period as a result of the sequence of flowers on individual inflorescences. The seed sheds very easily and it is therefore important that the decision to harvest is taken when the maximum potential quantity of high quality seed will be secured. Samples of pods are collected from the middle of the stems from representative plants in different parts of the field. When the seeds in the majority of the overall sample are light brown in colour and firm to the finger and thumb pressing test, the crop can be cut. It is normal practice to cut the crop and leave in windrows for approximately two weeks. The windrows are not usually turned during this period in order to avoid loss from shedding. At the end of the windrowing period the dried crop is picked up with a combine. There is usually too much green material for efficient direct combining, although in some dense crops of relatively short cultivars it may be possible to set the cutter bar high and only cut the pod-bearing branches.

Drying

It is usually necessary to dry the seed immediately after harvesting, and prior to any further processing, in order to ensure that its potential germination is maintained. The seed moisture content should be below 10% for storage in sacks, or 8% for bulk storage. When the moisture content is above 18% the air temperature for drying should not exceed 27°C; when below 18% it should not exceed 38°C. When using on-floor or bin driers the seed must not be too deep as the small seed restricts airflow.

Cleaning

The seed lot can be further upgraded by processing in an air/screen cleaner. Light or shrivelled seeds can be removed by a gravity separator or possibly with a spiral separator which is also suitable for separating split seeds.

Production of Hybrid Cultivars

Some cultivars are produced as triple-cross hybrids; the system is illustrated in Figure 7.1. The 11 year period can be reduced either by vernalising seed in the early stages or by multiplication of some of the stages in the other hemisphere.

In the production of hybrids, seed is normally harvested separately from each parent and then

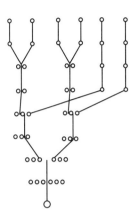

Year

1 – 4 Maintain inbred lines

   5 Sow double cross

   6 Harvest double cross

   7 Sow three-way cross

   8 Harvest three-way cross

   9 Sow triple cross

  10 Harvest triple cross

  11 Sell certified seed

FIGURE 7.1 System for production of triple-cross hybrid kale

mixed. This can be achieved either by mixing the seed prior to sowing, or sowing the parents in separate rows so that the mixture occurs during and after harvest. It is generally preferable to adopt the latter method, especially if the parent plants are of a different size and vigour. The seed grower should ensure that the established plant number is the same in each row. In practice, alternate rows provide a better seed set and have the added advantage of enabling seed crop inspectors to distinguish the parents' characters.

Cabbage, Cauliflower, Sprouting broccoli, Brussels sprouts, Vegetable kale, Kohlrabi

Classification of Cabbage Cultivars

   Method of seed production: open-pollinated or hybrid

   Season and use: fresh market, processing, storage, stock feed

   Morphological type: savoy or cabbage type, head shape

   Outer leaves: size, shape, colour pigmentation, degree of waxy cuticle, texture

   Resistance to high temperatures, splitting and early bolting

   Resistance to specific pathogens

Classification of Cauliflower Cultivars

Season and use, suitability for overwintering
Leaves: shape, midrib and vein character, inner leaf protection to curd
Curd: colour: green, white, pink or purple; relative size and shape, pattern of florets
Flower colour: white, yellow or intermediate
Resistance to specific pathogens

Classification of Brussels Sprouts Cultivars

Open-pollinated or hybrid
Season and use, suitability for mechanical harvesting
General vegetative characters: height of plant at first pick
Leaf: shape, colour, petiole length
Sprouts: relative size, colour (including pigmentation)
Resistance to specific pathogens.

Classification of Kohlrabi Cultivars

Season, uses, field or protected production.
Characters of 'bulb': colour, extent of pigmentation, shape, internal quality
Foliage: short, tall, pose of leaves, relative length of petioles

Environmental Requirements for Seed Production

The optimum pH for the vegetable crops is 6.0–6.5. Soils known to be deficient in boron, manganese or molybdenum should be avoided. The optimum NPK ratio is 1:2:2 and this should be taken into account during soil preparations.

Previous Cropping and Isolation

The previous cropping requirements, the need to control cruciferous weeds and links with seed-borne pathogens are as discussed for the fodder kales (*Brassica oleracea* var *acephala*). Cabbage, cauliflower, sprouting broccoli, vegetable kales and Brussels sprouts are usually established from transplants, therefore the previous cropping requirements are especially important for seed beds. The only possible exception is when selected plants are transferred to protected structures in which the soil or substrate has been partially sterilised.

The minimum isolation requirements between cultivars is 1000 m, with at least 1500 m between the different types. However, some seed regulations include a zoning scheme for *Brassica* spp. and the minimum distances may exceed those quoted here.

Seed Crop Establishment

Kohlrabi is normally drilled direct into the field as a seed-to-seed crop. Seed is sown in drills 75 cm apart, at a rate of 1.5–3 kg/ha.

The other vegetable brassicas are sown in seed beds; 60 g of seed sown in drills will produce approximately 10 000 plants, assuming satisfactory germination and young plant stand. Plants of some summer cauliflower cultivars are produced in greenhouses prior to transplanting.

The young plants are transferred to the field when they have approximately six true leaves. They are planted 30–60 cm apart with up to 90 cm between the rows; the optimum plant density within the rows and distances between the rows depends on the vigour of the cultivar and system of inter-row cultivations.

The taller cultivars of Brussels sprout and some of the heavier types of cabbage tend to topple as they reach the end of the vegetative stage, and as a prevention some seed producers earth-up the plants during earlier cultivations.

Seed Crop Management

The leaves of cabbages which form the heart are usually restrictive to the emerging inflorescence and it is usual practice to make a cross cut on the top of the cabbage head to reduce the restriction of the inflorescence, but care must be taken not to damage the internal growing point during this operation.

The terminal growing points of Brussels sprout plants are usually removed at the end of winter (during the final roguing); this stimulates the production of inflorescences from the secondary shoots which originate from the 'sprouts' in the leaf axils.

Summer cauliflowers are left to flower *in situ* provided that the seed crop is being produced in an area where the seed maturity can be completed before the onset of inclement weather, otherwise selected plants are lifted and replanted under protective structures, or protective structures are erected in the field.

The same technique of transferring selected plants into a protected environment can also be done for the other brassicas.

The Kohlrabi crop is usually left in the ground but either covered with protecting material or earthed-up. As with the other vegetable brassicas, selected mother plants can be lifted and transferred to a protected environment before the onset of inclement autumn weather.

Pollen beetle and seed weevil are important pests, as discussed for fodder kale (*Brassica oleracea* var *acephala*). Additional pests are *Brevicoryne brassicae* Linn. (mealy cabbage aphis) and the caterpillars of the cabbage white butterflies; these should be controlled with a suitable insecticide with due care given to pollinating insect populations, especially in protected structures.

Additional bee hives can be placed adjacent to *Brassica* seed crops at the start of anthesis. The use of insects for pollinating *Brassica* crops has been reviewed by Weiring (1964). *Diptera* spp. have been found to be especially useful pollinating insects in confined spaces such as greenhouses (Faulkner 1962).

Seed Crop Roguing

All of these crops are checked for trueness to type during the planting-out operation. Additional specific roguing requirements for individual crops are as follows.

Cabbage Before heading, the plants are checked for general foliage characters. After heading, individual plants are checked for timeliness and head characters. Early bolters and rosette plants are discarded.

Cauliflower and sprouting broccoli The plants are examined before the normal curding or floret period for vegetative characters including leaf type and leaf number; plants with precocious or button curds are discarded. When the mother plants are at market maturity they are checked for curd or floret colour and form, and shape and protection of curd.

Brussels sprouts After plant establishment the mother plants are checked for foliage characters, vigour and pigmentation. The second roguing is made when the lower axillary sprouts reach market maturity; at this time sprout quality and maturity period are assessed.

Vegetable kale The general plant vigour, size and leaf characters are checked at the end of the summer growing period and again in the spring, especially if winter hardiness is an important cultivar character.

Kohlrabi The initial roguing is at the final thinning of the direct sown crop when the general plant morphology is checked, including pigmentation. The second stage is at normal market maturity of the kohlrabi when shape, vigour, pigmentation and foliage characters are checked.

Harvesting

The harvest criteria are as described for fodder kale (*Brassica oleracea* var *acephala*). However, on small-scale production, plants or individual dried flowering shoots may be hand harvested as they reach maturity and placed on tarpaulins to finish ripening.

Threshing and Cleaning

The seeds are relatively brittle and a slow cylinder speed must be used. Further cleaning can be accomplished as described for fodder kale (*Brassica oleracea* var *acephala*).

The vegetable brassicas are dried to 5% moisture content for storage in vapour-proof containers.

Raphanus sativus L. Common Names: Radish, Fodder radish

The development of the four present-day radish types has been described by Banga (1976):

> *Raphanus sativus* L. var. *radicula*: the widely grown salad type
> *Raphanus sativus* L. var. *niger*: the large rooted type of Asia and Germany
> *Raphanus sativus* L. var. *mougri*: grown in Asia for its edible leaves and pods
> *Raphanus sativus* L. var. *oleifera*: fodder radish.

These types cross-pollinate freely with each other and with R. *raphanistrum* (the wild radish). The genus is pollinated by bees and other insects. The seed crops are usually produced as annuals and are not frost hardy, although there are some biennial types within the R. *sativus* var. *niger* group; these can act as annuals if grown in long days (Banga 1976).

Classification of Cultivars

Season of production, and use
Foliage characters: number of leaves at maturity, leaf shape in relation to position on the plant, other leaf characters
Root: shape, relative size, colour including proportion of bicolour if typical of the cultivar
Flower colour: white, yellow, pink

Environmental Requirements for Seed Production

Radish can tolerate soils with a pH from 5.5 to 6.8. The NPK ratio applied as a base-dressing is 1:3:4; higher proportions of nitrogen are given to fodder radish, depending on the soil's nutrient status.

Previous Cropping and Isolation

Ideally, fields designated for radish seed production should not have produced the crop for a minimum of five years, with wild radish eradicated as far as possible in previous crops. Other weeds which should have been eradicated as far as

possible are charlock and the species listed for fodder kale (*B. oleracea* vas. *acephala*).

The minimum isolation distance between seed crops is 1000 m, although care must be taken to ensure that ware crops are also taken into account as there may be early bolting plants in vegetable radish crops before they are cleared and fodder radish will also be in flower.

Seed Crop Establishment

The majority of radish seed crops are produced on a seed-to-seed basis and this is the method described here.

Fodder radish The seed crop is sown at a rate of 6 kg/ha in rows 45 cm apart.

Vegetable radish The seed crop is also sown at a rate of 6 kg/ha in rows from 50 to 90 cm apart to provide a distance of 5–15 cm between single plants. The lower final plant densities are used for the larger-rooted cultivars.

Seed Crop Management

The young plants of the vegetable radish are singled in order that the root characters can be determined in the subsequent roguing. The weed population should be controlled by cultivations early in the life of the seed crop.

The major pests are flea beetles (*Phyllotreta* spp.), and the pests listed for fodder kale seed crops should be controlled with appropriate insecticides when they become serious.

Seed Crop Roguing

Fodder radish The seed crop is normally rogued once, at the start of anthesis. At this stage the crop is checked for flower colour and any obvious off-types including wild radish.

Vegetable radish The seed crop is first checked when the stems start to elongate; at this stage early bolters and plants displaying off-type stem colour are removed, in addition to any wild radish plants remaining in the crop. The second roguing stage is immediately at the start of anthesis when the same points are checked as described for the first stage. Although flower colour can be observed at the start of anthesis, it is not always considered an important factor in the roguing of the final generation of vegetable radish unless obvious off-type flower colour occurs.

Root-to-seed Production of Vegetable Radish

Some stock seed of radish is produced on a root-to-seed system. In this case the radish crops are usually produced on a smaller scale than the seed-to-seed method and are lifted when they reach normal market maturity with the object of checking that the root character is true to the cultivar's description. Crops which have reached this stage by early summer are immediately replanted at the same density as described for seed-to-seed crops. The mother plants are rogued again at the two stages also described for seed-to-seed crops. Late-sown biennial types of *R. sativus* var. *niger* are normally overwintered in clamps before replanting.

Harvesting

The seed pods become brown and parchment-like as the seeds mature. The fodder radish crop is cut and windrowed when the seeds are observed to be turning brown. Further drying in windrows can take up to three weeks before the crop is harvested under dry conditions with a combine. For vegetable radish seed crops which have been finished under dry conditions, combines with a roller attachment or some types of bean threshers are considered most suitable, with the rollers adjusted to crack, but not crush, the pods; any unthreshed pods are returned to the threshing cylinder. If the seed crop is cut and left to dry in windrows, it is then passed through a thresher or a combine.

Drying, Processing and Storage

After threshing, the seed is immediately dried to approximately 10% moisture content. Further upgrading can be done on an air/screen cleaner.

Seed is reduced to 5% moisture content for vapour-proof storage.

Brassica campestris L. subsp. *rapifera* Metz. Common Name: True (vegetable) turnip

The vegetable turnip, which originated in Europe, is now widely cultivated elsewhere in the world. It is grown for both its edible swollen root and leaves, which are used as a cooked green vegetable. The root does not have a 'neck' as does the swede. It is mainly cross-pollinated by insects.

Classification of Cultivars

Season of maturity: suitability for early production (including protected cropping) or storage
Root: shape, colour of skin and flesh, colour of root crown
Leaf: shape, pose and relative size
Resistance to specific pathogen: *Plasmodiophora brassicae* Woron. (club root)

Environmental Requirements for Seed Production

The crop requires a soil pH of between 5.5 and 6.8. An NPK ratio of 1:1:1 should be applied during preparation, taking into account existing nutrient levels. If boron is deficient then boronated fertilisers should be used.

Previous Cropping and Isolation

A rotation which has excluded cruciferous crops within the previous five years should be adopted, especially if there has been any incidence of soil-borne pathogens. The crops in the previous two years should have been managed in a way that ensures the exclusion of cruciferous weeds.

The minimum isolation distance between flowering crops is 1000 m. The vegetable turnip readily cross-pollinates with swede, rape-type kales and turnip rape.

Seed Crop Establishment

Both seed-to-seed and root-to-seed systems are used for the production of vegetable turnip seed.

Root-to-seed This method is mainly used for the production of Basic Seed, and is therefore done on a relatively small scale. The mother plants are produced in exactly the same way as for the market crop produced for an autumn harvest. When the roots have reached maturity they are lifted and selected according to optimum cultivar characters. The selected roots are stored over winter and replanted in the early spring 30 cm apart, with 75 cm between the rows. This method can be very successful if the roots are replanted in polythene tunnels. The top 10 cm of each main flowering shoot is removed when it is approximately 30 cm high. This encourages the production of secondary shoots from the inflorescences and also condenses the overall seed ripening period.

Seed-to-seed With this method the seed is sown in the late summer at a thicker rate than the commercial root crop in order to achieve a dense plant stand. It is done in areas where the plants will be able to survive the winter, as the young plants remain in the field during this period. In exposed areas the young plants are covered, and to facilitate this they can be grown on a bed system. The following spring the shoots are reduced as described for the root-to-seed crop.

Seed Crop Roguing

Root-to-seed method The first roguing is done before the young plants have started to form swollen roots. At this stage they are checked for leaf type, colour and relative height. They are next checked immediately after lifting the roots; at this stage all the root characters are checked before they are stored. The third stage is at the start of anthesis when inflorescence characters are checked, including flower colour.

Seed-to-seed method Because of the high plant density this system does not offer a very detailed level of roguing. The plants are first rogued in the early vegetative stage when the crop is examined for leaf characters, colour and also relative height

and vigour. The second stage is at the start of anthesis when the crop is examined to check inflorescence details.

Harvesting and Threshing

The plants turn from green to a parchment colour as the seed ripens and it is usually preferable to cut the crop at this stage and leave it in windrows until the seed is more mature. Maturity is reached when typical samples of seed are seen to be turning brown and readily separate from their pods. Care must be taken with this assessment as the crop shatters very readily. At this stage the crop is either picked up with a combine or fed into a thresher.

Drying and Cleaning

The crop may require further drying as described for fodder kale (*B. oleracea* var *acephala*). Seed is further upgraded by an air.screen cleaner. Further improvement of the seed lot can be achieved by the methods described for fodder kale.

The seed is reduced to 5% moisture content for vapour-proof storage.

Brassica campestris L. subsp. *chinensis* Jusl. Common Names: Pak-choi, Chinese mustard

Brassica campestris L. subsp. *pekinensis* (Lous) Rups. Common Names: Pe-tsai, Chinese cabbage

These two species provide the majority of cultivars marketed under the general common name of 'Chinese cabbage', although in Asia it is usually only *B. c. pekinensis* which is referred to as the Chinese cabbage. The subspecies *B. c. chinensis* has a more open heart than *B. c. pekinensis*. Both subspecies are cultivated for their edible leaves and are important green vegetables in Asia; they have also become popular with urban populations in Africa, and more recently in northern Europe. *B. c. pekinensis* remains an important vegetable for winter storage and pickling in areas of Asia which experience severe winter conditions.

Classification of B. c. *subsp.* chinensis *Cultivars*

Open-pollinated or F1 hybrid: vernalisation requirement

Days to market maturity, days to flowering, heat tolerance

Leaf characters: relative shape, degree of wrinkling, intensity of green, character of midrib, character of rosette formed by outer leaves

Specific pathogen resistance

Classification of B. c. *subsp.* pekinensis *Cultivars*

Open-pollinated or F1 hybrid: vernalisation requirement

Days to market maturity, days to flowering, heat tolerance, bolting resistance

Shape and relative size of heart, solidity

Leaf characters: relative shape, degree of wrinkling, intensity of green, midrib character of outer leaves, pattern of serrations on leaf edge, degree of pubescence

Specific pathogen resistance

Resistance to tip-burn

Environmental Requirements for Seed Production

The seed production requirements and technology are similar for both subspecies and they are therefore dealt with together.

The crops grow best on soils which have a high water holding capacity, a pH between 6.0 and 7.5 and adequate available calcium levels. An NPK fertiliser ratio of 2:1:1 applied during site preparation will ensure sufficient levels of nutrients unless severe leaching takes place during the early stages of crop establishment. However, nutrient residues from previous crops should first be checked and taken into account as the crop is susceptible to high soluble-salt levels which can contribute to tip-burn if soil conditions become dry. If the boron status is low, boronated fertilisers should be used.

Previous Cropping and Isolation

Ideally, fields for seed production should not have grown cruciferous crops for at least two years, partly to minimise soil-borne pests and pathogens and partly to avoid carryover of dormant seeds of related crops and weeds. However, in some areas and communities where Chinese cabbage is a major crop, this requirement may not always be respected; particular attention should be given to the prevailing regulations.

Both subspecies are cross-pollinated, mainly by bees, although other insects also assist with pollination. The minimum isolation distance is 1000 m, although some authorities may stipulate greater distances than this, especially for Basic Seed production. Cultivars of the two subspecies readily cross-pollinate with each other and also to some extent with *Brassica juncea, B. nigra, B. napus, B. rapa*, their subspecies and cultivars.

F1 hybrid seed should be produced according to the maintenance breeder's instructions. The male and female parents are usually planted in a ratio of 1:1 or 1:2, usually in straight rows, although some breeders recommend a 1:1 ratio with an alternate plant pattern within the rows. In the latter case, clear instructions and careful supervision are required at harvesting.

Seed Crop Establishment and Management

Both head-to-seed and seed-to-seed methods are used, the former more usually for stock seed production and the latter for the final seed category. It is important to know the vernalisation requirement of the cultivar in relation to the local environment when planning the seed crop.

Head-to-seed method This system enables the seed producer to confirm morphological characters of individual plants before anthesis. The seeds are sown at a time when no vernalisation will occur until after head formation. The crop is grown either on the flat or on ridges, depending on local conditions and method of irrigation. If on the flat, seed is sown in drills 60 cm apart; for a ridge and furrow system the rows are up to 1.5 m apart. The young plants are thinned to 25–30 cm apart. An alternative, especially for small-scale production or when there is pressure on land use, is raising young plants in modules and then transferring them to the field when they have approximately five leaves.

Seed-to-seed method   This method is usually used in areas where the seed is sown at a time when the young plants are vernalised and the crop therefore does nor form the characteristic heart or 'head' of the cultivar before anthesis. The crop is drilled in rows 60 cm apart and thinned within the rows to approximately 30 cm apart. The seed-to-seed method is not usually practised where the crop is irrigated by a furrow system, but if this is necessary then the inter-row spacings are increased to 1.5 m. However, in areas where there is insufficient rainfall, furrow irrigation is preferable to overhead irrigation in order to minimise the risk of foliar disease and to provide better ripening conditions for the subsequent seed crop.

Seed Crop Management

Early weed control is essential to minimise competition and for efficient control of cruciferous weeds which act as alternative hosts to pests and pathogens of the seed crop.

In head-to-seed crops, the enveloping leaves of each Chinese cabbage head are removed to expose the undamaged interior growing point; this is not usually necessary for *Brassica campestris* subsp. *chinensis* cultivars.

Important diseases which should be controlled include *Sclerotinia sclerotiorum* (Lib.) de Bary (soft rot), *Peronospora parasitica* (Fr.) Tul. (downy mildew) and mosaic virus. Pests include *Pieris* spp. (cabbage white butterflies) and *Plutella* spp. (diamond back moth); the caterpillars of these two genera are often responsible for serious leaf damage. *Phyllotreta* spp. (flea beetles) can cause severe foliar damage to emerging seedlings.

Seed Crop Roguing

The first opportunity for roguing is provided for those crops which are raised in modules; during this operation transplants should be checked for leaf characters appropriate for the cultivar. Seed crops being produced on the seed-to-seed system should be checked for foliage characters and vigour when they have formed a rosette. The head-to-seed crops are checked for foliage characters, head size and head shape when the normal market head has been formed.

Harvesting

The mature seed pods turn a light brown colour as the seeds start to mature. The seed stalks can be cut and left in windrows if production is on a large scale. When produced on a relatively small scale or where there is adequate hand labour, selective hand harvesting of individual seed stalks is often practised. In some areas the whole plant is cut off and hung up for further drying. The seed pods shatter very readily and provision must be made for the cut material to be placed on or over tarpaulins. When ripening is completed the material is threshed.

Cleaning, Drying and Storage

The threshed seed can be upgraded by passing it through an air/screen machine. Seed is usually sun dried down to about 10% moisture content, or further dried to 5% for vapour-proof storage.

Lepidium sativum L. Common Name: Garden cress

Cress is grown as a salad crop which is mainly consumed as a raw salad while the plant is still in the young seedling stage. It is an annual when allowed to grow to anthesis and seed maturity.

Classification of Cultivars

There are two main morphological types, namely 'curled' and 'plain'; most seed companies do not distinguish the types further than this.

Seed Production

The seed production methods are as described for cruciferous oilseed crops and mustard.

Rorippa nasturtium-aquaticum (L.) Hayek. Common Name: Watercress

Watercress is produced as an aquatic crop, although the species can survive in soil without being immersed in water. Although many commercial crops are still propagated vegetatively from selected material, the main advantage of stock produced from seed is that it is virus free.

Environmental Requirements and Seed Production

R. nasturtium-aquaticum is a long-day plant; seed is generally produced by specialist growers of the commercial crop in temperate areas. There are no cultivars described, although the larger producers recognise the value of selected clonal material. The later flowering clones have a higher economic value when grown for the fresh market because of their longer cropping season. Therefore seed production techniques have been based on the use of protection for later flowering clonal material.

Clonal material which is seen to be vigorous and virus free is transferred either to aquatic beds covered with a polythene structure or to a soil-based substrate under protection. Aquatic beds for seed production are cleared of any previous crop and cleaned prior to receiving the vegetative material selected for seed production. The selection process of material from existing beds is usually done in late winter.

When the selected material is to be planted in a soil-based substrate, a pH of 7.0 is desirable and an NPK base fertiliser ratio of 2:1:1 is applied during preparation.

The flowering crop is partly cross-pollinated and some self-pollination takes place. The main pollinating agents are bees and *Diptera* spp. therefore an isolation distance of 1000 m from other flowering watercress crops is recommended.

Plant material for seed production is examined for virus symptoms and general morphological characters throughout the season until anthesis starts.

The seed pods of watercress shatter very readily, therefore the ripening seed heads are cut as soon as the maturing seeds start to turn a pale yellow colour. The cut material is placed in trays lined with paper to dry off. The trays are placed in the protective structure and sun dried.

The seeds are extracted by hand; this is achieved by hand rubbing the seed heads in dishes or trays. The seeds and chaff are then separated by passing the material through a series of fine sieves.

References

Banga, O. (1976). *Radish*. In *Evolution of Crop Plants* (Ed. N.W. Simmonds). Longman, London and New York.
Faulkner, G.J. (1962). Blowflies as pollinators of brassica crops. *Commercial Grower*, **3457**, 807–809.
Nieuwhof, M. (1969). *Cole Crops*. Leonard Hill, London.
Thompson, K.F. (1976). Cabbages, Kales etc. In *Evolution of Crop Plants* (Ed. N.W. Simmonds). Longman, London and New York.
UPOV (1984). *Guidelines for the Conduct of Tests for Distinctness, Homogeneity and Stability, Swede*. Doc. TG/89/3, International Union for the Protection of New Varieties of Plants, Geneva.
Watts, L.E. (1980). *Flower and Vegetable Plant Breeding*. Grower Books, London.
Weiring, D. (1964). The use of insects for pollinating brassica crops. *Euphytica*, **13**, 24–28.

Further Reading

Bleasedale, J.K.A. (1964). The flowering and growth of watercress (*Nasturtium officinale* N. Br.). *Journal of Horticultural Science*, **39**(4).
Crisp, P., Gray, A.R., James, H., Ives, S.J. and Angell, S.M. (1985). Improving purple sprouting broccoli by breeding. *Journal of Horticultural Science*, **60**(3), 325–333.
Opena, R.T., Kuo, C.G. and Yoon, J.Y. (1988). *Breeding and Seed Production of Chinese Cabbage in the Tropics and Subtropics*. Technical Bulletin No. 17, Asian Vegetable Research and Development Center, Taiwan.
UPOV (1976a). *Guidelines for the Conduct of Tests for Distinctness, Homogeneity and Stability, Cauliflower*. Doc. TG/45/3, International Union for the Protection of New Varieties of Plants, Geneva.
UPOV (1976b). *Guidelines for the Conduct of Tests for Distinctness, Homogeneity and Stability, Turnip*. Doc. TG/37/3, International Union for the Protection of New Varieties of Plants, Geneva.
UPOV (1976c). *Guidelines for the Conduct of Tests for Distinctness, Homogeneity and Stability, Cabbage (White cabbage, red cabbage and savoy cabbage)*. Doc. TG/48/3, International Union for the Protection of New Varieties of Plants, Geneva.

UPOV (1977). *Guidelines for the Conduct of Tests for Distinctness, Homogeneity and Stability, Brussels sprouts.* Doc. TG/54/3, International Union for the Protection of New Varieties of Plants, Geneva.

UPOV (1980a). *Guidelines for the Conduct of Tests for Distinctness, Homogeneity and Stability, Kohlrabi.* Doc. TG/65/3, International Union for the Protection of New Varieties of Plants, Geneva.

UPOV (1980b). *Guidelines for the Conduct of Tests for Distinctness, Homogeneity and Stability, Black radish.* Doc. TG/63/3, International Union for the Protection of New Varieties of Plants, Geneva.

UPOV (1980c). *Guidelines for the Conduct of Tests for Distinctness, Homogeneity and Stability, Radish.* Doc. TG/64/3, International Union for the Protection of New Varieties of Plants, Geneva.

CUCURBITACEAE: VEGETABLE CROPS

Citrullus lanatus (Thunb.) Mantsum et Nakai. Syn.
Citrullus vulgaris Schrader. Common Name:
Watermelon

Watermelon is widely grown in the tropics and subtropics, and because of its deep root system is also important in arid regions. Although usually regarded as a vegetable, the edible fruits are generally consumed as a dessert. The roasted seeds are eaten as a snack in some parts of the world, while in parts of Africa dried seeds are ground and used in cooked dishes.

There are open-pollinated, F1 hybrid and some seedless cultivars. The species is cross-pollinated by insects.

Classification of Cultivars

Season of fruit production, including suitability
    for protected cropping
Main type of market outlet for the fruit
Foliage cover, with special reference to protection for sun-scald
Vigour of vine, resistance to anthracnose, *Fusarium* wilt and sun-scald
Fruit
    External shape: spherical, oval or long
    Size and relative weight
    External rind colour or colours, and patterning
    Internal thickness of rind
    Flesh colour and intensity with reference to its intensity towards the centre
    Central cavity: absent or present
Seed: relative size, colour and patterning on testa

Environmental Requirements for Seed Production

Watermelon seed production is usually confined to areas where the ambient temperature does not fall significantly below 25°C.

The crop responds to bulky organic applications of up to 25 t/ha incorporated during preparation, otherwise a base-dressing of NPK in the ratio 1:1:1 is given. Excessive applications of nitrogen will produce too much vegetative growth without increasing potential seed yield.

Previous Cropping and Isolation

Satisfactory rotations should be adhered to, especially in fields where there is a history of *Fusarium oxysporum* Schlecht ex f. sp. *niveum* (E. F. Sm.) Snyder and Hansen (wilt) or *Colletotrichum lagenarium* (Pass.) Ell. and Halst. (anthracnose).

The usual minimum isolation distance is 1000 m, although 1500 m or greater is often stipulated for Basic Seed stocks.

Seed Crop Establishment

The seed crop is normally established from a direct sowing in the field. The seed is sown either on the flat or on ridges, depending on the method of irrigation. In some areas, mounds are used in preference to ridges. Furrow irrigation is used in conjunction with both mounds and ridge systems.

Two or three seeds are sown at each station, which are approximately 100 cm apart, with rows 120–180 cm apart. The sowing rate is 1–3 kg/ha depending on the seed size of the cultivar. The seedlings are thinned to one plant per station following emergence of the first true leaf.

Seed Crop Management

Irrigation frequency depends on soil type and rainfall. In arid areas the soil is brought up to field capacity prior to sowing and irrigation applied through the channels at approximately weekly intervals. Excessive overhead irrigation predisposes the crop to foliar pathogens. There is a moisture-sensitive stage during anthesis which extends through to young fruit development.

Seed Crop Roguing

There are four possible roguing stages. The first is before flowering when the vegetative characters are checked. The second stage is at early flowering when the morphology of undeveloped fruits is checked. The third stage is when the developing fruits are checked for trueness to type, and the final roguing is confirming the external morphological characters of the fruit to be harvested. Crops for Basic Seed production should be examined and rogued at all of these stages, but crops for the final seed category are usually only rogued at the first and third stages unless otherwise stipulated by legislation. It is important to remove off-type plants completely from the field to ensure that they cannot contribute pollen to the retained plants. Any plants showing symptoms of squash mosaic virus should be rogued out as soon as identified at any stage.

Harvesting and Seed Extraction

Mature fruits are harvested when the tendrils on the shoot bearing the fruit have withered; this is approximately one week later than harvesting for the fresh fruit market. In areas where there is sufficient hand labour, each fruit is hand harvested prior to wet seed extraction, otherwise in large-scale production the fruit is mechanically harvested and extracted. The methods for mechanical harvesting and seed extraction for fleshy fruits are described in Chapter 5. Careful attention to washing each seed lot after extraction will remove most of the fruit debris and minimise the amount remaining in the batch.

Drying and Processing

Immediately after the wet extraction process has been completed, the drying of the wet seed lot commences. The initial drying is done in either rotary driers or batch driers. Rotary driers, with a paddle which continuously turns the bed of seed, are usually preferred as they dry the seed batch more uniformly. The air temperature at the start of drying is 38–41°C. When no free moisture can be detected in pieces of fruit debris, the drying temperature is reduced to 32–35°C. The drying process is continued until the seed moisture content is 10%. The moisture content is reduced to 6% for storage in moisture-proof containers.

The seed lot can be further upgraded, if necessary, by processing with air/screen cleaners and indented cylinders.

Cucumis melo L. Common Names: Cantaloupe, Melon, Musk melon

The fruit of canteloupe or melon are mainly used as a dessert, although some of the oriental types are used for culinary purposes. Purseglove (1984) has classified melons into four types: the European cateloupe melon; musk melon; Casaba or winter melon; and the oriental melons. All the types freely cross-pollinate and there is therefore a range of intermediate types.

Classification of Cultivars

Season and market: outdoor or protected cultivation, suitability for shipping, storage potential
Fruit characters
Shape: length to width ratio
Rind: external texture and colour
Flesh: relative width of seed cavity and flesh, degree of sweetness, internal colour
Specific pest resistance
Specific pathogen resistance

Environmental Requirements for Seed Production

The crop grows best on soils with a pH between 6.0 and 6.8. Bulky organic manures can be applied during preparation, although this is not essential, especially when the seed is produced in rotations which include forage crops. The NPK ratio of 1:1.5:2 is applied as a base-dressing in the final stages of seed bed preparation. Supplementary nitrogen is applied at the start of anthesis, although later applications are likely to delay maturity and fruit ripening.

Previous Cropping and Isolation

Satisfactory rotations should be adhered to, especially where there is a history of *Fusarium* wilt or anthracnose (as described for watermelon *Citrullus lanatus*).

The minimum isolation distance is 1000 m, although some authorities may stipulate greater distances.

Seed Crop Establishment

Melon seed crops are grown on the flat, or on raised beds or ridges. The choice of system depends on the irrigation system to be used. Seeds are sown in groups of two or three at stations in the rows or on the ridges which are 1.25 to 2.0 m apart. Plant stations are 90 cm apart at the narrower row widths or down to 30 cm apart when the wider row widths are used. Two kilograms of seed will sow approximately 1 ha, although more seed will usually be required if it is drilled. The seedlings are singled following development of the first true leaves.

Seed Crop Management

When there is an average of four developing fruits per plant, the leaders and laterals are stopped by pinching them out.

The irrigation requirements and the need to avoid excessive overhead irrigation are the same as described for watermelon (*Citrullus lanatus*).

Seed Crop Roguing

The roguing stages and characters to be observed are the same as described for watermelon (*Citrullus lanatus*).

In addition to anthracnose and *Fusarium* wilt, listed for watermelon, the following viruses are seed-borne and infected plants should be removed during roguing when identified: cucumber mosaic virus (syn. *Cucumis* virus), melon mosaic virus, musk-melon mosaic virus (syn. *Marmor melonis*), squash mosaic virus and tobacco ringspot virus.

Harvesting

The ripe fruit of canteloupe and musk melons separate from the mother plant by an abscission layer. In large-scale seed production the crop is mechanically harvested when all maturing fruit have reached this stage. Alternatively, individual fruit are hand harvested. The fruit maturity of the winter and oriental types is detected by rind colour changes, e.g. green to yellow or yellow to white, depending on the cultivar. An additional indication of maturity is a softening of the fruit at the blossom end and an increase of aroma and waxiness of the rind's exterior.

The methods for mechanical harvesting and seed extraction for fleshy fruit are described in Chapter 5. Careful attention to washing each seed lot after extraction will remove most of the fruit debris and minimise the amount remaining in the batch.

Drying, Processing and Storage

These are the same as described for watermelon (*Citrullus lanatus*).

Cucumis sativus L. Common Name: Cucumber

Cucumbers are mainly cultivated for use as a salad or pickle. The salad crops are produced as either a field or protected crop, while those intended for pickling are normally only grown as a field crop.

Classification of Cultivars

Open-pollinated or hybrid

Season and market: pickling or salad; field or protected cropping

Fruit characters: relative size, shape, includes transverse section; rind colour at market and seed maturity; spines: degree of spininess, colour, relative size of warts at base of spines; retention of fruit shape if harvesting for market delayed; degree of bitterness

Vegetative characters: plant habit, leaf shape, production of lateral shoots

Resistance to cucumber leaf blotch and eelworm.

Environmental Requirements for Seed Production

The crop requires a soil with a pH of 6.5 and responds to base bulky organic manures applied at rates of up to 80 t/ha during preparations. Nutrients applied during the final seed bed preparations should be in the ratio of NPK 1:2:2, although the nutrient values of base-dressings should be taken into account. Where leaching of nitrogen is likely to take place, the NPK ratio is 2:1:1 with approximately half of the nitrogen applied as a top-dressing some four weeks after seedling emergence. However, as young cucumber plants are very susceptible to chemical leaf scorch, care must be exercised during the application of top-dressings.

Previous Cropping and Isolation

There should be a rotation of at least three years between a cucumber seed crop and any other cucurbit. Longer term rotations should be observed when there has been an incidence of root knot eelworm.

The minimum isolation distance should be 1000 m although some authorities may require greater distances than this, especially for the production of Basic Seed. These isolation distances are reduced when the seed crop is produced in pro-tected structures.

Seed Crop Establishment

The seeds of open-pollinated cultivars are sown two or three seeds per station 10–12 cm apart, in rows up to 2 m apart. The seedlings are thinned to a single plant per station after emergence. The seeds are sown on flat-topped ridges which are up to 30 cm high if a channel irrigation system is to be used.

For hybrid seed production there are usually six to eight female parent rows to two male (polli-nator) rows and the inter-row spacing is reduced to 50 cm.

Seed Crop Management

The main leaders on each plant are stopped when they have developed approximately five leaves and two main laterals are subsequently retained; other laterals are removed.

Seed Crop Roguing

There are four possible roguing stages. The first stage is before the first flowers open when the vegetative morphology and vigour are checked. The second stage is at the start of anthesis, when the same characters as described for the first stage are checked and in addition, if the seed crop is intended for the production of Basic Seed, the morphology of the developing fruits is checked. The third stage is when fruit is setting; in addition to the characters listed for the second stage, the time of production and fruit characters are con-firmed. The fourth roguing stage is when the fruits are ripe; at this stage the fruits are examined to determine that their morphological characters are in accordance with the cultivar's description.

Seed crops intended for the final seed category are usually only rogued at the third and fourth stages. In addition to any off-type plants identified at any of the four stages, plants which show symptoms of *Fusarium* wilt, cucumber angular leaf spot virus, cucumber green mottle mosaic virus (syn. Cucumis virus 2) or cucumber mosaic virus 1 (syn. cucumis mosaic virus 1) should be removed from the crop as these viruses are seed-transmitted.

Harvesting

The fruit should remain attached to the mother plant until fully mature. Fruit maturity for seed production purposes is indicated by development of a rind colour which is characteristic of the cultivar. This can be confirmed by cutting a sample of fruit open; when the seeds are mature they separate easily from the interior fruit flesh.

The methods for mechanical harvesting and seed extraction are described in Chapter 5. The individual fruits are cut longitudinally and the seeds scooped out when extracted by hand.

The freshly extracted seeds usually have a gelatinous layer around them which is removed by natural fermentation. When there is no gelatinous layer (as is often the case when the fruit has become very ripe before harvesting), the extracted seed is only washed.

When fermentation is necessary, the wet seed lot is placed in containers with some extra water to form a slurry. The rate of fermentation depends on the ambient temperature; the wet seed lot should be stirred every three or four hours and observations made to determine when the breakdown of the seeds' gelatinous coating has been completed. The process can take from one to two days. The containers should be covered with muslin during the process in order to minimise fritfly activity. On completion of the fermentation, the seed lot is further washed.

Drying, Processing and Storage

These are the same as described for watermelon (*Citrullus lanatus*).

Cucurbita species

Scientific name	Common name
Cucurbita maxima Duchesne	Pumpkin, Winter squash, Crookneck squash, Squash, Gourd
Cucurbita moschata (Duchesne) Duchesne ex Poiret	Pumpkin, Winter squash
Cucurbita mixta Pang.	Pumpkin, Winter squash
Cucurbita pepo L.	Marrow, Vegetable marrow, Courgette

The various types of *Cucurbita* spp. are widely grown in the tropics, arid regions and as summer season crops in the temperate regions. Although mainly cultivated for their fruits which are used as vegetables, the young leaves are cooked as a green vegetable in some tropical areas.

There is a very close similarity in seed production methods for all of the *Cucurbita* spp. listed above and they are therefore dealt with together.

Classification of Cultivars

Open-pollinated or F1 hybrid
Season of production
Culinary use: e.g. immature fruit, storage
Vegetative habit: trailing ('vine'), bush
Leaf characters
Fruit characters
   External: shape, relative size, rind colour (single colour or bicolour), striped, rind texture, prominent ribs
   Internal: flesh colour, shape and relative size of central cavity
Resistance to specific pathogens

Environmental Requirements for Seed Production

These species are not frost-tolerant and seed production is best achieved in the tropics or subtropics when there is either low or moderate rainfall. Wet or humid conditions predispose the plants to foliar pathogens.

This group of species is more tolerant of acid soil conditions than other members of Cucurbitaceae and will succeed on soils with a pH of 5.5 to 6.8. They respond to bulky organic base-dressings of up to 30 t/ha, although the equivalent amounts of nutrients can be applied if bulky manures are not available. The optimum NPK ratio is 1:2:2 applied during site preparation. A further top-dressing of NPK is applied when the fruit has started to set.

Previous Cropping and Isolation

Satisfactory rotations should be followed, especially in areas where *Fusarium* spp. have been recorded. The usual minimum isolation distance between any cultivars of these species is 1000 m. Crops for Basic Seed production should be isolated by 1500 m from other species in this group. Individual authorities may stipulate greater distances than these.

Seed Crop Establishment

The seed crop is either grown on ridges or on the flat, depending on the irrigation system. The interrow spacing depends on the vigour of the cultivar; for bush types it is up to 90 cm while for the trailing ('vine') types the rows are up to 3.5 m apart.

The distance between plants within the rows is usually approximately the same as for the cultivar's inter-row spacing. Approximately three seeds are sown per station and the young plants are subsequently singled.

F1 Hybrid Seed Production

The ratio of male to female rows is usually 1:5, although the breeder's recommendations should be followed if different. The male flowers on the plants in the female rows are suppressed by three applications of a 250 ppm ethrel solution: the first at first true leaf, the second at third true leaf, and a third application when the fifth true leaf is showing. The effect of the ethrel does not persist beyond the time that two to three fruits have developed; therefore, as further ethrel applications would not be effective, the development of later male flowers is prevented by manual removal of the plants' growing points.

Seed Crop Management

Irrigation should be applied at approximately weekly intervals, but if overhead systems are used, care must be taken to ensure that the foliage does not remain wet during the night otherwise foliar pathogens will become prevalent.

Seed Crop Roguing

The first roguing should be done in the early vegetative stage to check the vegetative characters; at this time non-trailing types are removed from trailing types (or *vice versa*). The second stage is just before the first flowers are open when the vegetative characters are again confirmed; in addition, the morphological characters of the undeveloped fruit below the female flowers are also checked. The third stage is when the first fruits are setting; at this time the developing fruit is checked for trueness to type in addition to the vegetative characters given for the previous stages. The fourth stage is to check that the more mature fruit is true to the cultivar's description.

In addition to the above characters, any plants found to have the following seed-transmitted viruses must be removed from the crop as soon as observed: cucumber mosaic virus (syn. Cucumis virus 1), musk-melon mosaic virus, Prunus necrotic ringspot virus (syn. peach ringspot virus).

Harvesting

The rind of mature fruit becomes hard at maturity and this is usually associated with a colour change. The fruit are mechanically or hand harvested as described for watermelon (*Citrullus lanatus*). The seed is extracted by the method described in Chapter 5. In the case of the winter squashes, which tend to have a relatively dry fruit texture, extra water is usually added during seed extraction to assist separation of the seed.

Drying, Processing and Storage

These are the same as described for watermelon (*Citrullus lanatus*).

Reference

Purseglove, J.W. (1984). *Tropical Crops, Dicotyledons.* Longman, London, 110.

Further Reading

UPOV (1978). *Guidelines for the Conduct of Tests for Distinctness, Homogeneity and Stability: Cucumber, Gherkin.* TG/61/3, International Union for the Protection of New Varieties of Plants, Geneva.

EUPHORBIACEAE: AGRICULTURAL CROP

Ricinus communis L. Common Name: Castor

The castor bean is a well established oilseed crop grown throughout the tropics. Although it originated in northeast Africa, the main producing countries are now India and Brazil with production increasing in China.

Descriptors for cultivars have not yet been published by the international organisations. Older cultivars are generally tall and unsuited to mechanical handling; newer cultivars, most of which are hybrids, are short and can be handled with suitable equipment. Atsmon (1989) states that plants may be pure red or pure green.

Environmental Requirements for Seed Production

Castor is not frost hardy and requires a temperature between 20 and 30°C during the growing season. At temperatures above 40°C it does not set seed satisfactorily. The plant is deep rooted and requires a deep, well drained but moisture retentive soil. The crop requires 350–600 mm of rain during the growing season, which lasts between 140 and 180 days (Cobley 1977).

Previous Cropping and Isolation

Seeds of castor can survive in the soil for three or more years and therefore previous crops other than castor should be grown for two to three years, during which time any volunteer plants should be eliminated.

The plants are monoecious. The male flowers occur at the base of the inflorescence and female in the upper part. The female flowers are receptive before the males release pollen, so that cross-pollination is usual. However, inbred lines have been produced and have been released as cultivars. Pollination is usually by wind, but insects also visit the plants, attracted by nectar released by glands at the junction of the leaf lamina and petiole. Isolation as for a cross-pollinated crop is normally required: 400 m for early generations and 200 m for Certified Seed production.

Seed Crop Establishment

Except for hybrid cultivars, establishment is the same as for an oilseed crop. Purseglove (1984) suggests rows 96–102 cm apart with 20–25 cm between plants in the row. Seed rates are 12–14 kg/ha.

For hybrid cultivars the appropriate proportions of female to male parents must be established. Production of hybrid seed has been based on a nuclear male sterility (NMS) system and is normally under the control of the breeder.

Seed Crop Management

During the growing period, management is the same as for an oilseed crop.

Seed Crop Inspection and Roguing

Off-types can usually be identified during flowering, which is the best time for both roguing and inspection.

In the production of hybrids by the NMS system it is necessary to rogue out the hermaphrodite plants in the female line before flowering. The Indian certification standards allow for no more than 1% deviants in the female line for Foundation Seed and 2% for Certified Seed; no monoecious plants are permitted in the female line when producing Foundation Seed (Tunwar and Singh 1988).

Harvesting

Older cultivars shed seed easily and are usually harvested by hand picking on several occasions.

Modern cultivars and hybrids shed seed less freely and are suitable for mechanical harvesting. Specialised harvesting equipment has been produced which will remove seed with a beating action. During harvest it is essential to avoid damage to the testa as this causes oil to exude which can go rancid very quickly, damaging the seed bulk.

Capsules must be dry when removed from the plant. A desiccant can be used to hasten drying, or where a frost can be expected as the seeds reach ripeness, this will also hasten drying.

Drying, Storage and Seed Cleaning

The seed is normally dry enough after harvest but may require hulling and will usually contain broken pieces of capsule and damaged seed. These impurities must be removed immediately as they can cause serious loss of germination if allowed to remain in the bulk. An air/screen cleaner will normally achieve this.

References

Atsmon, D. (1989). Castor. In *Oil Crops of the World* (Eds G. Robbelen, A.K. Downey and A. Ashri). McGraw-Hill, New York, Chapter 24.

Cobley, L.S. (1977). *An Introduction to the Botany of Tropical Crops* (revised by W.M. Steele). Longman, London, 302–306.

Purseglove, J.W. (1984). *Tropical Crops, Dicotyledons*. Longman, Harlow, 180–185.

Tunwar, N.S. and Singh, S.V. (1988). *Indian Minimum Seed Certification Standards*. The Central Seed Certification Board, New Delhi, 95–100.

GRAMINEAE: CEREAL GRAIN CROPS

There are 15 main species which are grown widely to produce grain; this is mainly used either as food for humans or as animal feed. In all of these there are conventional cultivars which are either self- or cross-fertilised. In several species, however, hybrid cultivars have been produced in both self- and cross-fertilised species. Hybrids may be *Single Cross* (F1 between two Inbred Lines); *Double Cross* (F1 between two Single Crosses); *Three-way Cross* (F1 between an Inbred Line and a Single Cross hybrid); or *Top Cross* (F1 between either an Inbred Line or a Single Cross and an open-pollinated cultivar).

To achieve hybridity it is necessary to ensure that the female parent is emasculated either by hand, by a chemical hybridisation agent (CHA), or by the use of genetic or cytoplasmic male sterility. The latter requires the normal sequence of production using a restorer (b) line to maintain the male sterile (a) line.

For quality control it is necessary to ensure that emasculation has taken place and that hybrid seed has been produced. In the normally cross-fertilising species this can usually be checked visually during field inspection.

However, in normally self-fertilising species it is not generally possible to check the degree of emasculation by simple observation in this way and special measures have been suggested: sample areas of the female parent could be protected in pollen-proof bags or tents (when CHAs are used, after treatment but before anthesis). This enables calculation of the number of seeds produced by the female parent in the absence of pollen from outside and this can be compared with the number of seeds produced by the female parent when exposed to pollen from the male parent. At least 98% sterility is usually required (OECD 1995a).

Tests of the hybrid seed using electrophoresis are also available to estimate the proportion of non-hybrid seed. A standard of 95% hybrid seed is usually applied.

Triticum species

Wheat is the most widely grown small-grain cereal and accounts for some 30% of production in the world. There are three species of *Triticum* used for food production, one of which, *T. spelta*, is on a small scale, mostly confined to mountainous areas in Europe. The main species used for bread-making around the world is *T. aestivum*. The species used for pasta and some bread is *T. durum*. Wheat grain is also an important livestock feed.

T. durum is tetraploid whereas *T. spelta* and *T. aestivum* are hexaploid. *T. spelta* has the speltoid characteristics of a fragile rachis and seed remaining enclosed in the glumes after threshing. Wheat is a temperate crop but is also grown at high altitude in the tropics and is being extended to lower altitudes by breeding adapted cultivars. There are two main types: winter wheat is a long-day plant and also requires vernalisation; spring wheat is either long-day or day-neutral and does not require vernalisation.

Hybrid cultivars of *T. aestivum* have been produced, mainly by use of chemical hybridising agents. They are used increasingly in North America. A problem for seed production is the need to provide suitable isolation in which to grow the seed crops because of the very wide distribution of conventional wheat crops (see, for example, Pickett and Galway 1997).

Triticum aestivum L. emend Fiori et Paol. Common Names: Wheat, Bread wheat

Classification of Cultivars (based on UPOV 1981a)

Intended use: bread flour, biscuit flour, animal feed
Seasonal type: winter, alternative, spring
Straw section below ear: pith thin, pith thick
Ear colour at maturity: white, coloured
Presence of awns or scurs: absent, scurs present, awns present
Grain colour: white, red

Distinguishing between Cultivars

Main features used for distinguishing cultivars in the field:

> Flag leaf attitude: rectilinear, recurved
> Time of ear emergence: early, late
> Glaucosity
>   of flag leaf sheath: weak, strong
>   of flag leaf blade: weak, strong
>   of ear: weak, strong
>   of neck of culm: weak, strong
> Plant height (stem + ear): short, long
> Ear shape: tapering, parallel, fusiform, clavate
> Ear density: lax, dense

Other features used for distinguishing between cultivars:

> Plant growth habit: erect, prostrate
> Distribution of awns or scurs: tip only, upper half, whole length
> Length of scurs at tip of ear: short, long
> Length of awns at tip of ear: short, long
> Anthocyanin colouration
>   of coleoptile: absent, present
>   of flag leaf auricles: absent, present
>   of anthers: absent, present
> Hairiness
>   of uppermost stem node: absent, weak, strong
>   of apical rachis segment (convex surface): absent, weak, strong
>   of inner surface of lower glume: weak, strong
> Lower glume
>   shoulder width: narrow, broad
>   shoulder shape: sloping, rounded, straight, elevated
>   beak length: short, long
>   beak shape: straight, curved, geniculate
>   internal imprint: absent, small, large
> Lowest lemma beak shape: straight, curved, geniculate
> Grain shape: round, ovoid, elongated
> Grain brush hair: short, long

Laboratory tests:

> Grain colouration with phenol (ISTA 1996)
> Electrophoresis of grain proteins (ISTA 1996)

Environmental Requirements for Growing the Crop

Feldman (1976) states that wheat is grown from 67°N in Norway, Finland and Russia to 45°S in Argentina, although in the tropics it is largely confined to the higher (cooler) altitudes. He gives the centre of origin as the Fertile Cresent, Anatolia and the Balkans. Wheat is thus very tolerant of a wide range of conditions, although it is most widely grown in southern Russia, the central plains of North America (USA and Canada), north-central China, India, Argentina and southwestern Australia. In Europe it is widely distributed, but the highest yields are obtained on the more fertile clays of northern Europe. The cultivars have almost all been bred for the conditions in which they are grown. For example, those grown in northern Europe are different from those grown in the Mediterranean area.

Previous Cropping and Isolation

To avoid ground-keepers in a seed crop it is necessary to allow an interval of two years free from wheat before the seed crop is sown. However, if the seed crop is of the same cultivar and is to produce the same grade of seed as previously, this interval can be reduced. Crops of other small-grained cereals should also be avoided prior to sowing a seed crop of wheat. Barley in particular grows readily from shed seed and cannot be eliminated from the growing wheat crop; when harvested with the wheat it is very difficult to remove from the wheat seed without high cleaning losses. Intervening crops are to be preferred which allow the field to be cleaned of weeds such as wild oats (*Avena fatua*, *A. ludoviciana*) or black grass (*Alepocurus mysuroides*).

T. aestivum is self-fertilising and so only requires isolation sufficient to avoid mixture with adjoining crops as provided by a physical barrier (e.g. hedge, fence or ditch) or a clear gap of 2 m.

However, when producing hybrid cultivars it is necessary to provide a greater isolation distance to ensure that the only pollen reaching the crop is that from the desired male parent. Wheat pollen is very buoyant and Edwards (1987) reports that it can travel at least 60 m. In the OECD Cereal Seed Scheme (OECD 1995b) an isolation distance of 25 m is required from all sources of pollen except the male parent.

Seed Crop Establishment

For conventional cultivars there are no particular measures to be taken for a seed crop which differ from those used to establish a commercial crop, apart from ensuring that the seed used to sow the crop is kept completely separate and that all equipment is cleaned before use. The seed rate varies between regions from 80 to 180 kg/ha. Lower seed rates in the earlier stages of multiplication will increase the multiplication rate and around 30 kg/ha has proved satisfactory. Crops are normally drilled in narrow-spaced rows and it is usual to leave 'tramlines' at suitable intervals to permit access for roguing and crop inspection.

For hybrid cultivars the appropriate ratio of female to male rows is planted and will normally be specified by the breeder. Ratios of 1.5:1 and 3:1 have been reported as satisfactory, but efficiency of different ratios varies with location. It is usual to sow by seed drill and varying the width of the drill from 2.1 to 3.7 m had little effect on the efficiency of pollination (Stroike 1987).

Seed Crop Management

In general, seed crops do not require management different from that of a commercial crop. However, it is important to give careful attention to detail to ensure that weed and disease problems are dealt with promptly and that all steps are taken to ensure that there is no possibility of mixture with other cultivars.

Fertiliser use is similar to that adopted in the area where the seed crop is growing, but care should be exercised in the application of nitrogen;

it is important that the crop does not lodge. Straw shorteners can be used, but it is advisable to avoid them if possible since they may distort the morphology of the plants, so making cultivar identification more difficult.

The same considerations apply to the use of herbicides, especially those based on growth hormones which may cause distortion of the ears. Weeds which have seeds difficult to separate from the wheat seed should be controlled. These include: *Avena fatua*, *A. sterilis* and *A. ludoviciana*, *Agrostemma githago*, *Asphodelus* spp., *Brassica* spp., *Convolvulus arvensis*, *Carthamus oxycantha*, *Cardaria* spp., *Centaurea repens*, *Euphorbia esula*, *Helogeton glomeratus*, *Lolium temulentum*, *Solanum carolinse*, *Senecio jacobea* and *Raphanus raphanistrum*. Ergot is not normally a problem in wheat, but care should be taken to avoid the inclusion of any infected grass heads with the harvest. Other weeds which can be troublesome in harvested seed are *Galium aparine* and *Allium vineale*; the latter is particularly objectionable as it creates a smell of onions in the seed, which is very persistent. There are herbicides available which will deal with these weeds, but their use is often controlled by legislation and their effectiveness may vary in different locations; it is therefore essential to seek advice from the manufacturers.

Disease control by the application of fungicides should follow normal practice. It is essential to control foliar diseases which may cause the seed to be poorly filled, so resulting in higher than normal seed loss during cleaning. Some seed-borne diseases are controlled by seed dressings; these include *Fusarium* spp., *Septoria nodorum*, *Tilletia caries* and *Ustilago nuda*. As with herbicides, it is necessary to consult with the manufacturers to secure the most effective fungicide treatment.

There are some pests which attack wheat crops, for example aphids. These should be controlled in a seed crop in the same way as in commercial crops.

Seed Crop Inspection and Roguing

Roguing is normally undertaken after ear emergence when the grain is still in the doughy stage.

However, there may be other times when particular features of rogue plants are more easily seen; for example, flag leaf attitude may be more visible before ear emergence.

Some cultivars tend to produce abnormal plants (for example, tall plants in a dwarf cultivar or speltoid ears). These should be removed during roguing. (See, for example, Smith and Whittington 1991; Laverack and Turner 1995.)

Because of the large production areas involved, roguing is usually confined to the earlier stages of multiplication (up to and including Basic Seed production).

Seed crops are usually inspected at about the same stage – after anthesis but before the grain is ripe – as this is the time when most of the characteristics which identify the cultivar can be seen. Cultivar purity standards of 99.9% for Basic Seed and 99.7% for Certified Seed are usual.

For the production of hybrid seed it is necessary to assess the effectiveness of pollination by electrophoresis of the seed proteins or by using the method of enclosing a sample of the female parent plants in pollen-proof bags or tents (see the introduction to this section). A 'hybridity standard' of 95% is normally required.

Harvesting

Most wheat crops are harvested by combine. Small machines are available for small plots in the earlier stages of multiplication.

In addition to conventional combines, which cut the straw more or less near the ground, there is now available a 'stripper header' which removes only the heads by rotary teeth. It is claimed that this method is faster and involves less separation in the combine.

An alternative is to swath the seed crop and pick up later, either by combine or by transporting the crop to a stationary thresher. Cutting by binder or by hand, stooking in the field and subsequent threshing is no longer practised to any extent.

The seed is ripe for combining when moisture content is between 8 and 20%. Below 8% and above 20% there is the danger that seed may be damaged during threshing. Swathing or cutting the crop can be done at an earlier stage, with moisture content up to 30%. At this stage the seed is still doughy whereas for combining it is hard and difficult to mark with the thumb-nail.

In some areas, where relative humidity may be very high as seed ripens, there is danger of premature germination or sprouting in the ear with some cultivars. Generally, cultivars with white grain are more prone to this than those with red grain. Plant breeders normally attempt to avoid this problem by breeding resistant cultivars. However, if growing a susceptible cultivar it is necessary to monitor the crop and to harvest earlier rather than later. Prematurely germinated seed may appear sound on visual examination but will have reduced germination capacity.

Drying and Storage

Seed must be carefully handled during threshing and subsequent storage. Rough treatment in the threshing drum or in bulk handling into store can cause damage (hair cracks in the seed coat) which can reduce germination. In store the seed should be reduced to 14% moisture content for short-term storage until the next sowing season; in higher latitudes where winter wheat is sown a few weeks after harvest, 16% moisture is satisfactory. For longer storage, moisture content should be reduced below 10% and below 8% for storage of samples in vapour-proof packages.

Seed which is above the moisture contents stated above must be dried immediately otherwise germination may suffer. This can be achieved in a commercial drier or on a ventilated floor. The temperature of the drying air should not exceed 44°C when initial moisture content is high, but this can be increased to 49°C as the seed dries. After drying, the relative humidity of the ambient air should not exceed 75% otherwise the moisture content of the seed will increase again (see, for example, Copeland and McDonald 1995).

Seed Cleaning

An efficient air/screen cleaner is the first requirement for cleaning wheat seed. Typical screen sizes for a three-screen cleaner are: top 8 mm round hole, middle 6.50 mm round hole, bottom 2.40 × 2.40 mm slot. Afterwards indented cylinders and/or a gravity separator may be used.

Triticum durum Desf. Common Name: Durum wheat

All of the points listed under *T. aestivum* apply equally to *T. durum* but the following additional points should be noted.

Additional characteristics for distinguishing cultivars (based on UPOV 1988):

Hairiness of middle third section of rachis: weak, strong

Lower glume shape (mid-third of ear): round, ovoid, elongated

Lower glume keel spicules (mid-third of ear): few, numerous

Colour of awns: white, red, brown/black

Anthocyanin colouration of awns: weak, strong

Divergence of awns: parallel, divergent

Grain colour: white, red, yellowish, brown/black

At harvest, particular attention must be paid to the danger of premature germination or sprouting in the ear, to which durum wheat is very prone in high humidity conditions. As a consequence, it is advisable to harvest the crop before the seed moisture content falls below 18% in those areas where wetting is likely after further maturation has proceeded.

Triticum spelta L. Common Name: Spelt

Federmann et al. (1992) report that spelt wheat has increased in popularity in southern Germany in recent years and that the flour is sought after; they describe an electrophoretic method whereby this flour can be distinguished from that of common or bread wheat.

The seed crops are grown in the same manner as those of *T. aestivum*. The seed remains enclosed in the glumes after threshing.

Oryza species

There are two species of *Oryza* (rice) which are cultivated, the main one being *O. sativa* which is widely distributed throughout the tropics and subtropics. It is grown in the areas approximately between latitudes 40°S and 45°N. *O. glaberrima* is largely confined to West Africa and is said to be more tolerant of less well controlled water regimes.

There are two main types of rice: upland rice is grown in areas where rain-fed conditions are suitable; irrigated rice is grown under various water regimes, varying from the controlled flooding of fields to the deep-water rice of Southeast Asia.

O. sativa comprises three subspecies: *indica* is a short-day plant, grown mainly in the warm humid tropics; *japonica* is mostly day-neutral, although there are some short-day cultivars, and it is grown outside the tropics; *javonica* is day-neutral and is largely confined to the equatorial climates of Java and other parts of Indonesia. Hybrids between these three subspecies are largely infertile, but modern cultivars based on such hybrids are available. Early work at the International Rice Research Institute produced the semi-dwarf cultivar IR8 which provided the basis for the production of higher yielding cultivars, notably cv. Tongril in South Korea.

There has been considerable progress in the production of F1 hybrid rice cultivars, particularly in China. The hybrids are mainly produced on the basis of cytoplasmic male sterility, with less progress in the use of chemical hybridising agents. Production of hybrid rice seed is labour intensive because it requires considerable attention during pollination. Rice flowers usually open in the morning and are closed by mid-day; pollen is most plentiful when the temperature of the air is 24–28°C. It is necessary to shake the crop (either by beating or by dragging a rope across it) to ensure the release of pollen from the male parent, and this may be necessary several times each morning during pollination. Production has therefore made less progress in countries where labour is more expensive.

Oryza sativa L. Common Names: Rice, Paddy

Classification of Cultivars (based on UPOV 1985)

Photoperiod sensitivity: insensitive, intermediate, sensitive

Water regime: upland, shallow, intermediate, deep (some cultivars may be adapted to more than one regime)

Plant type: semi-dwarf, tall, floating

Starchy endosperm: not glutinous, intermediate, glutinous, mixed

Time of maturity: early, late

Distinguishing between Cultivars

Main features used for distinguishing between cultivars in the field:

Pubescence of blade of penultimate leaf: weak, strong

Anthocyanin colouration of auricle: absent, present

Stem length (excluding panicle): short, long

Colour of stigma: white, light green, yellow, light purple, purple

Distribution of awns: tip, upper quarter, upper half, upper two-thirds, whole length

Other features used for distinguishing between cultivars:

Distribution of anthocyanin colouration on leaf: absent, on tips, on margins, in blotches, uniform

Curvature of flag leaf blade: absent, weak, strong

Time of heading: early, late

Anthocyanin colouration of lemma: absent, on keel, below apex, apex

Thickness of stem: thin, thick

Anthocyanin colouration of nodes: absent, weak, strong

Panicle length: short, long

Curvature of main axis of panicle: absent, weak, strong

Hairs on lemma: absent, short, long

Colour of tip of lemma: white, yellow, brown, red, purple, black

Length of longest awns: absent, short, long

1000 grain weight

Length and width of grain (both undecorticated and decorticated)

Shape of decorticated grain (front view): round, spindle-shaped

Colour of decorticated grain: white, brown, red, purple

Size of white core of polished grain: absent, small, large

Laboratory tests:

Phenol test. Jaiswas and Agrawal (1995) report that a modified test, which includes the addition of copper sulphate, iron sulphate, sodium hydroxide or sodium carbonate to the soak water, enhances the colouration of the grain.

High performance liquid chromatography and various methods using electrophoresis of seed proteins (see, for example, Cooke 1995).

Environmental Requirements for Seed Production

Upland rice is handled in the same way as other rain-fed cereal crops (wheat or barley). It is relatively tolerant of soil conditions but requires adequate rainfall, being very susceptible to dry periods; the growing period is three to four months.

However, most rice is produced under irrigation. It can be grown on a wide range of soils, but it is essential that adequate water is available to give controlled irrigation throughout the growing period. As a general rule, about 2 m of water are required during the growing season. The majority of the crops are grown in enclosed bunds – fields surrounded by banks which retain the flood irrigation water (sprinkler irrigation is only used on upland rice). Water depth, which can be between 15 and 30 cm, should be maintained during flowering and the early stages of seed development. When the seed is mature, the bund is drained and the soil allowed to dry before harvest.

Floating rice is grown in areas which flood naturally, generally in the river basins of Southeast Asia. The crop is planted before the monsoon rains and grows as the flood waters rise up to 5 m in depth. Chaudhary et al. (1995) describe the growth of floating rice in Cambodia and the rehabilitation of suitable cultivars after the prolonged war in that country.

Cultivars of subspecies *indica* are less tolerant of nitrogen because they tend to produce excessive vegetative growth. On the other hand, *japonica* can utilise more nitrogen. Modern cultivars based on hybrids between *indica* and *japonica* and which contain the semi-dwarf gene are designed to utilise higher nitrogen application.

Phosphorus is essential for optimum growth and is best applied in water-soluble form (e.g. superphosphate). Rice is relatively tolerant of soil pH and will grow in acid soils. Potash is normally adequate in the soil and irrigation water.

Previous Cropping and Isolation

Rice is often grown continuously on the same field for long periods. For the seed grower there are two points to consider: first, when changing from one cultivar to another it is essential to ensure that there are no ground-keepers in the field before the new cultivar is planted; and second, it is essential to ensure that the field is free from wild rice (*O. rufipogan* Griff.). Wild rice will normally increase unless measures are taken to combat it, since it is indistinguishable from cultivated rice except in the mature plant stage and is able to cross-fertilise with it. To avoid both these possibilities, an interval of at least two years free from all rice is desirable. However, it is also possible, by water management and cultivation, to minimise the occurrence of both ground-keepers and wild rice. Early ploughing will encourage the germination of dropped seed and the seedlings can be destroyed by cultivation or by flooding the field for two to three weeks. Alternatively, ploughing deep will bury the seeds and subsequent flooding will cause the buried seed to rot.

Rice is normally self-fertilised and needs only protection from other cultivars at harvest; for upland rice there should be a physical barrier (e.g. fence or hedge) or a gap of at least 2 m between crops. For irrigated rice the cultivar will normally occupy a whole field or bund.

For hybrid rice isolation of the female plants is necessary from sources of pollen other than the male parent and 40 m is usual.

It is also necessary to ensure that there is no wild rice in the immediate vicinity of the seed production field in order to avoid all possibility of either mixture with the crop seed or cross-fertilisation between the seed crop plants and the wild rice.

Seed Crop Establishment

Establishment of seed crops does not differ from the methods used for commercial crops. Upland rice is handled in the same way as other rain-fed cereal crops. Irrigated rice can be sown *in situ* or can be transplanted.

When seed is sown in the field the seed rate is 70–130 kg/ha. For upland rice it is preferable to drill in rows like other cereal crops, as this gives better access for inspection of the growing crop. When practical, irrigated rice should also be sown in rows for the same reason, but it is usually necessary to broadcast the seed after the field has been shallow flooded when pre-germinated seed is often used. Transplanting is more economical of seed in the early stages of multiplication as only about 25 kg/ha of seed crop is needed in the seed bed, but it is more costly.

For hybrid rice seed production it is necessary to intersperse the male and female parents; for example, five to six rows of female parent are alternated with one row of male. However, the exact layout is normally specified by the breeder.

Seed treatments are available against the main seed-borne diseases: blast (*Piricularia oryzae*), brown spot (*Cochliobolus miyabeanus*) and Bakianae disease (*Gibberella fujikuroi*).

Seed Crop Management

Seed crop management should follow the established good management practices for commercial crops. Irrigation must be maintained to ensure uniform crop growth.

Weed control is essential. Wild rice seed cannot be separated from rice seed and the plants can only be identified after the crop has reached maturity. Seed crops should therefore be inspected at this stage and any wild rice plants removed from the field. Other weeds are *Erinochloa* spp., *Panicum* spp., *Sorghum halepense, Paspalum distichum, Fimbristylis milacea, Scirpus supinus* and *Cyperus rotundus* (see, for example, Rao and Moody 1990).

Seed Crop Inspection and Roguing

Where possible, seed crops are inspected at anthesis and at maturity. However, when fields are flooded during growth the first inspection is not always practicable. Inspection at maturity is the only way to identify plants of wild rice. Except for upland rice, roguing is most practical after the field has been drained prior to harvest. Typical standards for varietal purity at inspection are 99.9% minimum for early generations and 99.7% for Certified Seed.

For hybrid rice it is necessary to determine the effectiveness of hybridisation. A standard of 95% hybridity is usual.

Harvest

The seed is usually ready for harvest four to six weeks after pollination. For upland rice the moisture content is normally about 15%; below this figure the seed is liable to damage during threshing. Irrigated rice is harvested at about 18% moisture and should not be above 25% or damage may occur.

Small areas are cut by hand; the plants may be cut at ground level in which case they can be tied in sheaves and stacked in the field to dry; alternatively only the ripe heads are removed and taken to a suitable drying floor. These smaller quantities are normally threshed by hand by beating the panicles on a prepared clean floor, or a small thresher can be used (there are models designed for threshing rice). Seed threshed by

hand will usually still have the husk (lemma and palea) attached.

For larger areas, combines are used; they are particularly suited to upland rice. When used for irrigated rice they are generally fitted with flotation tyres to prevent the machine sinking in the soft ground. Combines fitted with headers are also used.

Drying and Storage

In humid climates it is necessary to reduce the seed to a moisture content below 12%. In dry areas, storage at 14% between harvest and the next sowing season is satisfactory. Seed can be dried on any dryer designed for small cereal grains or on a suitable drying floor.

When it is desired to store samples for an extended period, moisture content should be reduced to 8%.

Seed Cleaning

For small quantities, winnowing, either by hand or in a small winnower, will remove the husk and chaff; hand picking is also possible, especially to remove seeds of wild rice which are coloured red.

For seed cleaning an air/screen cleaner is required. Screen sizes are dependent on the type of grain (long-grained or short-grained rice).

The air/screen cleaner can be coupled with an indented cylinder and/or a gravity separator.

Oryza glaberrima Steudel. Common Name: African Rice

This species is largely confined to West Africa and has now been replaced by *O. sativa* in most areas. There are a few cultivars, including some that are floating. Management for seed production does not differ from that for *O. sativa*.

Hordeum vulgare L. *sensu lato*. Common Name: Barley

H. vulgare is treated as one species with two, sometimes three, different forms, although some authorities have given these forms the status of botanical varieties or subspecies. The forms are distinguished by the number of rows of grain in the ear. The ear or spike of barley has the spikelets arranged in rows on opposite sides of the rachis, three on each side. In the 'two-row' form only the centre row on each side develops, and the remaining four rows remain sterile; this is stated by Harlan (1976) to be the earliest known form. The 'six-row' form develops all six rows of spikelets. Some authorities also recognise a 'four-row' form, but practically none of the modern cultivars can be so classified.

Barley is mainly used for animal feed, although a considerable quantity is used to provide malt for brewing and other purposes. A smaller amount is used directly for human food. Cultivars are largely bred specifically for one or other of these purposes. Most brewers using malt prefer a lower protein content in the grain and this is usually achieved in two-row barley. There are, however, some beers brewed from six-row malt. The grain of the six-row form may be less uniform in size than that of the two-row because the outside grains tend to twist as they develop; this can cause uneven grain size which is detrimental to malting, but in modern six-row cultivars this problem has been largely overcome.

For the seed grower it is particularly important to ensure that no mixture occurs between malting and non-malting types, as even a very small admixture of non-malting would be detrimental to the value of a malting barley.

There are a few cultivars which are classified as 'naked barley' in which the seed threshes from the lemma and palea; being free of husk, the grain has a higher feed value.

Some hybrid cultivars have been produced but have met with limited success.

Classification of Cultivars (based on UPOV 1981b)

Form: two-row, six-row
Time of sowing: winter, spring
Main use: animal feed; malting

Distinguishing between Cultivars

Main features used for distinguishing cultivars in the field:

Flag leaf attitude: erect, recurved
Hairiness of lower leaf sheath: absent, present
Anthocyanin colouration of flag leaf auricle: absent, present
Intensity of this colouration when present: weak, strong
Anthocyanin colouration of awn tips: absent, present
Intensity of this colouration when present: weak, strong
Glaucosity of the ear (spike): weak, strong
Ear density: lax, dense
Awn length relative to ear length: shorter, longer
Spiculation on awn margins: absent, present
Sterile spikelets: (two-row only): parallel, divergent

Other features used for distinguishing between cultivars:

Time of 50% ear emergence: early, late
Plant height: short, long
Rachilla hair type: short, long
Anthocyanin colouration of lemma nerves: weak, strong
Spiculation of inner lateral nerves of lemma: weak, strong
Hairiness of ventral furrow of grain: absent, present
Type of lodicules: frontal, clasping

Laboratory tests:

The colour of the grain can be enhanced by examination under ultra-violet light.
Electrophoresis of seed proteins (hordeins) (ISTA 1996).

Environmental Requirements for Seed Production

Barley is generally regarded as a temperate crop, although it can be grown at higher altitudes in the subtropics. It is reasonably tolerant of poorer soils but does best under fertile conditions and will not tolerate wet situations. Lighter soils are used to produce the best samples for malting, as it is more possible to restrict the amount of available nitrogen. White (1990) reported that seed source had a significant effect of 1000 grain weight and mineral content of seed, but that these differences had no effect on the yield of crops grown from these different seed lots; this confirms the tolerance of a wide range of growing conditions in respect of yield of seed, but also suggests that some areas may be better than others for producing seed of higher 1000 grain weight. However, the White data are not sufficiently comprehensive to identify the precise conditions required.

For seed growing, situations which reduce the danger of lodging are to be preferred since badly lodged crops will produce poor quality seed. In damp harvest weather, seed may germinate on the ear before it can be secured.

Previous Cropping and Isolation

For barley seed production it is particularly important to plan previous cropping carefully. Barley seed can survive for long periods in the soil and in some instances has become a weed in subsequent crops. The winter six-row form is especially difficult in this respect. There should therefore be at least a two year interval free of barley before sowing a seed crop, although this can be reduced when the same cultivar is to be grown in succeeding years. Intervening crops between barley seed crops should be such that volunteer cereal seedlings can be eliminated, preferably by an effective herbicide.

Barley is recognised as self-fertilising, but some cultivars (especially six-row forms) can cross-pollinate to an appreciable extent. Many cultivars have been selected to self-fertilise before the glumes open, but there are some which open their glumes during flowering and these are particularly susceptible to out-pollination. Such cultivars are also liable to be more susceptible to *Ustilago nuda* (loose smut). Thus, isolation requirements are normally those for a self-fertilised crop, i.e. a gap of 2 m or physical barrier (e.g. fence or hedge) but for some cultivars it is necessary to provide 250–300 m to guard against cross-fertilisation. When hybrid cultivars are produced, the seed production field should be isolated by at least 25 m from all other barley crops except those of the male parent. To protect against seed-borne diseases which are spread by air-borne spores, at least 50 m is normally required.

Seed Crop Establishment

There are no requirements for a seed crop which differ from those for a commercial crop. Seed rates of 80–200 kg/ha (depending on local conditions and time of sowing) are used except in the early generations of production, when lower rates achieve a higher multiplication rate (rates as low as 30 kg/ha have been used). The crops are sown with a normal seed drill and tramlines are left to provide access for roguing and other operations.

When hybrid cultivars are grown it is necessary to ensure that the correct proportion of male and female parents is grown; this is usually specified by the breeder of the cultivar.

Seed Crop Management

There are no special management requirements except that extra care must be taken to avoid all possibility of mixture with other cultivars and to eliminate particularly those weeds of which the seed will be difficult to remove from the harvested seed. Nitrogen application should be adjusted to avoid lodging, as lodged crops are vulnerable to wet weather at harvest. The use of straw shorteners should be avoided since they can cause growth habit which makes it difficult to assess cultivar purity.

The main diseases of barley for which chemical seed treatments are available are *Fusarium* spp. (foot rot), *Erysiphe graminis* (powdery mildew), *Puccinia striiformis* (yellow rust), *Septoria nodorum* (leaf and glume blotch), *Rhynchosporium secalis* (leaf blotch or scald), *Pyrenophera teres* (net blotch) and *Ustilago nuda* (loose smut). For appropriate treatments it is advisable to consult the manufacturers, as this is a developing area of research.

It is desirable to cultivate a strip of about 1 m around the seed crop to prevent ingress of weeds and to reduce the risk of infection from *Claviceps purpurea* (ergot) which can be carried on hedge-row grasses, although barley is not very susceptible to this disease. Most seed certification schemes include standards for the maximum number of weed plants to be tolerated per unit area. These weed species vary according to the situation in each country, but include, for example, *Avena fatua, A. ludoviciana, A. sterilis* (wild oats), *Lolium temulentum* (darnel), *Raphanus raphanistrum* (wild radish) and *Agrostemma githago* (corn cockle). Hormone weed killers should be used with care as they may cause ear distortions which mask cultivar identity.

Seed Crop Inspection and Roguing

Inspections should be timed to coincide with the maximum expression of the characteristics which distinguish the cultivar. Generally this will be after anthesis when the grain is starting to fill, and several of the characteristics listed above will be visible. Standards for cultivar purity are usually one off-type per 1000 for Basic Seed and three per 1000 for Certified Seed. For hybrid cultivars it is necessary to ensure that emasculation of the female parent has been achieved, and a standard of 95% is considered sufficient. In addition to the estimation of the effectiveness of emasculation, the inspector should also check that the proportions of male and female parents are correct.

Roguing will normally be most effective at this time also. However, it is also useful to rogue crops at early ear emergence if there is a problem with early emerging rogues.

Harvesting

Most crops are now direct combined. The seed should be hard, with the endosperm well past the doughy stage and with moisture content between 8 and 20%. Combining when the seed is above 20% moisture can cause injury to the seed and reduce its germination, and below 8% the seed is liable to be broken or cracked by the combine cylinder. Opoku and Gamble (1995a) reported that seed harvested by swathing, which was at a lower moisture content than that combined direct, suffered more damage. They also determined that slower cylinder speeds at threshing reduce damage and that damage was greater in naked barley than in the normal type. Opoku and Gamble (1995b) reported that the effects of damage caused by harvesting at too low a moisture content or using too high a cylinder speed could be observed in crops grown from such seed, both in seedling emergence and in eventual seed yield.

When hybrid cultivars are produced, it is essential to keep the male and female production separate.

Drying and Storage

For short-term storage, for example when winter barley seed is to be used within a few weeks of harvest, 16% moisture is considered suitable. For storage for sowing in the following spring, moisture content should not be more than 14%. For storage over longer periods, the moisture content should be reduced below 10% or 8% if samples are to be kept for several years.

Drying has to be undertaken promptly if seed moisture content is high and the temperature of the drying air should not exceed 44°C; at lower moisture contents the temperature of the drying air may be increased to 49°C. It is advisable to make routine inspections of seed in store and to take prompt action if any signs of deterioration are detected.

Seed Cleaning

An air/screen cleaner is the main equipment required for cleaning barley seed. For a three-screen cleaner typical screen sizes are: top, 8 mm round hole; middle, 7 mm round hole; bottom, 2 × 20 mm slotted. Air/screen cleaners are usually coupled with indented cylinders and/or a gravity separator. A de-awner is also necessary if the seed has not been de-awned in the combine or thresher; however, care should be exercised that the seed is not hit too hard as this could damage the seed coat and so affect germination.

Avena species

There are three species of *Avena* which are grown to produce grain. The most widely grown is *A. sativa* L. with the common name Oat or Common oat. It is suited to a wide range of conditions and is somewhat more tolerant of wet weather and of soil acidity than other small-grain cereals such as wheat and barley. *A. nuda* L., the naked oat, and *A. byzantina* K. Koch, the red oat, are more special-ised. The naked oat has a high feeding value because the grain kernel is free of the lemma and palea which are attached to the kernel of the common oat, but grain yield is lower than common oat. The red oat is confined to drier areas: the Mediterranean countries of Europe and North Africa, Australia, South Africa and South Amer-ica. Oats generally are less popular than wheat or barley because yields are less and the grain is not as versatile for food use. Oats are, however, generally a good feed for livestock (particularly horses) because of the higher fibre and oil content. The straw has a high feeding value for livestock.

Two other species of *Avena*, known as 'wild oats', are difficult weeds of cultivated land, although herbicides have been developed which are effective in crops of species other than oats.

A. fatua L. and *A. ludoviciana* are similar, the latter being a variety of subspecies of *A. sterilis* L. The former germinates mainly in the spring, whereas *A. ludoviciana* behaves as a winter cereal. They are difficult to distinguish from one another but differ in the way that seeds are shed: *A. fatua* seeds fall separately and the base of each has an abscission scar; seeds of *A. ludoviciana* fall as a complete spikelet and only the first seed has an abscission scar.

Avena sativa L. Common Name: Oat

Classification of Cultivars

The following characteristics are based on UPOV (1981c).

Seasonal type: winter, alternative, spring
Hairiness of uppermost stem node: absent, present
Colour of lemma of primary grain: white, yellow, brown, grey, black
Glaucosity of lemma of primary grain: absent, present

Distinguishing between Cultivars

Main features used for distinguishing cultivars in the field:

Hairiness of lower leaf sheaths: weak, strong
Hairiness of margins of leaf below flag leaf: weak, strong
Flag leaf attitude: rectilinear, recurved
Orientation of panicle branches: unilateral, equilateral
Attitude of panicle branches: erect, drooping
Attitude of spikelets: erect, pendulous
Glaucosity of glumes: weak strong
Intensity of glaucosity of lemma of primary grain: weak, strong
Tendency for awns on primary grain: weak, strong
Hairiness of back of lemma of primary grain: absent, present (not white or yellow oats)

Other features used for distinguishing between cultivars:

Plant growth habit: erect, prostrate
Intensity of hairiness of uppermost stem node: weak, strong
Length of glumes: short, long
Length of lemma of primary grain: short, long
Time of panicle emergence: early, late
Plant height (including panicle): short, long
Length of basal hairs of primary grain: short, long
Length of rachilla of primary grain: short, long
Width of rachilla of primary grain: narrow, wide
Grooves of rachilla of primary grain: weak, strong

Laboratory tests:

The colour of the grain can be enhanced by examination under ultra-violet light.

Electrophoresis of seed proteins (see, for example, Cooke 1989; Sharopova et al. 1997).

Environmental Requirements for Seed Production

The requirements for seed crops do not differ from those for crops grown for other purposes. The species is confined to the mainly temperate regions, and can be grown successfully in the wetter areas; it is not suited to very dry soils. It will tolerate a lower pH than wheat or barley. Generally, the straw of oats is weaker than that of wheat or barley and will lodge under excessive nitrogen regimes, although modern cultivars with short stiff straw are more tolerant in this respect.

Previous Cropping and Isolation

In quality control systems, standards for cultivar purity are normally one per 1000 or higher for earlier generations and three to 10 per 1000 for later generations. There are also usually similar standards for plants of other cereals (wheat and barley). It is essential to plan previous cropping carefully to avoid the risk of volunteer plants occurring. For crops to produce Pre-Basic or Basic Seed, a two year break from oat crops is recommended; other cereals should also be avoided since winter barley seeds in particular are relatively long-lived in the soil. For crops to produce later generations, two years free from oats are also desirable, but one year free from the other cereals is generally specified in quality control schemes. When it is necessary to grow an oat crop in the preceding two years, care must be taken to ensure that the crop is grown from authentic seed of the same cultivar as is to be grown for seed.

A. sativa is self-fertilised and natural cross-pollination is generally less than 1%. Therefore, provided there is a clear gap (generally a fence or hedge or clear ground of at least 2 m), around the crop, no other isolation is needed. However, Bicklemann (1993a), in a controlled experiment, reported that the frequency of intravarietal, intervarietal and interspecific crossing was 0–1.1% for six cultivars with the lowest frequencies, and 0–8.7% for the three with the highest frequencies. Although intravarietal crossing was the most frequent (and should not be important for seed production), *A. fatua* was found to cross with one of the cultivars with a frequency of up to 5%. It is therefore advisable to ensure that seed production fields are not surrounded by fields with a high concentration of wild oats.

Seed Crop Establishment

Seed rate for a seed crop is no different from that used for commercial crops, usually about 80–100 kg/ha. Oat crops can be sown broadcast, but are more usually sown by drill in close spaced rows (10–20 cm apart). For seed crops it is advisable to leave regular gaps or 'tramlines' when drilling; these provide guides for subsequent operations such as spraying or roguing. Irrigation is not usually required and is not generally economic.

Seed Crop Management

There are no requirements for management of a seed crop which differ from those for commercial crops, except that it is necessary to have regard to the measures required to safeguard cultivar purity. All equipment brought onto or through the field must be cleaned to ensure that no unwanted seed is dropped.

There are several weeds of which the seeds are difficult to separate from oat seed. Wild oats are impossible to remove completely from oat seed and there is no herbicide which can be used safely on a growing crop; therefore the only way to ensure seed free from wild oats is to use only fields with no history of wild oats. Wild oat seed can remain viable in the soil for long periods and if a field has become contaminated it is necessary to undertake a long-term, systematic programme of eradication; at least four years and preferably longer are required, during which other crop species are grown to enable herbicides to be applied and suitable cultivations carried out. In the final stages of a programme it will usually be necessary to remove the remaining wild oat plants by hand. Other weeds that should be controlled include *Lolium temulentum* L., *Raphanus raphanistrum* L. and *Agrostemma githago* L.; control measures do not differ from those used for commercial crops.

There are systemic fungicides available as seed treatments against *Erysiphi graminis* (powdery mildew), although this is not seed-borne. Seed-borne diseases of most importance are *Fusarium* spp. (brown foot rot and ear blight), *Ustilago hordei* (covered smut), *Pyrenophora avenae* (leaf stripe or spot), *Ustilago avenae* (loose smut). As far as possible, disease-free seed should be used to sow the seed crop; chemical treatments are available and advice should be sought from the manufacturers as this is a rapidly developing technology. There are no generally distributed pests which are seed-borne.

Seed Crop Inspection and Roguing

The best time for inspection and roguing is after anthesis when the grain is starting to fill; at this stage several key characteristics can be observed: panicle shape, hairs at uppermost node of stem, characteristics of the spikelet, awnedness.

A feature of oat crops is the occurrence of 'fatuoids' which exhibit the characteristics of the cultivar in which they originate except for those of the floret, which resemble wild oats. In a detailed study of fatuoids, Bicklemann (1993a, b, c) points out that there are two possible causes: natural crossing or mutation. Of these two, natural crossing, which is interspecific as opposed to intervarietal or intravarietal, would be the most objectionable for seed production since these would exhibit most of the characteristics of wild oats. Bicklemann (1993c) describes an electrophoretic test in which some fatuoids could be distinguished as arising from intravarietal crossing (because the prolamin pattern was similar to that of the parent cultivar); however, fatuoids with patterns dissimilar from those of the parent cul-

tivar could have arisen either from intervarietal or interspecific crossing. Whereas the former might not be considered as cause for concern in a seed lot, the latter could cause rejection if occurring at a frequency above the standard required by the quality control system.

Harvesting

Most seed crops are harvested by combine. The oat panicle tends to ripen less uniformly than the spikes of wheat or barley so that the timing of harvest has to be judged to provide the maximum of fully matured seed. The straw is normally still slightly green but the seed is hard at this stage.

Drying and Storage

For short-term storage the seed should be below 16% moisture; when the seed is to be stored for more than six months, moisture content must be reduced below 14%. For long-term storage in bulk, moisture content below 10% is required. For the storage of samples in moisture-proof containers the moisture content should be below 8%.

When drying seed from a high moisture content (above 22%), the temperature of the drying air should not exceed 44°C. As moisture falls below 20% the air temperature can be increased to 49°C.

Seed Cleaning

For cleaning oat seed, an air/screen cleaner is essential. In three-screen cleaners, typical screen sizes are: top, 10 mm round hole; middle, 9 mm round hole; and bottom 2 × 20 mm slotted.

Indented cylinders or gravity separators are also used for particular cleaning problems. It is sometimes advantageous to remove the awns from seed to allow it to flow more freely when sowing; however, de-awners must be used with great care as mechanical damage can be caused and this may reduce the germination capacity.

Seed Treatment

Chemical seed treatments against the main fungal diseases are available (see section on 'Seed Crop Management').

Avena nuda L. Common Name: Naked oat

Naked oats are used for specialist production of high feeding value grain; the fibre content is much reduced because the grain threshes free of the lemma and palea (Doyle and Valentine 1988). Seed crops are grown in the same way as those of *A. sativa*. However, great care must be exercised in handling the seed of naked oats as the lack of protection of the lemma and palea make it very vulnerable to mechanical damage. Valentine and Hale (1990) report no difference between the germination of seed harvested by a conventional combine and that harvested by a stripper-header. Reducing the drum speed had little effect in their experiments, although they quote earlier work (Thornton and Lawes 1981; Thornton 1986) as indication that a reduced drum speed of 850 rpm would give a general improvement in germination and that increasing the concave clearance would also be advisable.

Avena byzantina K. Koch. Common Name: Red oat

Red oats are grown to a very limited extent in the warmer, drier areas. Hybrids between *A. sativa* and *A. byzantina* are also available. There is no difference in the requirements for seed growing from those outlined for *A. sativa*.

X *Triticosecale* Wittmack. Common Name: Triticale

Triticale is a new species derived from a cross between wheat (mainly *Triticum durum*) and rye (*Secale cereale*). Durum wheat is tetraploid and rye diploid. The cross produces an octoploid after chromosome doubling (Pickett 1988). Triticales derived from crosses between durum wheat and rye have been further crossed with bread wheat (*Triticum aestivum*) in the production of many of the modern triticale cultivars (National Research Council 1989). The objective has been to combine the best features of rye (suitability for poorer soils,

winter hardiness, good disease resistance) with those of wheat (enhanced nutritional properties, higher grain yield, more easily managed plant type) and to eliminate the poorer qualities of each species.

Rye is cross-fertilising and wheat self-fertilising. Thus although triticale is normally regarded as self-fertilising, it is more liable to out-crossing than wheat.

Classification of Cultivars (based on UPOV 1989a)

Seasonal type: winter, spring
Length of stem (including ears and awns): short, long
Time of ear emergence: early, late

Distinguishing between Cultivars

Main features used for distinguishing cultivars in the field:

Flag-leaf
    Attitude: rectilinear, recurved
    Length: short, long
    Width: narrow, wide
    Glaucosity of sheath: weak, strong
Ear
    Glaucosity: weak, strong
    Density: lax, dense
    Length: short, long
    Shape: tapering, parallel
Anthers – anthocyanin colouration: weak, strong

Other features used for distinguishing between cultivars:

Stem neck – hairiness: sparse, dense
Auricles – anthocyanin colouration: weak, strong
Lower glume
    Length of first beak: short, long
    Hairiness (external): weak, strong

Laboratory tests:

> Colouration of the grain with phenol (ISTA 1996)
> Electrophoresis of seed proteins (ISTA 1996)

Environmental Requirements for Seed Production

Triticale is generally more tolerant of less fertile soils than wheat, although many modern cultivars are also able to take advantage of fertile conditions. It can therefore take advantage of a wider range of soil types that wheat. Gawronska and Grzelak (1993) report that genotypes differ significantly in their response to drought stress during seedling emergence. Cultivars also differ markedly in their ability to resist lodging.

Cultivars may be either long-day or day-neutral. Some therefore are not suitable for growing at lower latitudes.

There are considerable differences between cultivars in their ability to respond to different environmental conditions, and the advice of the breeder should be sought before placing a seed crop in any particular location.

Previous Cropping and Isolation

The requirements for previous cropping are not different from those for wheat (*T. aestivum*). There should normally be two years free from other small-grained cereals, including triticale. Although wheat grains are somewhat smaller than those of triticale, they are very difficult to remove from triticale seed.

Pickett (1988) reports variation in the extent to which cultivars are liable to out-pollination when exposed to other pollen; therefore the presence of ground-keepers in a crop for seed will be a source of possible out-pollination as well as of mechanical mixture at harvest.

The variability between cultivars in the extent to which they are liable to out-pollination also presents difficulty for deciding on suitable isolation for a seed crop. The OECD Cereal Seed Scheme (OECD 1995b) specifies that isolation of a triticale seed crop should differ, depending on the type of cultivar, as follows.

1. Cross-pollinating cultivars of triticale should be isolated from crops of rye and other crops of triticale by 300 m for Basic Seed production and by 250 m for Certified Seed production.
2. Self-pollinating cultivars of triticale should be isolated from other triticale crops by 50 m for Basic Seed production and by 20 m for Certified Seed production.

An additional hazard for triticale seed crops is the susceptibility of most cultivars to infection by ergot (*Claviceps purpurea*) in the temperate regions, although resistance has been incorporated into many modern cultivars. Ergot can infect some hedgerow grasses and it is advisable to provide a sterile strip around the crop to prevent its ingress.

Seed Crop Establishment

Seed crops of triticale are established in a similar manner to those of wheat. The crop is sown in rows and tramlines are left at suitable intervals to facilitate subsequent operations. Seed rate varies between 80 and 180 kg/ha depending on the custom for wheat in different areas.

Weed and disease control measures are similar to those employed for wheat. Triticale cultivars are less susceptible to diseases (including seed-borne diseases) than wheat, and control measures can generally be scaled down. The problem of ergot has been mentioned above: there are no effective seed treatments although ergots can be removed from small quantities of infected seed by floating them off in a 20% solution of common salt; the triticale seed must be washed and dried immediately and some loss of germination is likely. Care should be taken to control weed grasses which might be infected with ergot before sowing the crop.

Seed Crop Management

There are no major differences from the management described for wheat. Formerly, cultivars were rather weak in the straw and it was necessary to use less nitrogen, but modern cultivars are more resistant to lodging.

Seed Crop Inspection and Roguing

The timing of inspection and roguing is similar to that described for wheat (*T. aestivum*).

Harvesting

Crops are normally harvested by combine, either direct or following swathing. When ripe, the straw loses its green colour and the grain is hard and difficult to mark with the thumb-nail. Shedding of seed prior to harvest is not normally a problem but there can be a tendency in some cultivars for the head to break off at the neck if harvest is delayed. Moisture contents at harvest should be similar to those for wheat (*T. aestivum*).

Drying and Storage

The seed should be handled in the same way as wheat (*T. aestivum*).

Seed Cleaning

Cleaning losses may be higher in some cultivars because of the tendency to produce shrivelled or poorly developed seed. The seed can also be larger than that of wheat. Otherwise there are no differences from the methods described for wheat (*T. aestivum*).

Secale cereale L. Common Name: Rye

Rye is an annual, the most commonly grown cultivars of which are winter sown, and requires vernalisation. Although most cultivars are diploid, some tetraploid cultivars have been developed (Evans 1976). The crop is cross-pollinated and there are some synthetic and hybrid cultivars available. Because it grows rapidly, rye is also used as a fodder crop for livestock, either as a catch crop in the autumn or as an early spring crop.

Classification of Cultivars (based on UPOV 1978)

Reproductive system: open-pollinated, synthetic, hybrid
Seasonal type: winter, alternative, spring
Ploidy: diploid, tetraploid

Distinguishing between Cultivars

Main features used for distinguishing cultivars in the field:

Glaucosity of flag leaf sheath: weak, strong
Time of ear emergence: early, late
Glaucosity of ear: weak, strong
Hairiness of stem below ear: weak, strong
Length of leaf below flag leaf: short, long
Attitude of ear: erect, recurved
Ear length: short, long
Ear density: lax, dense
Plant height (stem and ear): short, long

Other features used for distinguishing between cultivars:

Plant growth habit: erect, prostrate

Laboratory tests:

Colour of aleurone layer of seed: light, dark
Anthocyanin colouration of coleoptile: absent, present
Electrophoresis of seed proteins (see, for example, Cooke 1995).

Environmental Requirements for Seed Production

Rye is a crop for temperate regions and the main centres of production are in Poland and the western areas of the former Soviet Union. It is suitable for growing on light acid sandy soils and is often treated as preferring these conditions. However, it produces the best crops on more fertile soils although under these conditions it cannot usually compete economically with other cereals. The winter cultivars which require vernalisation are generally very winter hardy and can be grown in areas where winter conditions are severe (Bell 1965).

The crop requires the same fertiliser treatment as barley. Nitrogen should be used sparingly as the straw is generally weaker than that of other cereals.

Previous Cropping and Isolation

An interval of two years free from rye and other small-grained cereals is required before sowing a rye seed crop. When mature, rye sheds seed very easily and this can cause the appearance of volunteer plants in subsequent crops. Volunteer plants are doubly undesirable in a seed crop as they may release pollen which will pollinate some plants of the seed crop and lead to the production of off-type plants when the seed is used. Although seed crops of the same cultivar may be grown consecutively in many seed certification schemes, the practice is not to be recommended; volunteer plants will be of a different generation of a cross-pollinated crop.

Ergot (*Claviceps purpurea*) is a particular problem for rye production and must be avoided in seed crops, since ergots in the seed will infect the following crop. Ergots will not normally live for more than a year so the two year gap between rye crops should eliminate any shed from previous crops. Attention should also be paid to hedgerow grasses, many of which can also be infected and can pass on the infection to rye.

As a cross-pollinated crop, rye requires a greater isolation distance than other small-grained cereals. For crops which are to produce Basic Seed or earlier generations, a separation of at least 300 m is usually required. This distance can be reduced to 250 m when later generations are to be produced. Diploid and tetraploid cultivars must be separated from one another since infertility can result from pollination by a parent of different ploidy.

Seed Crop Establishment

Seed crops are normally drilled with a seed rate of 80–150 kg/ha. It is advisable to leave wider spaces between the rows at suitable intervals to allow access for subsequent operations and for roguing or seed crop inspection. For the production of hybrid cultivars it is necessary to sow the correct proportions of male to female parents (specified by the breeder). For the production of synthetics, the seed should be supplied with the correct proportions of the constituent lines.

Seed Crop Management

Apart from the attention to detail required for quality control, seed crops do not require management different from that of a crop for food production. Weed control follows the same pattern as for other small-grained cereals. Wild oats (*Avena fatua*, *A. ludoviciana*, *A. sterilis*), wild radish (*Raphanus raphanistrum*), corncockle (*Agrostemma githago*) and darnel (*Lolium temulentum*) are some of the most undesirable weeds.

Seed Crop Inspection and Roguing

The best time for roguing is soon after ear emergence when most of the field characteristics can be seen. If ergots appear later in small numbers of plants, it may be worth removing and destroying those plants.

Field inspection is normally undertaken at or soon after anthesis. For hybrid cultivar production it will be necessary to check the efficiency of emasculation of the female parent. A sterility level of 98% is considered satisfactory.

Harvesting

Rye is normally harvested by combine. The seed should be hard and difficult to mark with the thumb-nail. Moisture content should be below 20% as combining at a moisture content higher than this can damage germination. Rye straw retains more green colour when the seed is ripe than that of wheat or barley; the straw is tough and can be a problem to combine if at all damp. Very dry seed is also vulnerable to damage during threshing.

Drying and Storage

Immediately after harvest the seed must be dried to below 16% moisture. For safe keeping until the next growing season, moisture content must be below 14%. If seed is to be kept for a year (for example, to eliminate ergots), moisture content must be below 10%; for longer periods below 8% is necessary.

Slow drying is preferable. In dry areas this may be achieved by spreading on a clean floor and turning frequently. When forced-air drying is used at seed moisture contents of 16% or above, the temperature of the drying air must not exceed 44°C. At lower moisture contents 49°C is satisfactory.

Seed Cleaning

Rye seed can be damaged by rough handling in the cleaning plant and this will impair germination. An air/screen cleaner will usually be all that is required, but an indented cylinder or gravity separator may also be used. Typical screen sizes for a three-screen cleaner are: top, 13 mm round hole; middle, 8 or 8.5 mm round hole; bottom, 1.5 × 2.0 mm slot.

When small quantities of seed contain ergot, it is possible to float the ergots out of the rye in a 20% solution of common salt; however, the rye seed must be washed and dried immediately after treatment and some loss of germination is likely.

Sorghum species. Common Name: Sorghum

There is some confusion about the species of sorghum. Doggett (1965) states that 'the species boundaries are vague, and there are all sorts of intermediate forms'. However, the accepted designation for cultivated sorghum for grain production is *Sorghum bicolor* (L.) Moench. Harlen and de Wet (1972) described five main 'races' which they distinguished by spikelet characteristics as follows.

(i)   Bicolor: grain elongate, sometimes slightly obovate, nearly symmetrical dorso-ventrally; glumes clasping the grain, which may be completely covered or exposed as much as one-quarter of its length at the tip; spikelets persistent.

(ii) Guinea: grain flattened dorso-ventrally, sublenticular in outline, twisting at maturity 90° between gaping involute glumes that are nearly as long or longer than the grain.

(iii) Caudatum: grain markedly asymmetrical, the side next to the lower glume flat or even somewhat concave, the opposite side rounded and bulging; the persistent style often at the tip of a beak pointing towards the lower glume; glumes half the length of the grain or less.

(iv) Kafir: grain approximately symmetrical, more of less spherical, not twisting; glumes clasping and variable in length.

(v)   Durra: grain rounded obovate, wedge-shaped at base and broadest slightly above the middle; glumes very wide, the tip of a different texture from the base and often with a transverse crease across the middle.

There is considerable variability in many of the older cultivars and the above classification is not absolute. The races are interfertile so that intermediate types may be expected. In India and the USA, F1 hybrids have been developed, generally based on inbred seed-bearing lines from Durra, and pollinators from Kafir.

For practical purposes of seed growing, the grain sorghums can be grouped together under the species name *S. bicolor*.

There are three other species which are used as fodder (cut or grazed) for livestock. They are *S. almum* Parodi, *S. sudanense* Stapf and *S. bicolor X sudanense*. These are described in the section on 'Gramineae: Tropical Grasses'.

Classification of Cultivars

These characteristics are taken from IBPGR (1984), House (1985) and UPOV (1989b).

Breeding system: open-pollinated, synthetic, inbred line, hybrid

Time of panicle emergence: early, medium, late

Plant height (at panicle emergence): short, long

Photoperiod sensitivity: insensitive, neutral, sensitive

Stalk juiciness: dry, juicy

Juice sweetness: insipid, sweet

Distinguishing between Cultivars

Main features used for distinguishing cultivars in the field:

Plant colour: pigmented, tan

Leaf colour: pale green, dark green

Leaf midrib colour: colourless (white), dull green, yellow, brown

Glume colour: white, yellow, brown, red, purple, black, grey

Panicle shape: elliptic, oval, broomcorn

Panicle density: lax, dense

Awning at maturity: awnless, awned

Other features used for distinguishing between cultivars:

Anthocyanin colouration of stigma: absent, present

Yellow colouration of stigma: absent, present

Length of stigma: short, long

Length of flower and pedicel: short, long

Seed covering by glumes: uncovered, covered

Glume length: short, long

Twinning of seeds: single, twin

Laboratory tests:

Seed coat colour: white, yellow, red, brown, buff

Seed coat lustre: not lustrous, lustrous

Seed subcoat: absent, present

Seed form: dimpled, not dimpled (plump)

Endosperm texture: corneous, intermediate, starchy

Environmental Requirements for Seed Production

Sorghum is grown in the tropics and subtropics and requires warm temperatures throughout growth. It is killed by frost in the field and even seed with 16–19% moisture is killed when stored at –29°C (Copeland and McDonald 1995). Sorghum generally can tolerate dry conditions but is also grown successfully in wetter areas and is sometimes irrigated. However, it is susceptible to sporophytic moulds which develop on the ripening heads and discolour the seed, causing loss of germination; it is thus desirable for seed production to avoid those areas where rain is likely to occur during maturation before harvest. The main centres of production are in Africa and India, where it is regarded as more reliable than maize (*Zea mays*) in the drier areas. Hybrid cultivars are grown in India and also in the USA, where it is mainly grown to provide grain for feeding to livestock. Some hybrid seed is also produced in Italy, France and Argentina.

Sorghum is grown on a wide range of soils from clays to light sands. In many of the areas where sorghum is grown, little fertiliser or organic manure is used. However, where it is grown under better conditions it responds well to moderate applications of complete fertiliser backed up by top-dressing with nitrogen.

Previous Cropping and Isolation

Usually, one season free from sorghum will be satisfactory before sowing a seed crop. It is important to ensure that the field is free from weeds and trash which might harbour disease inoculum. Of particular importance is the parasitic weed striga (*Striga* spp.) which produces large numbers of seeds that can remain viable in the soil for long periods. Control in previous crops to the seed crop is essential. Some resistant sorghum cultivars have been reported and may help to reduce striga populations (Okonkwo and Garba 1993; Carsky et al. 1996).

Although mostly self-pollinated, sorghum will cross-pollinate to about 5–10%. It is therefore usual to provide isolation. Indian seed certification standards require 200 m for Basic Seed and 100 m for Certified Seed production for open-pollinated cultivars; these distances are increased to 400 m for isolation from Johnson grass

(*S. halepense*) or other forage sorghums. Requirements for the production of hybrid seed are the same except that the Basic Seed requirement is increased to 300 m (Tunwar and Singh 1988a). Chopra (1987) states that for the production of hybrid seed it is desirable to produce the seed in areas where sorghum is not a traditional crop; if traditional areas have to be used the seed should be grown in the off-season for normal cropping.

Seed Crop Establishment

For seed production it is preferable to plant in rows. Seed rates are relatively low, about 8–16 kg/ha or only 3 kg/ha in dry areas. For hand sowing the rows may be 46–90 cm apart with the plants spaced 15–60 cm apart in the row, depending on the availability of water. For sowing by seed drill, about 75 cm between the rows and 10 cm between plants in the row is generally satisfactory.

For the production of hybrid seed it is necessary to plant the correct proportion of female to male plants. Usually this is 6:2 but the breeder may specify other proportions. Hybrid cultivars are produced on the basis of cytoplasmic male sterility. The crop should be surrounded by four rows of the male line to improve isolation from other sources of pollen. As male and female lines do not always flower at the same time, it is often necessary to plant them at different times; male lines usually have a shorter period between sowing and flowering and when this is the case are sown 10–14 days after the female to ensure pollen supply when the female flowers are receptive. Flowering can also be influenced by application of nitrogen or irrigation water; if applied to one of the lines, flowering date can be brought forward by up to seven days. This requires forward planning to ensure that the water or nitrogen can be applied to one or the other line separately. Date of flowering can be estimated by taking plants from each line and examining the stage of development of the floral initials about four weeks after seedling emergence; this is done by carefully slitting the stems of the plants. If development is further advanced in one line than in the other, action should be taken.

In very dry areas there is little response to fertiliser, but under wetter conditions the crop will respond well to nitrogen; the application should be split between the seed bed and a top-dressing applied as the stem is beginning to elongate. The latter is best applied as a side-dressing about 5 cm away from the base of the plants.

Seed Crop Management

The seed crop should be managed in the same way as other crops. Crops may be earthed up to encourage the formation of prop roots and so reduce the risk of lodging. Weed control is especially important in the early stages of growth, but as the plants grow taller they will compete well. Apart from hand weeding and inter-row cultivation there are no easy options for control; therefore it is essential to choose clean fields for seed production, particularly in areas where striga is common.

Sorghum is very attractive to birds as the crop ripens, and in Sub-Saharan Africa the *Quelea quelea* causes much damage. Some resistance to bird damage has been recorded in some cultivars by Okonkwo and Kilishi (1990). For seed crops it is probably worthwhile to employ bird scarers around the crop or to prevent the birds from roosting in nearby trees by various measures.

Seed Crop Inspection and Roguing

It is usual to inspect sorghum seed crops at least twice and those of hybrid cultivars three times. First inspection should be during flowering to check isolation; for hybrid cultivars two inspections are usual while flowering continues, and it is additionally necessary to ensure that none of the female line plants are shedding pollen. Plants in the female line which are shedding pollen should not exceed 0.1%. Final inspection should be at maturity, just before harvest, to check cultivar identity and purity by reference to the seed colour.

Roguing should precede crop inspection. For hybrid cultivars it is necessary to remove any female plants which are shedding pollen as soon as possible.

Harvesting

Seed is physiologically mature when seed moisture content falls below 25–30%. However, seed at this moisture content is still soft and can easily be damaged during harvesting operations. Harvest should therefore be delayed until moisture content falls below 15%. At this stage the plants will have lost their lower leaves but the upper leaves may still be green or may have turned yellow, depending on cultivar. Sorghum is very prone to sprouting in the ear, so harvest should not be delayed if the weather is favourable.

Many of the crops are harvested by hand where fields are small. If only the heads are cut it is possible to go over the crop two or three times, which is an advantage for some older cultivars which do not mature uniformly. The cut heads are left to dry before threshing, either in the field or on a clean drying floor.

Larger areas are harvested by combine and this method is particularly suited to the modern, more uniform and short-strawed cultivars (often hybrids). When harvesting hybrid cultivars, the male rows must be removed first and the heads or seed removed from the field before the hybrid seed is harvested. If the male rows are combined, it is necessary to check that no heads have been missed and to remove any found before continuing with the female line after cleaning the combine.

Drying and Storage

Heads which have been harvested by hand are dried to 12% moisture before threshing. Heads dried on a drying floor should not be spread more than 20 cm deep and should be turned frequently. Combined seed may be dried using forced air, but the temperature of the drying air should not exceed 40°C. Seed moisture should be below 12% for short-term storage, but should be reduced below 10% for longer periods. Sorghum is very susceptible to storage insect pests and fumigation of seed in store is often worthwhile. All storage areas must be kept scrupulously clean and an insecticide used on the floors. In humid areas it is preferable to store the heads and to thresh only when the seed is needed.

Seed Cleaning

Pre-cleaning of the threshed or combined seed is necessary to remove trash such as broken straw. Thereafter an efficient air/screen cleaner will usually finish cleaning, sometimes supplemented by a gravity separator.

Pennisetum glaucum (L.) R. Br. emend Stuntz Syns. *P. americanum* (L.) Leeke and *P. typhoides* (Burm. f.) Staph et C.E. Hubb. Common Names: Pearl millet, Bulrush millet

Pearl millet is generally more tolerant of dry conditions than sorghum and is grown mainly in Sub-Saharan Africa and in India. Both synthetic and hybrid cultivars have been introduced in India.

Classification of Cultivars

The following are based on Purseglove (1985a), Tunwar and Singh (1988b) and Wanous (1990).

Breeding system: open-pollinated, synthetic, inbred line, hybrid
Maturity: early, medium, late
Photoperiod sensitivity: insensitive, sensitive (short-day)
Main use: grain production, animal fodder

Distinguishing between Cultivars

Main features used for distinguishing cultivars in the field:

Shape of inflorescence: cylindrical, conical, spindle, club, dumb-bell, lanceolate, oblan-ciolate, globose
Exposure of seed: seed exposed in spikelet at maturity, not exposed
Bristle length: shorter than seed, longer than seed

Other features used for distinguishing between cultivars:

Seed shape: obovate, lanceolate, elliptical, hexagonal, globular
Seed colour: ivory, cream, yellow, grey, brown, purple, black

Environmental Requirements for Seed Production

Pearl millet is grown on lighter soils than sorghum (*Sorghum bicolor*) in areas with low rainfall. Cobley (1976) and Purseglove (1985a) state that it can be grown with as little as 250 mm rainfall per year. However, it is sensitive to drought after sowing and may fail if there is not sufficient moisture for seedling growth. Earlier maturing cultivars are more suitable in the drier areas. It is also sensitive to excessive rainfall when mature and awaiting harvest.

Pearl millet is best suited to fertile soils and responds well to nitrogen and phosphate, particularly in the early growth stages.

Previous Cropping and Isolation

To reduce the risk of volunteer plants occurring, there should be an interval of one year free from millet before a seed crop of pearl millet. Remaining seed dormant in the soil is less of a problem than with sorghum.

Pearl millet is reasonably tolerant of weed competition but clean fields are to be preferred. *Striga* spp. will attack pearl millet, but it is less susceptible than sorghum.

The main soil-borne disease is downy mildew (*Sclerospera graminicola*). Air-borne diseases which may also be seed-borne are smut (*Tolyposporium penicillariae*) and ergot (*Claviceps microcephala*); the former can be controlled by an appropriate seed dressing and the latter may be removed from the seed by flotation in a brine solution. Other control measures include attention to the elimination of ground-keepers in preceding crops and isolation from infected crops or plants.

The species is cross-pollinating and there are two types of flowers. The earlier-maturing, upper flowers are hermaphrodite. The later-emerging, lower flowers are usually male and reach anthesis later. Isolation is therefore required for seed crops. For open-pollinated and synthetic cultivars,

200 m is required for the final generation and 400 m for earlier generations. For the production of hybrid cultivars, 200 m is satisfactory for the final production year, but for the earlier years during production of inbred lines up to 1000 m is preferred. Because pollen release takes place over a long period, isolation in time is not possible.

Seed Crop Establishment

The crop can be broadcast, but for seed production it is better to sow in rows as this makes subsequent operations, including field inspection, easier. Pearl millet generally tillers profusely and therefore rows should be spaced 60–100 cm apart with 10–20 cm between plants in the row. The relatively small seed is sown at a seed rate of between 3 and 9 kg/ha and the crop may be thinned. Seed crops can also be transplanted to the desired spacing and this is specially effective for the earlier generations. Weed control during early growth is necessary.

For the production of hybrid cultivars the breeder will usually specify the required proportion of female to male rows to be established; generally this is 6:2. The four outer rows of the field should form a male surround.

Seed Crop Management

Pearl millet requires little attention during growth, although it will respond to irrigation in exceptionally dry areas. It is particularly attractive to seed-eating birds (*Quelea* spp.) and it is essential to guard crops carefully as they mature; some resistance to bird attack has been recorded for some cultivars.

Seed Crop Inspection and Roguing

The best time to determine cultivar identity and purity is during anthesis. However, roguing earlier will enable the removal of plants infected by downy mildew. It is therefore best to rogue fields at least twice. Smutted ears and those infected by ergot will appear during the grain filling period and a third visit to remove these is necessary.

Inspection for quality control is most effective during anthesis and a second inspection to check on disease incidence prior to harvest is desirable.

Hybrid cultivars are produced on the basis of cytoplasmic male sterility and inspection should verify its effectiveness.

Harvesting

Larger fields may be harvested by combine. For small areas, hand harvesting is usual and the ears are cut on more than one occasion as the seed matures. The seed is physiologically mature when the moisture content falls to 25%. However, it is necessary that seed be dried to 12% moisture for safe storage and therefore it is advisable to allow moisture content to fall to 15% before harvest, if possible, so as to reduce drying time. However, pearl millet is susceptible to sprouting in the ear and harvest should not be delayed if the weather is likely to be unsettled.

Drying and Storage

Combined seed should not be stored with moisture content above 12%. When using forced-air drying the temperature of the drying air should not exceed 40°C.

Hand harvested ears are dried to 12% moisture before threshing, either in the field or on a clean drying floor; however, exposure to very hot sunlight should be avoided. Seed heads should not be spread on a drying floor to greater than 20 cm deep.

Seed Cleaning

Seed stores must be kept clean to avoid storage pests. The threshed seed usually contains trash which is best removed by winnowing or by pre-cleaning before the main cleaning operation. Subsequently an efficient air/screen cleaner is all that is required, but occasionally a gravity separator may be used in addition.

Eleusine coracana **(L.) Gaertn. Common Name: Finger millet**

Finger millet is grown in the tropics in East Africa and in India. Some improved cultivars have been produced in India, but the species has received less attention than the more widely grown cereals.

Distinguishing between Cultivars

Mehra (1963, quoted in Purseglove 1985b) describes two groups of finger millet cultivars:

1. African highland types with long spikelets, long glumes, long lemmas and with grains enclosed within the florets;
2. Afro-Asiatic types with short spikelets, short glumes, short lemmas and with mature grains exposed in the florets.

Main features used for distinguishing between cultivars in the field (based on Purseglove 1985b):

>   Plant height: dwarf, tall
>   Colour of vegetative organs: green, purple
>   Type of inflorescences: spikes straight and open, spikes incurved and closed, spikes branched, resembling a cockscomb
>   Length of spikes: short, long
>   Number of spikelets per spike: few, many
>   Length of spikelet: short, long
>   Density of spike: lax, dense
>   Seed colour: white, orange red, deep brown, purple/black

Other features used for distinguishing between cultivars:

>   Tillering: few tillers, many tillers

Environmental Requirements for Seed Production

Finger millet is more restricted in the areas where it can be grown than other species such as maize or sorghum. It requires a well distributed rainfall which is not excessive (between 800 and 1250 mm/year). In Africa it is generally grown between 1000 and 2000 m altitude (Purseglove 1985b).

Average maximum temperature above 27°C and average minimum above 18°C are said to be required.

Free-draining soils which do not waterlog in excessive rainfall are required. Soils should be fertile with a good nutrient balance.

Dry weather is necessary when the grain matures and at harvest because the spikes are difficult to dry.

Previous Cropping and Isolation

Seed crops should be preceded by two years free from finger millet, during which crops are grown which permit weed control. Wild millet (*E. africana*), which is very similar to finger millet, is particularly objectionable; striga (*Striga* spp.), a parasitic weed, is not usually a serious problem in finger millet but is difficult to remove when it does occur. Weeding is usually done by hand in those areas where the crop is grown.

Finger millet is generally self-fertilising but out-pollination does occur to about 1%. When particular cultivars are being grown it is therefore advisable to provide some isolation (40 m from other sources of pollen). Otherwise crops should be separated by a physical barrier or a gap of 2 m.

Seed Crop Establishment

Prior to sowing, fertiliser or farmyard manure should be applied; finger millet requires fertile conditions. A firm fine tilth should be prepared as the species will not tolerate a seed bed which is not properly compacted. The crop may be sown broadcast or in narrowly spaced rows. Seed rates vary between 5 and 20 kg/ha, depending on region and method of sowing. In some places the crop is transplanted and this is more economical of seed, giving a higher multiplication rate.

Hand weeding is usually done when the crop seedlings are about 5 cm high. The crop is very susceptible to weed competition in the young stages and it is essential to keep it clean at this time. If the crop is in more widely spaced rows (25 cm apart) the plants are thinned so that they are 10 to 12 cm apart.

Seed Crop Management

Once established, the crop requires little attention until harvest. It is reasonably resistant to diseases and pests. The main disease is blast (*Piricularia grisea*), to which some cultivars are resistant.

Seed Crop Inspection and Roguing

In India, two inspections are made, one during flowering and the other at maturity. The maximum permitted off-types at final inspection are 0.05% for early generations and 0.1% for Certified Seed (Tunwar and Singh 1988c). Roguing is not generally practised, but the best times are the same as for inspections.

Harvesting

According to Purseglove (1985b), cultivars vary in their ability to resist shattering. Some therefore will need close attention as the crop matures to ensure that it is harvested before there is too great a loss of seed.

In Africa the crop is often harvested by hand and this allows for harvesting on consecutive occasions: the individual heads are cut with a few centimetres of straw and taken to a clean floor for drying, only the ripe heads being cut on each occasion. Otherwise the crop may be harvested as other cereals when the majority of the heads are mature.

Drying and Storage

Seed heads are usually dried to 12% moisture content before threshing, otherwise threshing can be difficult. Drying in the sun is usual, but if forced-air drying is used the temperature of the air should not exceed 40°C.

In very humid climates it may be necessary to store seed in controlled conditions, although generally finger millet stores well and is not as susceptible to pests as other millets.

Seed Cleaning

It is generally necessary to pre-clean the threshed seed to remove debris; this involves winnowing the seed either by hand or in a winnowing machine. Subsequently an air/screen cleaner should complete the operation, although in some instances a gravity separator is also used.

Other Minor Cereal Grain Crops

The following grain crops are grown in parts of Africa and Asia. All except foxtail millet (*Setaria italica*) are cross-pollinating, at least in some degree. Some are also used as forage grasses in the tropics.

Scientific name	Common name
Brachiaria deflexa C.E. Hubb	Fonio
Coix lachryma-jobi L.	Job's tears
Digitaria exilis Staph	Fonio
Digitaria ibura Staph	Black fonio
Echinochloa colonum (L.) Link	Barnyard millet, Jungle rice
Echinochloa crus-gali (L) P. Beauv	Barnyard millet
Echinochloa frumentacea (Roxb.) Link	Japanese barnyard millet
Eragrostis tef (Zuccagni) Trotter	Teff
Panicum miliaceum L.	Proso millet
Panicum sumatrense Roth ex Roem & Schult	Little millet
Paspalum scrobiculatum L.	Kodo millet
Phalaris canariensis L.	Canary grass
Setaria italica (L) P. Beauv	Foxtail millet

Species for which cultivar descriptors have been published in OECD (1995c) are *Panicum miliaceum, Phalaris canariensis* and *Setaria italica*.

See also Langer and Hill (1991), Purseglove (1985c) and Wanous (1990) for further details.

References

Bell, G.D.H. (1965). The comparative phylogeny of the temperate cereals. In *Essays on Crop Plant Evolution* (Ed. Sir J. Hutchinson). The Syndics of the Cambridge University Press, London, 70–102.

Bicklemann, U. (1993a). On fatuoids of *Avena* spp. and their significance for seed production: I. Flowering behaviour of oats and origin of fatuoids and hybrids. *Plant Varieties and Seeds*, **6**, 21–26.

Bicklemann, U. (1993b). On fatuoids of *Avena* spp. and their significance for seed production: II. Description by floret morphological characters. *Plant Varieties and Seeds*, **6**, 27–32.

Bicklemann, U (1993c). On fatuoids of *Avena* spp. and their significance for seed production: III. Description of prolamin electrophoretic characters. *Plant Varieties and Seeds*, **6**, 65–74.

Carsky, R.J., Ndikawa, R., Kenga, R., Singh, L., Fobasso, M. and Kamuanga, M. (1996). Effect of sorghum variety on *Striga hermonthica* parasitism and reproduction. *Plant Varieties and Seeds*, **9**, 111–118.

Chaudhary, R.C., Hillerislambers, D. and Puckridge, D.W. (1995). Improvement of deepwater rice for Cambodia: a vertical and lateral support model. *Plant Varieties and Seeds*, **8**, 175–185.

Chopra, K.R. (1987). Technical and economic aspects of seed production of hybrid varieties of sorghum. In *Hybrid Seed Production of Selected Cereal Oil and Vegetable Crops* (Eds W.P. Feistritzer and A.F. Kelly). FAO Plant Production and Protection Paper 82, Food and Agriculture Organisation of the United Nations, Rome, 193–216.

Cobley, L.S. (1976). *An Introduction to the Botany of Tropical Crops* (revised by W.M. Steele). Longman, London, 49–54.

Cooke, R.J. (1989). The use of electrophoresis for the distinctness testing of varieties of autogamous species. *Plant Varieties and Seeds*, **2**, 3–14.

Cooke, R.J. (1995). Variety identification: modern techniques and applications. In *Seed Quality: Basic Mechanisms and Agricultural Implications* (Ed. A.S. Basra). The Haworth Press, New York, 279–318.

Copeland, L.O. and McDonald, M.B. (1995). *Principles of Seed Science and Technology*. Chapman and Hall, New York, Chapter 8.

Doggett, H. (1965). The development of the cultivated sorghums. In *Essays on Crop Plant Evolution* (Ed. Sir J. Hutchinson). The Syndics of the Cambridge University Press, London, 50–69.

Doyle, C.J. and Valentine, J. (1988). Naked oats: an assessment of the economic potential for livestock feed in the United Kingdom. *Plant Varieties and Seeds*, **1**, 99–108.

Edwards, I.B. (1987). Breeding of hybrid varieties of wheat. In *Hybrid Seed Production of Selected Cereal Oil and Vegetable Crops* (Eds W.P. Feistritzer and A.F. Kelly). FAO Plant Production and Protection Paper 82, Food and Agriculture Organisation of the United Nations, Rome, 3–34.

Evans, G.M. (1976). Rye. In *Evolution of Crop Plants*

(Ed. N.W. Simmonds). Longman Group, Harlow, 108–111.

Federmann, G.R., Goecke, E.U. and Steiner, A.M. (1992). Research Note: Detection of adulteration of spelt (*Triticum spelta* L.) with flour of wheat (*Triticum aestivum* L. emend Fiori et Paol) by electrophoresis. *Plant Varieties and Seeds*, **5**, 123–125.

Harlan, J.R. (1976). Barley. In *Evolution of Crop Plants* (Ed. W.W. Simmonds). Longman, Harlow, 93–98.

Harlan, J.R. and de Wet, J.M.J. (1972). A simplified classification of cultivated sorghum. *Crop Science*, **12**, 172–176.

House, L.R. (1985). *A Guide to Sorghum Breeding*, 2nd edition. International Crop Research Institute for the Semi-arid Tropics, Patancheru AP 502 324, India.

IBPGR (1984). *Revised Sorghum Descriptors*. International Board for Plant Genetic Resources, Rome.

ISTA (1996). International Rules for Seed Testing 1996. *Seed Science and Technology*, **24** (Supplement), 254–257.

Jaiswas, J.P. and Agrawal, R.L. (1995). Varietal purity determination in rice: modification of the phenol test. *Seed Science and Technology*, **23**, 33–42.

Langer, R.H.M. and Hill, G.D. (1991). *Agricultural Plants*. pp. 113, 126–135. Cambridge University Press, Cambridge.

Laverack, G.K. and Turner, M.R. (1995). Roguing seed crops for genetic purity: a review. *Plant Varieties and Seeds*, **8**, 29–46.

National Research Council (1989). *Triticale: a Promising Addition to the World's Cereal Grains*. National Academy Press, Washington DC.

OECD (1995a). *Up-to-date Version of the Seed Schemes as of 15 June 1995*. Organisation for Economic Co-operation and Development, Paris, 54–56.

OECD (1995b). *Up-to-date Version of the Seed Schemes as of 15 June 1995: Cereal Seed*. Organisation for Economic Co-operation and Development, Paris, 35–66.

OECD (1995c). *Control Plot and Field Inspection Manual*. pp. 154, 159, 165. Organisation for Economic Co-operation and Development, Paris.

Okonkwo, C.A.C. and Garba, A. (1993). Selection of striga-resistant sorghum germplasm in the northern Guinea savanna. *Plant Varieties and Seeds*, **6**, 197–205.

Okonkwo, C.A.C. and Kilishi, N.A. (1990). Bird resistant sorghum germplasm in the northern Guinea savanna. *Plant Varieties and Seeds*, **3**, 119–126.

Opoku, G. and Gamble, E.E. (1995a). Seed quality of normal and naked OAC Kippen barley. *Plant Varieties and Seeds*, **8**, 73–80.

Opoku, G. and Gamble, E.E. (1995b). Storability of seeds of normal and naked types of OAC Kippen barley. *Plant Varieties and Seeds*, **8**, 197–206.

Pickett, A.A. (1988). Factors affecting seed production of Triticale (*X Triticosecale* Wittm.). *Plant Varieties and Seeds*, **1**, 63–74.

Pickett, A.A. and Galway, N.W. (1997). A further evaluation of hybrid wheat. *Plant Varieties and Seeds*, **10**, 15–32.

Purseglove, J.W. (1985a). *Tropical Crops: Monocotyledons*. Longman, Harlow, 203–214.

Purseglove, J.W. (1985b). *Tropical Crops: Monocotyledons*. Longman, Harlow, 146–156.

Purseglove, J.W. (1985c). *Tropical Crops: Monocotyledons*. pp. 134, 142–145, 157, 199, 201–202, 256. Longman Group Ltd., Harlow.

Rao, A.N. and Moody, K. (1990). Weed seed contamination in rice seed. *Seed Science and Technology*, **18**, 139–146.

Sharopova, N.R., Portyanko, V.A. and Sozinov, A.A. (1997). Classification of European oat (*Avena sativa* L.) varieties using biochemical and morphological markers. *Plant Varieties and Seeds*, **10**, 33–38.

Smith, J.E. and Whittington, W.J. (1991). Effects of roguing frequency of atypical wheat (*Triticum aestivum* L.) Plants. *Plant Varieties and Seeds*, **4**, 77–86.

Stroike, J.E. (1987). Technical and economic aspects of hybrid wheat production. In *Hybrid Seed Production of Selected Cereal Oil and Vegetable Crops* (Eds W.P. Feistritzer and A.F. Kelly). FAO Plant Production and Protection Paper 82, Food and Agriculture Organisation of the United Nations, Rome, 176–186.

Thornton, M.S. (1986). *Investigations into the problems associated with the development of naked oats as a crop*. PhD Thesis, University of Wales, Aberystwyth (quoted in Valentine and Hale 1990).

Thornton, M.S. and Lawes, D.A. (1981). Improvement of nutrient quality and ease of processing in oats. *HGCA Progress Report on Research and Development 1979–1980*. HGCA, London, 58–60 (quoted in Valentine and Hale 1990).

Tunwar, N.S. and Singh, S.V. (1988a) *Indian Minimum Seed Certification Standards*. The Central Seed Certification Board, New Delhi, 44–50.

Tunwar, N.S. and Singh, S.V. (1988b). *Indian Minimum Seed Certification Standards*. The Central Seed Certification Board, New Delhi, 51–57.

Tunwar, N.S. and Singh, S.V. (1988c). *Indian Minimum Seed Certification Standards*. The Central Seed Certification Board, New Delhi, 62–63.

UPOV (1978). *Guidelines for the Conduct of Tests for Distinctness, Homogeneity and Stability: Rye*. The International Union for the Protection of New Varieties of Plants, Geneva.

UPOV (1981a). *Guidelines for the Conduct of Tests for Distinctness, Homogeneity and Stability: Wheat.* Doc. TG/01/8, International Union for the Protection of New Varieties of Plants, Geneva.

UPOV (1981b). *Guidelines for the Conduct of Tests for Distinctness, Homogeneity and Stability: Barley.* Doc. TG/19/7, International Union for the Protection of New Varieties of Plants, Geneva.

UPOV (1981c). *Guidelines for the Conduct of Tests for Distinctness, Homogeneity and Stability: Oat.* Doc. TG/20/7, International Union for the Protection of New Varieties of Plants, Geneva.

UPOV (1985). *Guidelines for the Conduct of Tests for Distinctness, Homogeneity and Stability: Rice.* Doc. TG/16/4, International Union for the Protection of New Varieties of Plants, Geneva.

UPOV (1988). *Guidelines for the Conduct of Tests for Distinctness, Homogeneity and Stability: Durum wheat.* Doc. TG/120/3, International Union for the Protection of New Varieties of Plants, Geneva.

UPOV (1989a). *Guidelines for the Conduct of Tests for Distinctness, Homogeneity and Stability: Triticale.* Doc. TG/121/3, International Union for the Protection of New Varieties of Plants, Geneva.

UPOV (1989b). *Guidelines for the Conduct of Tests for Distinctness, Uniformity and Stability: Sorghum.* Doc. TG/122/3, International Union for the Protection of New Varieties of Plants, Geneva.

Valentine, J. and Hale, O.D. (1990). Investigations into reduced germination of seed of naked oats. *Plant Varieties and Seeds*, **3**, 21–30.

Wanous, M.K. (1990). Origin, taxonomy and ploidy of the millets and minor cereals. *Plant Varieties and Seeds*, **3**, 99–112.

White, E.M. (1990). The effect of source of seed on seedlot characteristics and yield potential in spring barley cultivars. *Plant Varieties and Seeds*, **3**, 31–41.

Further Reading

Parry, D. (1990). *Plant Pathology in Agriculture.* Cambridge University Press, Cambridge.

GRAMINEAE: GRAIN AND VEGETABLE CROPS

Zea mays L. Common Names: Maize, Corn, Sweetcorn

Maize is grown in both tropical and temperate regions, although in the higher latitudes it becomes increasingly less likely to produce an economic grain yield. It is grown as a grain crop for human consumption and as an animal feed, it is cut as a green fodder in higher latitudes, and is an important vegetable for human consumption. There are also specialist uses for the grain such as oil and alcohol production.

The species is divided into the following main types:

(i)   Dent: seed with a dent in the apex caused by shrinkage of soft starch;
(ii) Flint: seed filled, no dent;
(iii) Flour: endosperm soft starch;
(iv) Sweet: seed contains a large proportion of sugar;
(v)   Popcorn: seed with a thick layer of hard starch surrounding soft starch in the centre of the endosperm;
(vi) Waxy: seed with starch composed entirely of amylopectin, giving a waxy appearance.

Sweet and popcorn are grown as specialist food crops for human consumption, the sweet type being used as a vegetable and the popcorn being prepared as a dry cereal. The main agricultural types are the dent and flint, with waxy being grown for special uses. Flour is now rarely grown and is largely confined to local communities where it has been grown traditionally.

Maize is cross-fertilised and is one of the first species used in the early development of hybrid and synthetic cultivars. Modern hybrid cultivars are based on cytoplasmic male sterility, although because the male and female inflorescences are separate on the same plant, it is possible to emasculate the male plants by hand or mechanically. Hybrids were developed in the USA in the 1920s and now account for most of the area grown in developed countries; increasingly they are being used in developing countries, although open-pollinated cultivars are still used extensively, including some composite and synthetic cultivars.

Classification of Cultivars

These characteristics are based on UPOV (1980) and OECD (1995a)

Breeding system: open-pollinated, composite/synthetic, inbred line, hybrid
For a hybrid cultivar: single-cross, double-cross, top-cross, other
Type of seed: dent, flint, flour, sweet, popcorn, waxy
Maturity: early, late
Photoperiod sensitivity: insensitive, neutral, sensitive

Distinguishing between Cultivars

Main features used for distinguishing between cultivars in the field:

Emergence of tassel (50% of plants): early, medium, late
Start of anthesis: early, medium, late
Anthocyanin colouration of anthers: weak, medium, strong
Emergence of silk (50% of plants): early, medium, late
Anthocyanin colouration of silk, of nodes, of middle internodes, of middle leaf sheaths, of tassel glumes: weak, medium, strong
Closed anthocyanin ring at base of tassel glume: absent, present
Anthocyanin colouration of cob glumes: weak, medium, strong
Intensity of anthocyanin colouration of cob glumes: weak, medium, strong
Plant height (including tassel): short, tall
Height of insertion of ear relative to plant height: low, high
Angle between main axis and lateral branches in lower third of tassel: small, large

Attitude of lateral branches in lower third of tassel: rectilinear, recurved

Number of primary lateral branches of tassel: few, many

Length of main axis of tassel above lowest side branch: short, long

Length of main axis of tassel above upper side branch: short, long

Ear coverage at maturity: tip visible, tip not visible

Length of peduncle of ear: short, long

Ear length: short, long

Diameter of middle of ear: small, large

Shape of ear: conical, cylindrical

Number of rows of grain in ear: few, many

Other features used for distinguishing between cultivars:

Anthocyanin colouration of first leaf sheath: weak, medium, strong

Length of first leaf blade: short, long

Width of first leaf blade: narrow, wide

Ratio of first leaf length to width: small, large

Shape of first leaf tip: pointed, obtuse, round

Leaf attitude in central third of plant: rectilinear, recurved

Hairs on margin of leaf sheath: few, many

Colour of tip of grain: white, yellow, orange, red, black

Colour of dorsal side of grain: white, yellow, orange, red, black

Laboratory tests:

Electrophoresis of seed proteins (see, for example, Brink et al. 1989).

Environmental Requirements for Seed Production

Although grown in some temperate regions, for seed production maize requires a warm climate with adequate moisture. It is not very drought resistant and will not withstand waterlogging. Rainfall of about 600 mm with good distribution throughout the growing season is most suitable. Irrigation can be used to supplement rainfall, especially in the period from silking to harvest.

For good growth leading to a good seed crop, the temperature from planting to harvest must be adequate; the seed will not germinate at temperatures below 10°C. Ideally temperature should be about 20°C for good germination and temperatures of around this level throughout the growing period are best for seed growing.

A cultivar is adapted to a particular latitude and cannot be grown satisfactorily in regions where the day length is not suited. Cultivars which are adapted to areas with short days will be much later to mature if grown in areas with long days.

Crops can be grown on a wide range of soils, but do best on well drained fertile loams with a good content of organic matter. Soils with adequate lime content are required (pH about 6.0) and should be well supplied with nutrients. Maize responds well to fertiliser application, particularly nitrogen.

Previous Cropping and Isolation

Maize plants arising from seed shed from earlier crops do not usually occur in fields being prepared for a seed crop, and seed can therefore be grown where the immediately preceding crop was also maize. However, when crops follow one another within the same year, volunteers may create difficulties and this practice should be avoided for seed growing.

Maize is cross-fertilised and seed crops require good isolation from other sources of pollen; 200 m is generally considered adequate in most certification schemes for the production of Certified Seed. Greater distances may be required for the production of earlier generations, although OECD (1995b) specifies the same distance throughout. In the Indian standards (Tunwar and Singh 1988) distances up to 400 m are specified, while in FAO (1982) 600 m is stated as necessary separation when the crops are of cultivars with different grain colour or texture. These greater distances are also needed to separate sweetcorn or popcorn

TABLE 7.1   Border rows to be discarded to achieve adequate isolation (source: FAO 1982)

Isolation distance available (m)	Number of rows to be discarded according to field size (ha)						
	Less than 1.6	1.6–2.4	2.4–3.2	32.–4.0	4.0–4.8	4.8–6.4	Over 6.4
175	3	3	2	2	2	1	1
150	5	5	4	4	3	3	2
125	7	7	7	6	5	5	4
100	9	9	8	8	7	7	6
75	11	11	10	10	9	9	8
50	13	13	12	12	11	11	10

from other types. Separation in time may also be possible and may be enhanced by adjusting the sowing dates of the two crops. The time difference must be sufficient to ensure that the female silks of the first crop are not receptive of pollen when pollen is being shed in the second crop; when crops to produce hybrid seed are grown in this way it is also necessary to ensure that the male tassels of the first crop have stopped shedding pollen when the female silks of the second are receptive.

In hybrid seed production, planting border rows of the male parent around the crop will allow shorter isolation distances in some quality control schemes. Table 7.1 shows such an arrangement as specified in FAO (1982); Tunwar and Singh (1988) specify for more field sizes and isolation distances in the Indian seed certification scheme.

For open-pollinated cultivars, isolation distances may also be reduced by discarding the seed from the border rows around the crop as specified in Table 7.1.

Seed Crop Establishment

Maize seed is sown in rows with the seed spaced in the rows. Seed rate depends on two factors.

1. Optimum plant population will vary with the growing conditions and the growth characteristics of the cultivar from 25 000 to 100 000 plants/ha, with 50 000 the most usual. The rows are spaced 50 to 100 cm apart, with 75

TABLE 7.2   Distance between rows and spacing of plants in the row for different plant populations (source: FAO 1982)

Desired no. of plants per ha	Spacing of plants in the row (cm) for distance between rows (cm)					
	50	60	70	75	80	90
35 000	57	48	41	38	36	32
40 000	50	41	36	33	31	28
45 000	44	37	32	30	28	25
50 000	40	33	29	26	25	22
55 000	36	30	26	24	23	20
60 000	33	28	24	22	21	18
65 000	31	26	22	20	19	17
70 000	29	23	20	19	18	16
75 000	27	22	19	18	17	15
80 000	25	21	18	17	16	14
85 000	23	20	17	16	15	13
90 000	22	19	16	15	14	12
100 000	20	17	14	13	12	10

cm the most usual. To achieve a plant population of 50 000/ha with 75 cm between the rows, the plants must be spaced at 26 cm in the row; Table 7.2 gives data for other population sizes.

2. The size of seed differs from one cultivar to another. Small-seeded cultivars require a lower seed rate than those with larger seeds to achieve the desired plant population per hectare. George (1985) suggests a seed rate of 30 kg/ha for cultivars of sweetcorn with large seed and 15 kg/ha for those with smaller seed. When a plate planter is used, it is adjusted to take account of seed size and seed which has been size-graded is used; details of seed-sizing

arrangements are given in the section on 'Seed Cleaning'.

When hybrid seed is to be produced it is necessary to plant the female and male parents in the correct proportions to achieve good pollination. The breeder of the cultivar should advise on this. For single-cross hybrids the usual proportion is four female:two male rows. For double-cross hybrids it is usual for the male parent to be more vigorous and the ratio can be widened to 6:2 or even 8:2.

For hybrid seed production it is essential to ensure that the male parent is producing pollen when the female silks are receptive. When there is a difference between the parents, some adjustment can be made by sowing one earlier than the other.

For better pollen distribution the male parent chosen should be taller than the female.

When cytoplasmic male sterility is used in the female parent it is necessary to use a restorer, otherwise the progeny will be sterile. This is done by using a male parent which contains one or more specific restorer lines. The provision of restorer lines is the responsibility of the breeder.

Seed Crop Management

The management of a seed crop does not differ from that of a crop for feed or other use, except in the case of hybrid seed production. It is necessary to maintain soil moisture, supplementing rainfall with irrigation as necessary. This is particularly important during the period between pollination and maturity, especially for sweetcorn seed production.

The objective of weed control is to reduce competition and there are no weeds which have seeds difficult to remove after harvest. Control by inter-row cultivation is possible when the plants are small, but it is usual to use herbicides which will give control throughout the life of the crop.

Maize is responsive to nitrogen provided that other nutrients are in adequate supply.

There are numerous pests and pathogens which attack maize. The main seed-borne pathogens according to George (1985) are listed below.

Pathogen	Common name
Acremonium strictum W. Gans. Syn. *Cephalosporium acremonium* Corda	Kernel rot
Cephalosporium maydis Samra, Subet & Hingorani	Late wilt, late blight, slow wilt
Cochliobolus carbonum R.R. Nelson Syns *Drechslera zeicola* (Stout) Subram & Jain, *Helminthosporium carbonum* Ullstrup	Charred ear mould, southern leaf spot
Cochliobus heterostrophus (Drechsl.) Drechsl. Syns *Drechslera maydis* (Nisik.) Subram & Jain *Helminthosporium maydis* Nisik.	Southern leaf spot or blight
Diplodia spp.	Dry ear rot, stalk rot, seedling blight, root rot, white ear rot
Gibberella fujikuroi (Saw.) Wollenw. Syn. *Fusarium moniliforme* Sheld.	*Gibberella* ear rot, kernel rot, stalk rot
G. f. var. *subglutinans* Edw. Syn. *F. m. subglutinans* Wollenw. & Reink.	Seedling blight
G. zeae (Schw.) Petch Syn. *F. graminearum* Schwabe.	Seedling blight, cob rot
Marasmius graminum (Lib.) Berk.	Seedling and foot rot
Sclerophthora macrospora (Sacc.) Thirum, Shaw & Naras. Syn. *Sclerospora macrospora* Sacc.	Crazy top
Sclerospora philippiensis Weston Syn. *S. indica* Butler	Philippine downy mildew, crazy top

Pathogen	Common name
Ustilaginoidea virens (Cooke) Tak.	False smut, green smut
Ustilago maydis (D.C.) Corda Syn. *U. zeae* (Schw.) Unger	Smut, blister smut, loose smut
Erwinia stewartii (E.F. Smith) Dye Syns *Bacterium stewartii* E.F. Smith *Xanthomonas stewartii* (E.F. Smith) Dowson	Bacterial wilt, bacterial leaf blight, Stewart's disease, white bacteriosis

Viruses that attack maize are maize leaf spot virus, maize mosaic virus, sugar cane mosaic virus, wheat streak mosaic virus and corn stunt

Seed Crop Inspection and Roguing

Any off-type plants occurring should be removed before the silks are receptive of pollen following the description of the cultivar. For the production of hybrid seed based on cytoplasmic male sterility, an important part of roguing is to remove any female plants which are producing pollen. This should be done frequently (at least every second day) over a period of about two weeks before the silks are receptive of pollen. Similarly, when reliance is placed on mechanical or hand emasculation, it is necessary to check that the work has been done satisfactorily and to remove any tassels which have been missed. These checks should be continued after anthesis of the male parent to ensure that only that pollen reaches the silks of the female parent.

Inspection of the seed crop for quality control follows the same pattern as for roguing. For open-pollinated cultivars, one visit to check on the isolation and on the cultivar purity is required. For the production of hybrid seed, more than one inspection is required and up to four may be necessary from the start of tasselling to the end of the pollen shedding period.

For open-pollinated cultivars, standards of cultivar purity are of the order of 99.5% minimum for Basic Seed and 99.0% for Certified Seed.

For hybrid cultivars the standards are higher, usually 99.9% for the production of the seed of the male and female parents and 99.0% for the female parent when producing hybrid seed. Standards are also set for the number of female plants which are found to be shedding pollen: the first inspection should take place when about 5% of the silks are receptive; at each inspection at least 99.5% of the female plants should not be shedding pollen and the total for all inspections should be at least 99%. To identify the plants which are shedding pollen, tassels, sucker tassels or portions of tassels are counted when 50 mm or more of the central axis, the side branches or a combination of the two have their anthers extended from their glumes and are shedding pollen.

Harvesting

For seed production, maize is harvested as intact cobs either by hand or by specialised 'corn picker'. The seed is physiologically mature at a moisture content of about 45% but is vulnerable to mechanical damage at this stage; machine harvesting should therefore be delayed until moisture content has fallen to below 25%. Harvest maturity is indicated by a 'black layer' at the base of the grain. The use of combines which thresh the seed from the cob is not recommended for seed production.

Drying and Storage

The harvested cobs usually still have the husks attached, although some corn pickers do partially remove the husks, and the moisture content of the whole is too high for immediate shelling of the cobs. The most usual way of drying is to place the cobs in cribs constructed with mesh sides to provide free flow of air, although drying on a clean floor protected from rain and intense direct sunlight is also possible. The size of cribs should be adjusted according to the humidity usually encountered in the area where the seed is to be stored, and suggested dimensions are given in Table 7.3.

If heated air is used for drying, the temperature should not exceed 42°C. There must be a good

TABLE 7.3   Crib dimensions for drying maize cobs (source: FAO 1982)

Dimensions of crib (m)			Volume (m^3)	Capacity (kg ears)	For use where air humidity is
Length	Width	Height			
35.00	1.35	3.00	142	70000	High
35.00	1.50	4.00	210	110000	Moderate
35.00	2.00	3.00	210	110000	Low

circulation of air through the ears, and when initial moisture content is high the drying must be done slowly.

After drying, the moisture content should not exceed 14%. The ears can then be shelled, but before doing so it is advisable to check that the ears conform to the characteristics of the cultivar when the quantities are small enough in the earlier stages of multiplication, and especially for sweetcorn.

Before shelling, the husks must be removed from the ears. Shelling can be done either by hand or in a mechanical sheller designed for seed. After shelling, the moisture content should be reduced to 10–12% for medium-term storage and 8–10% for long-term storage.

Seed Cleaning

After shelling, the seed should be pre-cleaned to remove broken cobs or husks and dirt. Subsequently an air/screen cleaner is used to separate the small, medium and large seed. Seed is usually planted with a plate planter and the seed is size-graded to match the plates fitted to the planter. Medium-sized seed is passed through a 12 mm but is retained by a 6 mm round-hole screen. The seed is further separated into 'flat' and 'round'; flat seed will pass through a slotted screen whereas round seed will be retained. Flats occur in the middle of the ear and rounds at either end, while the largest seeds occur towards the base. Once separated, the flats and rounds are each further graded into 'large', 'medium' and 'small' using round-hole screens or on a gravity separator.

***Euchlaena mexicana* Schrad. Syn. *Zea mexicana* (Schrad.) Kuntzer. Common Name: Teosinte**

Teosinte is closely related to maize and will cross readily with it. It occurs as a weed in Central America but is also grown as a forage crop and minor grain crop in India. It is similar to maize but produces tillers from the base.

References

Brink, D.E., Price, S.C. and Martinez, C. (1989). Monoclonal antibodies against zeins. *Seed Science and Technology*, **17**, 1–6.

FAO (1982). *Technical Guideline for Maize Seed Technology*. Food and Agriculture Organization of the United Nations, Rome, 41, 64, 98.

George, R.A.T. (1985). *Vegetable Seed Production*. Longman, Harlow, 288–293.

OECD (1995a). *Control Plot and Field Inspection Manual*. Organisation for Economic Co-operation and Development, Paris, 58–60.

OECD (1995b). *OECD Scheme for the Varietal Certification of Maize and Sorghum Seed Moving in International Trade: Appendices 11 and 11a*. Organisation for Economic Co-operation and Development, Paris.

Tunwar, N.S. and Singh, S.V. (1988). *Indian Minimum Seed Certification Standards*. The Central Seed Certification Board, New Delhi, 31–43.

UPOV (1980). *Guidelines for the Conduct of Tests for Distinctness, Homogeneity and Stability: Maize*. Dec. TG/02/4. International Union for the Protection of New Varieties of Plants, Geneva.

GRAMINEAE: AGRICULTURAL CROPS – TEMPERATE GRASSES

As the name implies, these species are grown in the temperate regions. Unlike the cereal grains they are grown for their leaves, either as fodder for livestock or for amenity purposes such as lawns or sports fields. Because of this, seed growing is a specialist operation, often requiring a different growing method from the fodder crops. There are annual and biennial species or subspecies, but most are perennial. The annuals are harvested for seed from a spring-sown crop whereas the biennials and perennials produce seed in the year after the crop is established; they will only produce seed as lengthening days and higher temperatures cause them to change from the vegetative to the reproductive phase. Tillers formed in the autumn produce the most seed and management is directed to encourage autumn tiller production and to avoid damaging them in the spring.

Lolium spp. are the most important and some cultivars are used both for fodder and for amenity purposes. To some extent, however, these two purposes are incompatible. For fodder, the maximum amount of leaf per unit area per year is required, whereas for amenity purposes slow growth with dense ground cover is needed to ensure minimum maintenance. Other species are generally more specialised. For example, cocksfoot or orchard grass (*Dactylis glomerata* L.) is used exclusively as a fodder and red fescue (*Festuca rubra* L. *sensu lato*) mainly for amenity purposes. As a consequence almost all seed crops of these species are grown with a definite market in mind and most are on contract between a grower and a seed company.

Most of the species are cross-fertilised and show more plant-to-plant variation within a cultivar than self-fertilised species. Cultivars are therefore influenced by the environment in which they are growing since some plants of the cultivar may be favoured and others suppressed. This is relevant when seed is produced in areas favourable for seed production which may be far removed from the areas where the seed is to be used to sow fodder crops. For example, much grass seed is produced in the west of the USA for use in the livestock-producing areas of the east. Experience and experiment indicate that it is essential to limit the number of generations that a cultivar is produced for seed in a different environment from that in which it was selected. The advice of the breeder should be followed on this point.

In some specialised seed growing areas where the climate is dry but suitable for seed maturation, irrigation is used. If irrigation is not used in such areas, seed yields may be reduced by drought stress. Irrigation can be used after sowing to establish strongly tillering plants, withholding irrigation later encourages the production of uniform flowering stems but it is required later to ensure that the seed fills well. Irrigation should be integrated with the application of nitrogen which may have to be divided into smaller, more frequent doses to prevent excessive leaching. The drought resistant species (for example, tall fescue, *Festuca arundinacea* Schreber) are generally less responsive to irrigation.

Recent work has suggested that endophytes (*Acremonium* spp.) assist seed production of meadow fescue, Italian ryegrass and perennial ryegrass. Another recent development is the use of growth regulators to enhance seed yield.

Lolium species. Common Name: Ryegrass

Lolium multiflorum and *L. perenne* are the two most important species, but there is an indistinct division between them and they will cross-fertilise. This has been used by plant breeders to create hybrid cultivars so that there is now a continuum between the annual Westerwolds of *L. multiflorum* through the hybrids of *Lolium* × *Boucheanum* to the perennial cultivars of *L. perenne*.

There are diploid and tetraploid cultivars in each species. Tetraploids have seeds approximately twice the size of diploids and whole plants contain 1–2% more moisture.

L. rigidum is an annual species which is used to a limited extent. It does not cross freely with either *L. multiflorum* or *L. perenne*.

Distinguishing morphological features of *L. multiflorum* and *L. perenne* are as follows.

Feature	*L. multiflorum*	*L. perenne*
Awns on outer palea	Generally present	Usually absent
Rachilla	Flattened	Oval
Shoots	Rounded	Flattened
Auricles	Prominent	Less prominent

Hybrid cultivars usually exhibit the features of the parent they resemble most closely. These morphological features, however, are not absolute in either species and from a practical point of view persistence in the sward is generally the most important feature. *L. multiflorum* is annual or biennial and lasts in the sward for a maximum of three years, whereas *L. perenne* is perennial, with the later flowering cultivars being generally the most persistent.

In the laboratory the 'fluorescence test' may distinguish shorter lived cultivars: seedlings produced in the dark on non-fluorescent paper are examined under ultra-violet light and a high proportion of fluorescent seedlings usually indicates a less persistent cultivar (for precise details of this test see ISTA 1996).

Biennial cultivars of *L. multiflorum* and perennials of *L. perenne* respond to lengthening days and temperature to change from the vegetative to the reproductive phase. Recent work in New Zealand has shown that it is possible to use heading date under combinations of vernalisation and day length to distinguish cultivars (Halligan et al. 1993): with 10 weeks of vernalisation at 5°C, followed by 14 h photoperiod (mean photosynthetic photon flux density 675 mol/m²/s) the authors could distinguish 14 of the 15 cultivars of perennial ryegrass tested in less than six months. Earlier work (Halligan et al. 1991) gave similar results for Westerwolds, Italian and Hybrid cultivars.

Alternative laboratory tests include the use of allele frequencies of isozyme systems as described in Chapter 3 (see also Lallemand et al. 1991; Booy et al. 1993).

Lolium multiflorum Lam. Common Name: Italian ryegrass (including Westerwolds)

The annual Westerwolds cultivars are taken for seed in the sowing year. Some biennial Italians will also produce seed in the sowing year if sown early, but this is not an economic yield and these cultivars are taken for seed in the second year. All crops provide only one seed harvest, as second harvests do not produce an economic yield and are usually contaminated by plants from shed seed.

Classification of Cultivars

Seeding in the sowing year: annual, biennial
Ploidy: diploid, tetraploid

Distinguishing between Cultivars

Main features used for distinguishing cultivars in the field (based on UPOV 1980a):

Time of inflorescence emergence: early, late
At inflorescence emergence
   Plant growth habit: erect, prostrate
   Flag leaf length: short, long
   Flag leaf width: narrow, wide
Stem length, including fully expanded inflorescence: short, long

Other features used for distinguishing between cultivars:

Leaf colour in autumn: light green, dark green
Plant growth habit (spring of second year): erect, prostrate
Natural height of plant (spring of second year): short, long
Inflorescence length: short, long

Laboratory tests:

Fluorescence test (ISTA 1996)
Vernalisation requirement (Halligan et al. 1991, 1993)
Determination of leaf colour (McMichael and Camlin 1994)
Electrophoresis (see Chapter 3; ISTA 1996; Lallemand et al. 1991; Booy et al. 1993).

Environmental Requirements for Seed Production

Biennial cultivars of Italian ryegrass require a temperate climate in latitudes with suitable day length and temperature to ensure vernalisation. A uniform distribution of rainfall and dry weather at harvest are required. In areas where there is high humidity, blind seed disease (*Gloeotina temulenta*) is frequently encountered and areas where it occurs should be avoided for seed production as there is no economic treatment against it.

Soils should be fertile, with pH above 5 and preferably about 6. Phosphorus (P) and potash (K) requirements are similar to those for wheat. Nitrogen (N) is used to stimulate growth at appropriate times. However, excessive use of N, which causes lodging, must be avoided; it is important that crops do not lodge before fertilisation occurs as this would result in some inflorescences remaining unfertilised.

Previous Cropping and Isolation

Fields should be chosen which have been free of sown grasses with similar seed size for at least two years. Grass weeds and volunteer plants from sown grass crops must be controlled in the preceding crops. The following are the main species to be avoided: *Avena fatua, A. sterilis, A. ludoviciana* (wild oats), *Alepecurus mysuroides* (black grass), *Agropyron repens* (couch grass), *Poa annua* (annual meadow grass), *Poa trivialis* (rough stalked meadow grass), *Holcus lanatus* (Yorkshire fog), *Bromus* spp. (bromes) and *Lolium temulentum* (darnel).

Grass plants in previous crops can be controlled by cultivations or the use of defoliants such as paraquat, either on the seed bed before sowing or on stubbles. Spring-sown crops provide the best opportunity to eliminate volunteer grass plants as they give more time for treatment before sowing. Glyphosphate can be applied to preceding crops and controls many broad-leaved weeds as well as grasses.

Italian ryegrass is cross-fertilised by wind-dispersed pollen. Isolation of seed crops is therefore essential. When producing seed for further multiplication, 200 m is usually specified for fields of 2 ha or less, and 100 m for larger fields; these distances can be halved with producing seed for the establishment of fodder crops. Diploids will not cross readily with tetraploids but pollen may cause infertility in the recipient inflorescence; isolation of 50 m is therefore recommended.

Seed Crop Establishment

Seed crops should be established in rows 10 to 20 cm apart; broadcasting is sometimes used when the seed bed is too wet to carry a seed drill, but is not recommended.

Annual (Westerwolds) cultivars are sown in early spring without a cover crop and are harvested for seed in the same year.

Biennials may be sown with a cover crop in spring, or in the autumn with no cover crop, and are harvested in the following year. Autumn sowing should be early enough to ensure that the seedlings become established before the onset of severe weather which is late August or early September in northern Europe.

Seed rate for diploids is 12–16 kg/ha and for tetraploids 16–22 kg/ha. The higher rates are used when seed bed conditions are poorer than desirable or when broadcasting. Leaving a sterile strip 1 m wide around the crop will reduce ingress of weed grasses from the field boundaries. P and K fertiliser (45–50 kg/ha P and 55–65 kg/ha K) will be needed as a seed bed dressing on most soils. N is also used at sowing time, but when seed crops are sown under a cover crop it should be applied sparingly to avoid causing lodging in the cover crop. Such seed crops benefit from nitrogen application in the autumn after the removal of the cover crop. Similarly, seed crops sown in the autumn benefit from an application of 20–50 kg/ha N. Autumn-applied N stimulates growth which can be grazed, preferably by sheep, to encourage tillering as a sound basis for seed development in the following year. If livestock are not available, excessive growth should be mown and removed from the field before the onset of winter.

After establishment, when plants are growing strongly and have at least two leaves, ethofumesate may be applied in late autumn/early winter when soil is moist to control many grass weeds and volunteer grasses, and cereals, cleavers (*Galium aparine*) and chickweed (*Stellaria media*). There are several herbicides which can be used to control broad-leaved weeds in the spring of the year after sowing, but timing is critical and the manufacturer's advice should be sought. In general, plants should have reached the four-leaved stage before application, but the crop should not be treated within four weeks of inflorescence emergence.

Seed Crop Management

The annual cultivars need little or no attention between establishment and harvest.

For biennial cultivars a further 100 to 175 kg/ha N (depending on field conditions) should be applied but it is important to avoid excessive early lodging; in areas where there is heavy spring rainfall, N application should not exceed 125 kg/ha. The N may be applied as a single dressing as growth is beginning (mid-March in northern Europe) or as a split application, one-third early and the remainder later after defoliation.

It is usual to defoliate Italian ryegrass seed crops in spring before shutting up the field for seed. Failure to do so may result in excessive lodging before anthesis leading to poor seed set and to excessive leaf growth at seed harvest. However, care must be taken to avoid removing elongating flowering stems so it is important to time defoliation correctly. Normally the field should be shut up for seed by mid-April in the UK. Defoliating too late will result in lower seed

yield because of fewer or smaller inflorescences. Grazing is preferable, although it may be necessary to top the field after the stock are removed to establish an even base for the seed harvest. Alternatively, an early light silage cut may be taken. Wet conditions, which might cause the soil to poach, must be avoided.

There are few diseases or pests which require treatment. Two which are seed-borne are blind seed disease and ergot (*Claviceps purpurea*). For the former there is no remedy except to use disease-free seed; some cultivars are resistant and the pathogen will not survive for more than two years in dry seed. Very small quantities of seed can be treated by soaking in hot water at 50°C for 30 min followed by immediate cooling and drying, but this is likely to reduce germination. Straw burning on the field has been effective in destroying the pathogen which might become soil-borne, but this is now generally prohibited on environmental grounds. There is no treatment for ergot; some ergots may be removed from seed during cleaning but the best prevention is to use only ergot-free seed. Ergots can establish on grasses in the field boundaries and then spread into the seed crop; such grasses should therefore be prevented from flowering.

Pests such as slugs, fritfly and aphids sometimes attack developing seedlings. They may be treated in the same manner as in cereal crops.

Seed Crop Inspection and Roguing

Roguing Italian ryegrass seed crops is not usually practicable as the crop is dense and swayed at the time that rogues can be seen. Sometimes it is possible to hand pull some obvious weed grasses which are taller than the crop.

The usual time for inspection is at the beginning of inflorescence emergence. This allows the main morphological characteristics to be observed and is early enough for isolation requirements to be corrected if necessary. Assessment of cultivar purity is difficult and is usually done by concentrating attention on a predetermined number of standard quadrats. However, because of the nature of the crop, it is difficult to distinguish individual plants, and standards are therefore sometimes expressed as number of off-types per unit area rather than in percentage terms.

Harvesting

Harvest may be by swathing and picking up later with a combine, or by combining direct. In areas where harvest weather is normally good, swathing is to be preferred, but in those areas where the seed must normally be dried immediately after harvest, direct combining is better. Mature seed sheds easily and in the nature of the crop there will be inflorescences at different stages of ripening, but the best seed will be from older, longer inflorescences. Harvest should therefore be timed to take the seed from the majority of these early tillers.

Ripeness is indicated by a fading of the green colour in the inflorescences and in some cultivars the anthocyanin colouration intensifies. The seed is doughy at this stage. Seed moisture content can also be used as a guide to harvest timing: the crop is sampled by taking inflorescences from several places at random and sealing them in a plastic bag for subsequent threshing and determination of moisture content. Frequent sampling will be needed because moisture content changes rapidly as the seed ripens. Temporary apparent increases in moisture content caused by rain are discounted. Moisture content can be expected to fall by 2 to 3% per day in good conditions. Moisture contents when harvest can begin are as follows:

Harvest method	Diploid	Tetraploid
Swathing	43%	45%
Combining direct	38%	40%

A swathed crop should be allowed to dry to 14% moisture, or less if possible, before picking up with the combine, but in damp weather this may have to be done when the seed still contains more moisture.

Drying and Storage

Seed threshed at more than 14% moisture content must be dried immediately. Germination can fall very quickly when damp seed is stored even for a few hours, particularly when the cultivar is tetraploid.

For on-floor drying systems, seed depth should be restricted to 55 cm when moisture content is above 35% and this requires a floor area of 8 m^2 per tonne of seed. Airflows required are:

Moisture content of seed (%)	35	40	45
Air flow (m^3/min)	16.2	21.5	27.8

No heat should be used at first. The drying air may be reduced to 65% humidity in the final stage and the top layer of seed must be reduced to 20% moisture content within five days. If necessary, the seed should be turned to achieve this.

Continuous-flow flat-bed dryers can also be used. Seed depth must be restricted to 15 cm to avoid moisture gradients and overheating the bottom layer. Maximum temperatures for the drying air are:

Moisture content of seed (%)	35	40	45
Maximum air temperature (°C)	54	49	38

Seed must be cooled before placing it in store. Seed may be stored in bags or in bulk. If the latter, it is essential to monitor the seed frequently and to take instant action if there is any sign of heating.

Seed Cleaning

An air/screen cleaner will usually achieve a sufficiently good sample. In some cases additional cleaning on a gravity separator or indented cylinder may be required. Cleaning ryegrass seed is a highly specialised operation and should be left to the contracting company. High cleaning losses can result from poor setting of the equipment.

Lolium perenne L. Common Name: Perennial ryegrass

All of the notes on seed production of Italian ryegrass (*L. multiflorum*) apply also to perennial ryegrass (including the features used for distinguishing between cultivars) except for the following points.

1. *Number of harvest years.* There is a range of heading dates from early to late. The early heading cultivars are normally taken for seed only once in the year after the sowing year. Later cultivars may be retained for a second or third harvest.
2. *Isolation.* In addition to the spatial isolation measures outlined for Italian ryegrass, isolation in time is possible for some cultivars of perennial ryegrass which differ widely in time of anthesis. There can be a month between anthesis of the early and late cultivars.
3. *Seed rates.* Slightly lower seed rates can be used for perennial ryegrass than for Italian.
4. *Spring defoliation.* Extra care is required when defoliating perennial ryegrass seed crops in the spring, since there is less recovery of any developing inflorescences which are removed when defoliating. Early perennials should be shut up for seed about a month earlier than suggested for Italian and if growth is not excessive, defoliation can be avoided. Late perennials are more tolerant and may be treated like Italians, although spring growth in these cultivars is usually less vigorous and later.
5. *Second or third harvest year.* Crops which are to be taken for seed in a second or third harvest year should be cleaned as soon as possible after the first harvest by removing all cut material and trimming those areas which were not cut close enough during harvest; this provides a uniform base for the next harvest. On soils low in P and K, further dressings may be needed. N is applied at the same times as in the first year, but rates can be increased by 30%. Autumn and spring defoliation follows the same pattern as in the first year.

On some soils it is advantageous to roll the field to press in any stones which have come to the surface so as to avoid damage to harvest machinery.

6. *Nitrogen application.* There is some evidence that the use of plant growth regulators in conjunction with nitrogen increases seed yield (Young et al. 1995).
7. *Blind seed disease.* This is more frequent in perennial than in Italian ryegrass and greater care is needed to avoid fields which have shown evidence of the disease in previous years and to use only disease-free seed.
8. *Evidence of harvest readiness.* Early perennials are similar in appearance to Italians when ready for harvest, but the later cultivars remain greener. Moisture content when ready for harvest is about 2% lower in perennials compared with Italians.

Lolium x Boucheanum Kunth. Common Name: Hybrid ryegrass

Hybrids are treated either as Italian or perennial according to which species they resemble most closely (see, for example, Hides et al. 1996; Kelly 1988).

Lolium rigidum Gaud. Common Name: Wimmera ryegrass

The main area of production of this species is southwestern Australia. It is annual and is grown for seed similarly to Westerwold Italian ryegrass.

Festuca species

There are two species which are used in agriculture for leys, for grazing or hay, *F. pratensis* Huds, and *F. arundinacea* Schreb. *F. pratensis* is not as tall as *F. arundinacea* and is altogether a more slender plant.

There are also three other species which are of limited use in agriculture but are used extensively for amenity purposes; these are *F. rubra* L., *F. ovina* L. *sensu lato* and *F. heterophylla* Lam.

Festuca pratensis **Hudson. Common Name: Meadow fescue**

Classification of Cultivars (based on UPOV 1984a)

Ploidy: diploid, tetraploid, hexaploid, octoploid, decaploid
Inflorescence emergence: early, late

Distinguishing between Cultivars

Main features for distinguishing cultivars in the field:

At inflorescence emergence in second year
    Plant growth habit: erect, prostrate
    Natural plant height: short, tall
    Flag leaf length: short, long
    Flag leaf width: narrow, wide

Other features used for distinguishing between cultivars:

In autumn of year of sowing
    Plant growth habit: erect, prostrate
    Leaf colour: light green, dark green
    Leaf width: narrow, wide
    Emerging inflorescences: absent, few, many
After inflorescence emergence in second year
    Length of longest stem: short, long
    Length of upper internode: short, long
    Inflorescence length: short, long

Laboratory tests:

Ultra-thin isoelectric focusing and isozyme electrophoresis are both effective according to Hahn and Schoberlein (1995).

Environmental Requirements for Seed Production

Previous Cropping and Isolation

Meadow fescue requires similar conditions to those described for *Lolium perenne*. It is cross-fertilised and requires isolation.

Seed Crop Establishment

Establishment procedures are similar to those described for *L. perenne*, except that the seed rate should be that noted for *L. multiflorum* (diploid), namely 12 kg/ha in good conditions or 16 kg/ha when conditions are less satisfactory.

Seed Crop Management, Inspection and Roguing

Management, inspection and roguing are similar to those described for *L. perenne*. Defoliation is beneficial in the autumn, but grazing or cutting must be completed early in the spring of a harvest year, and should be omitted altogether unless growth is excessive.

Harvesting

Seed is ready for swathing when it is turning brown but retains some of its green colour, and stems are beginning to lose their green colour. Early maturing cultivars are generally less leafy than those maturing later and are therefore more suitable for combining direct. There is no published information on seed moisture contents at harvest.

Drying, Storage and Seed Cleaning

The suggestions made for *L. perenne* may be followed for meadow fescue.

X Festulolium Braunii **(K. Richt.) A. Camus. Common Name: Festulolium**

These are hybrids between *Festuca* and *Lolium* which are known to occur in nature. Recent plant breeding in the Czech Republic and the Netherlands has resulted in one or two cultivars becoming available. Seed production by such hybrids in the past has been poor.

Festuca arundinacea **Schreb. Common Name: Tall fescue**

All of the points made concerning meadow fescue (*F. pratensis*), including the distinguishing features of cultivars, apply to tall fescue except for the following points of difference.

Environmental Requirements for Seed Production

Tall fescue generally tolerates drier conditions than meadow fescue and is somewhat more difficult to establish. It is usual to take three or more seed harvests from each crop. Therefore fields should be selected which can accommodate the seed crop for at least four years (a sowing year and three harvest years) or longer.

Seed Crop Establishment and Management

Tall fescue yields best if established in rows spaced widely apart. When moisture is not limiting the rows may be 45 cm apart, but in drier conditions up to 60 cm can be used; even wider spacing may be used in exceptionally dry conditions. The seed rate is about 8 kg/ha. Spring sowing is to be preferred and a cover crop may be used provided it is not too heavy. First harvest year yields are normally greater when the crop has been sown without a cover crop; the decision of whether to use one will thus depend on the relative returns which might result. Hare (1994) reported that in New Zealand, autumn early sowing is best; delay of three weeks markedly reduced seed yield in the following year.

Tall fescue differentiates inflorescences early and benefits from autumn nitrogen. Inter-row cultivation may be used to control weeds but should not be too close to the rows as this could damage the roots.

Defoliation should be practised with great care and must on no account be done too late in the spring as this will cause severe reductions in yield. After a seed harvest, the field should be cleaned up and all excess herbage removed. Nitrogen applied at stem elongation improves seed yields (Hare and Rolston 1990).

Harvesting, Drying and Storage, and Seed Cleaning

Harvest ripeness is similar to that for meadow fescue but harvest should not be delayed as the seed sheds very easily. Andrade et al. (1994) obtained the best yields by windrowing at moisture content of 350–410 g/kg.

Other *Festuca* Species

Scientific name	Common name
Festuca rubra L. sensu lato	Red fescue, Chewing's fescue, Creeping red fescue
Festuca ovina L. sensu lato.	Sheeps' fescue, Fine leaved sheeps' fescue, Hard fescue
Festuca heterophylla Lam.	Shade fescue

Festuca rubra has creeping rhizomes and surface runners, while *F. ovina* does not; *F. heterophylla* is also without runners and is distinguished from *F. ovina* by flat stem-leaves, whereas all leaves of *F. ovina* are bristle-like (Rose 1989).

The following procedures apply to all three species except where indicated.

Classification of Cultivars, Excluding Shade Fescue

Ploidy: diploid, tetraploid, hexaploid, octoploid
Leaf blade folding: open, closed (year of sowing)
Rhizomes: absent, present
Leaf glaucosity: absent, present (autumn of year of sowing)
Awns: absent, present
Lemma hairiness: absent, present

Distinguishing between Cultivars, Excluding Shade Fescue

The following information is based on UPOV (1980b).

Main features used for distinguishing cultivars in the field:

Time of inflorescence emergence: early, late
Flag leaf length: short, long
Stem length: short, long
Inflorescence length: short, long
Red fescue only – flag leaf width: narrow, wide

Other features used for distinguishing between cultivars:

In the year of sowing
Anthocyanin colouration of leaf sheath: weak, strong
Plant growth habit: erect, prostrate
Leaf width: narrow, wide
Leaf colour: pale green, dark green

Characteristics of Shade Fescue

There are no UPOV guidelines for shade fescue; the following characteristics are listed in OECD (1995a, p. 139).

In second year at time of heading
Plant growth habit
Leaf colour
Time of inflorescence emergence
Leaf length
Leaf width
Inflorescence length
Inflorescence density
Inflorescence glume anthocyanin
Flag leaf attitude
Distribution of main foliage relative to height of the plant

Environmental Requirements for Seed Production

The turf fescues require fertile conditions for seed production. P and K requirements are similar to wheat and pH should be around 6. A regular rainfall to maintain soil moisture throughout the growing season is necessary, with dry weather at harvest. Moderate summer temperatures are desirable.

Previous Cropping and Isolation

For turf production it is essential to have seed free of all other grass seed. Therefore careful preparation of the proposed field to eliminate all grass plants (whether cultivated or weeds) is essential. At least two years will be required for this process, during which crops should be chosen which permit grass-killing herbicides to be used. There are several herbicides suitable for this purpose and the advice of manufacturers should be sought (see also Canode 1980).

The three species are cross-fertilised and adequate isolation is required: for fields of 2 ha or less, 200 m for crops to produce seed for further multiplication and 100 m for other fields; for fields larger than 2 ha these distances can be halved.

Seed Crop Establishment

The best seed yields are obtained from crops sown in rows spaced 30 cm apart with seed rate of 1 to 1.5 kg/ha. It is desirable to sow in spring without a cover crop. If a cover crop is used it is essential to use a crop species which can be harvested and removed early (for example, oilseed rape). Establishment is slow and this is a critical time for weed control; it is essential to eliminate both grass weeds and broad-leaved weeds at this stage. Treatment of the seed bed before sowing to kill all weeds, and then sowing on an undisturbed soil is one method. An alternative is to use a special drill (developed in the USA) which deposits a protective covering of charcoal 2.5 cm wide over the sown seed; this allows the use of overall spraying of herbicides which are absorbed by the charcoal so protecting the developing fescue seedlings (Copeland and McDonald 1995).

Application of N to the seed bed is desirable, although if a cover crop is used the application of N should be delayed until it is removed.

Seed Crop Management

Autumn defoliation should be early and light as the early-formed tillers will provide the best seed. An autumn application of N will stimulate tillering.

In a seed harvest year, N should be applied in the spring. The amount must be adjusted according to where the crop is situated; for example, in northern Europe 40–60 kg/ha N has been recommended while in the USA up to 112 kg/ha N has been suggested, especially for older stands. There is no advantage in splitting the application.

The fescues are early to change from the vegetative to the reproductive phase and therefore spring defoliation should be avoided unless vegetative growth is excessive, in which case it must be removed as early as possible.

Several harvests can be obtained from an established crop provided it does not become sod-bound.

Seed Crop Inspection and Roguing

The removal of off-types is not generally practicable but hand pulling of small numbers of grass weeds is desirable. However, higher numbers of grass weeds should be treated with herbicides and if necessary the field should be grazed or used for fodder for a year to allow time for their elimination.

Inspection usually takes place after inflorescence emergence but before anthesis. This allows isolation to be checked in good time.

Harvesting

Direct combining is possible but it is more usual to swath and pick up later. The crop is relatively easy to thresh and handle; the seed does not shed readily before being ready for cutting, and swathing can proceed when seed moisture content has fallen to 25%. Moisture content of the seed should be allowed to fall to 14% or below before picking the crop up.

Drying and Storage

Seed which is threshed at 14% moisture content or below can be stored for some time after pre-cleaning to remove trash. At moisture content above 14%, however, immediate drying is essential. This can be either by spreading thinly of a well ventilated floor and turning frequently or in a forced-air drier. The temperature of the drying air must not exceed 49°C when seed moisture content is 30% or above; this can be increased gradually to 54°C as the seed dries.

Seed Cleaning

Cleaning seed for use in the establishment of fine turf is a highly specialised task and is usually done by the contracting seed company. An air/screen cleaner is the main equipment used.

Alopecurus pratensis L. Common Name: Meadow foxtail

There are very few cultivars of this species and very limited areas are grown for seed, mainly in central Europe and northwestern USA.

Distinguishing between Cultivars

The following information is based on OECD (1995a, p. 113).

Distinguishing between Cultivars in the Field

> In year after sowing year at inflorescence emergence
> Time of inflorescence emergence: early, late
> Length of longest stem: short, long
> Flag leaf length: short, long
> Flag leaf width: narrow, wide
> Flag leaf attitude: rectilinear, recurved
> Distribution of foliage: well below flowering stems; not well below flowering stems

Other features used for distinguishing between cultivars:

> In year of sowing
> Plant height: short, long
> Plant growth habit: erect, prostrate
> Tendency to produce inflorescence: weak, strong
> In seed harvest year at end of flowering
> Stem length: short, long

Growing the Seed Crop

Seed crops are grown in a similar manner to those of *L. perenne*. Meadow foxtail is cross-fertilised and requires adequate isolation. The seed heads tend to ripen over an extended period and much seed may be lost through shedding. Harvest is usually by direct combining with the cutter bar adjusted so that it works above the dense foliage and takes only the seed heads. This lower growth must then be removed to allow new tillers to develop for the next year's harvest.

Dactylis glomerata L. Common Names: Cocksfoot, Orchard grass

Classification of Cultivars

The following information is based on UPOV (1984b):

Ploidy: diploid, tetraploid
Time of inflorescence emergence (second year): early, late
Silicate teeth on leaves: absent, present

Distinguishing between Cultivars

Main features used for distinguishing cultivars in the field:

At inflorescence emergence
Growth habit: erect, prostrate
Length of flag leaf: short, long
Width of flag leaf: narrow, wide
Length of longest stem: short, long
Length of upper internode: short, long
Inflorescence length: short, long

Other features used for distinguishing between cultivars:

In year of sowing
Leaf colour: light green, dark green
Leaf width: narrow, wide
Growth habit: erect, prostrate
Tendency to form inflorescences: weak, strong

Environmental Requirements for Seed Production

Cocksfoot can withstand drier conditions than ryegrass (*Lolium* spp.) but otherwise the requirements of the species are similar.

Previous Cropping and Isolation

It is usual to take up to three harvests from a cocksfoot seed crop, so it is necessary to plan cropping so that the field will be available for four years. Preceding crops should be such that there is ample opportunity to eliminate grass plants, both weeds and sown grasses.

Cocksfoot is cross-fertilised by wind-borne pollen and seed crops therefore require isolation. This is usually 100 m for fields of 2 ha or less and 50 m for larger fields when producing the final generation; for earlier generations these distances should be doubled.

Seed Crop Establishment

Inter-row spacing of between 45 and 60 cm is usual, with the wider spacing in drier areas; when seed crops are grown in very dry areas it may be advantageous to space the rows wider than 60 cm apart. The seed rate is approximately 8 kg/ha.

The best seed yields are obtained when crops are sown in the spring without a cover crop. If a cover crop is used it must not be too heavy and should preferably be one which can be harvested early, to give the seedlings time to establish strongly before the onset of winter.

A moderate application of N to the seed bed for crops sown without a cover crop will assist in the quick establishment of the seedlings. Application of N to a cover crop should not be too heavy as this may encourage a heavy crop which could retard the establishment of the cocksfoot.

Seed Crop Management

Inter-row cultivations, especially after a cover crop has been removed, are beneficial in suppressing weeds and in preventing soil capping. However, they must not be too deep nor too close to the cocksfoot plants or the roots may be damaged. An alternative method of suppressing weeds is to use herbicides, which may be applied as overall sprays or as inter-row treatments. This is usually cheaper, but does not provide the advantage of improving the condition of the soil.

Defoliation is not usually necessary in the autumn of the sowing year. However, if there is excessive leaf growth it may cause loss of seed yield if it is killed by frost; it should be removed towards the end of the year and in any case before the end of January in the following year (or equivalent in the southern hemisphere).

Spring cultivations are not normally needed but a spring top-dressing of N fertiliser in February/March (or equivalent in the southern hemisphere) should be given (about 100–150 kg/ha N in the UK). Cocksfoot does not normally lodge and responds well to N application.

After each seed harvest the field should be cleared of excess herbage and debris; if the soil is very compacted it is beneficial to cultivate between the rows, but avoiding damage to the roots of the cocksfoot plants. This work should be completed before the end of the year and no later than January of the subsequent year in the northern hemisphere. If the amount of debris is excessive it must not be allowed to smother the cocksfoot plants and must be removed as soon as possible after harvest; fields treated in this way may require a second visit.

N should be applied in the same manner as in the first year, although for very early flowering cultivars there may be some advantage in splitting the dressing and applying half in autumn and half in spring.

Seed Crop Inspection and Roguing

Roguing is not usually practicable in cocksfoot seed crops. However, there may be occasions when it is worthwhile to remove by hand small numbers of weed grasses or obviously rogue plants. The latter are normally only visible as early heading plants in a late flowering cultivar and therefore can only be rogued for a very limited time.

Inspection should take place at the start of inflorescence emergence. For crops sown in widely spaced rows, the sample areas for detailed examination should be a predetermined length of row with half the inter-row space on either side.

Harvesting

Cocksfoot sheds ripe seed readily and therefore it is necessary to watch the crop carefully as it ripens. In areas where the weather is reliable, the crop can be swathed and picked up later; however, where it is intended to use this method the rows should not be too wide apart, as this may create conditions in which the swath cannot be kept off the ground. In other areas it is better to combine direct.

Swathing should start when the seed moisture content is about 44%; most of the seed will be light brown but some still greenish and the stems below the inflorescences will be yellow to brown. Some shedding may already have occurred. It is preferable to pick up the swaths when seed moisture content has fallen to 14%, but depending on the weather it may be necessary to do this earlier and to complete the drying of the seed in store.

Direct combining can start when the seed moisture content has fallen to 30% or below. This will normally be about 10 days after the swathing stage, but combining may have to be started earlier if seed shedding becomes excessive.

Drying and Storage

Seed harvested above 14% moisture must be dried immediately, as even a few hours in store at higher moisture content can damage germination. The advice given for *Lolium multiflorum* is valid also for cocksfoot.

Seed in store must be monitored to ensure that it remains in good condition. Immediate action must be taken if there are any signs of heating or if moisture content rises.

Seed Cleaning

An air/screen cleaner will do most of the work. Specialist equipment may be needed in the final stages and this is best left to the contracting seed company.

Phleum species. Common Name: Timothy

There are two species of *Phleum* used in agriculture. *P. pratense* L. is the most widely grown in leys. *P. bertilonii* DC. is diploid and is a smaller plant; it is less used in leys but can be used in turf, especially in shade conditions. Both are grown for seed in the same manner.

Classification of Cultivars

This information is based on UPOV (1984c).

  Ploidy: diploid, hexaploid (there are currently
    no tetraploids)
  Time of inflorescence emergence: early, late (in
    year after sowing year)

Distinguishing between Cultivars

Main features used for distinguishing cultivars in the field:

  In second year at inflorescence emergence:
    Flag leaf length: short, long
    Flag leaf width: narrow, wide
    Stem length (inflorescence included): short,
      long
    Length of upper internode: short, long
    Length of inflorescence: short, long

Other features used for distinguishing between cultivars:

  In year of sowing
    Speed of inflorescence emergence: slow, fast
  In spring of second year
    Leaf colour: light green, dark green
    Leaf width: narrow, wide
    Plant growth habit: erect, prostrate

Environmental Requirements for Seed Production

Timothy prefers moister conditions and in this respect resembles ryegrass (*Lolium* spp.) rather than cocksfoot *Dactylis glomerata*. Otherwise the conditions described for cocksfoot apply equally to timothy.

Previous Cropping and Isolation

Timothy is not as easy to establish as ryegrass or cocksfoot and fields are usually maintained to produce seed for several years. The young seedlings are easily smothered by weed growth and the chosen field must therefore be clean and free from all weeds. This may involve planning the preceding crops so that herbicides and other measures can be used to eliminate weeds and volunteer plants.

  Isolation is required as timothy is cross-fertilised: for fields of 2 ha or less, 200 m when the seed to be produced is for further multiplication or 100 m for production of the final generation. These distances can be halved for larger fields.

Seed Crop Establishment

Timothy can be grown in narrowly spaced rows and will form a solid stand from inter-row spacing of 10–20 cm; however, it is often preferable to establish inter-row spaces which can be cultivated to reduce weeds, and 45–60 cm is satisfactory for this purpose.

  The seed rate is about 6 kg/ha for the wider spaced rows and 8 kg/ha when they are closer together. A cover crop can be used but it is generally better to sow in spring without. In the latter case it is advantageous to mix seed of a quick-growing plant such as lettuce or mustard with the timothy seed, sufficient to give about one plant every metre; this will show the rows earlier and enable inter-row cultivations to proceed; the young plants of lettuce or mustard can be destroyed by use of a selective herbicide.

Seed Crop Management

Seed crops can be grazed in the autumn and in the spring of a harvest year. Defoliation in spring can be about two weeks later than for cocksfoot (*Dactylis glomerata*), particularly for those cultivars which are late in inflorescence emergence. Otherwise management is similar to that described for *D. glomerata*.

Seed Crop Inspection and Roguing

Roguing is only practical when the impurities to be removed are few in number. Weed and volunteer grasses of other species can be hand pulled as soon as they can be distinguished. When growing a late cultivar it is sometimes worthwhile to remove plants which are showing inflorescences markedly earlier than the bulk of the crop.

Seed crops should be inspected at the time of inflorescence emergence.

Harvesting

It is usual to swath timothy and to pick up later with a combine. The seed threshes more easily if there are alternating moist (dew or light showers) and dry (sun and wind) conditions during the time in swath.

The crop is ready for swathing when the seed has changed colour to grey with a brownish tinge and the inflorescence is losing seed from the tip. Seed may shed very easily after this stage and it is sometimes preferable to use the combine as a swather with the concave set wide; this threshes only the ripest seed and the remainder can be picked up from the swath after a suitable ripening period.

Drying Storage and Seed Cleaning

The requirements are similar to those outlined for Italian ryegrass. The seed of timothy is very small and must be handled with care as it can easily be lost from gaps through which larger seeds would not pass.

Bromus species

There are several species of *Bromus* which are used in agriculture in leys. The two with several cultivars listed in OECD (1995b, p. 7) are:

> *Bromus catharticus* Vahl. (syn. *B. willdenowii* Kunth.). Common name: Rescue grass, Prairie grass
> *Bromus inermis* Leysser. Common name: Smooth brome, Soft brome, Bromegrass.

The OECD list also includes the following, but each has only very few or no cultivars listed:

Scientific name	Common name
Bromus arvensis L.	Field brome
Bromus biebersteinii Roem et Schult.	Meadow brome grass
Bromus carinatus Hook et Arn.	California brome
Bromus erectus Hudson.	Erect brome
Bromus sitchinensis Trin.	Alaska bromegrass
Bromus stamineus Desv.	Southern brome

Bromus inermis Leyss

Soft brome is somewhat more widely grown than rescue grass, and cultivars have been developed in North America, Japan, New Zealand and Europe. In the USA the southern group of cultivars has more vigorous underground root stocks then the northern group, and starts growth earlier in the spring. OECD (1995a, pp. 118–27) gives the following as the primary characteristics for distinguishing between cultivars.

> In second year at time of flowering
> Plant growth habit
> Plant height
> Anthocyanin at base of plant
> Tiller diameter above first node
> Tiller density
> Blistering of plant parts
> Length of uppermost internode
> Length and width of longest leaf
> Leaf colour
> Length and width of flag leaf
> Attitude of flag leaf
> Pubescence of flag leaf

Seed crops are grown in a similar manner to those of *Festuca arundinacea*. The seed is chaffy and consequently is difficult to thresh and to clean and may present difficulties at sowing time; it does not store satisfactorily for more than a year.

Bromus catharticus Vahl.

Cultivars of rescue grass have been developed in Europe, South America and New Zealand. Rescue grass prefers warmer conditions than soft brome and is listed as a 'warm season, southern grass' by Copeland and McDonald (1995). OECD (1995a, pp. 118–27) gives the same characteristics as for soft brome for distinguishing between cultivars.

Seed crops are grown in a similar manner to those of soft brome (*B. inermis*), but there is some evidence that spacing between the rows can be less: in New Zealand spacing from 15 to 30 cm between rows made no difference to seed yield, although yield was reduced with a spacing of 60 cm and by broadcasting (Hampton 1989). There is some evidence that growth regulators can be used to increase seed yield (Hampton et al. 1989).

Arrhenatherum elatius (L.) Beauve. ex J.S. et K.B. Presl. Common Name: Tall oat grass; False oat grass

Tall oat grass is mainly grown in central and eastern Europe. There are few cultivars.

Distinguishing between Cultivars

The following characteristics are listed in OECD (1995a, p. 114).

  In the sowing year
    Plant growth habit
    Plant natural height
    Leaf length
    Leaf width
    Leaf colour
    Tendency to form inflorescences
  In the second year
    Leaf colour
    Time of inflorescence emergence
    Length of longest stem
    Flag leaf length
    Flag leaf width
    Inflorescence length
    Plant growth habit

Environmental Requirements for Seed Production

The wild plant is very common in road verges and can tolerate a wide range of conditions. In agriculture it is used mainly on the lighter, less fertile soils.

Previous Cropping and Isolation

Previous crops which enable the eradication of grass weeds and volunteer grasses should be planned.

The species is cross-fertilised and requires spatial isolation.

Seed Crop Establishment and Management

Seed yields are higher from crops sown in rows wide apart (up to 1 m). Tall oat grass responds well to nitrogen. Seed crop management is similar to that described for *Dactylis glomerata*.

Seed Crop Inspection and Roguing

Roguing and inspection are similar to those described for *Dactylis glomerata*.

Harvesting

The seed sheds easily although there are now some cultivars which are less prone to this. The crop should be harvested when the stem turns yellow and before the top half of the panicle has shed. Direct combining is usual. The crop is difficult to thresh and the long twisted awns on the dorsal pales make it difficult to handle.

Drying, Storage and Seed Cleaning

Seed should be dried immediately after harvest in a manner similar to that described for cocksfoot. The seed should be handled carefully; if the pales are removed the germination will suffer.

Other Species of Grass for Poorer Soils

There are two other species from which cultivars have been bred. These are:

Cynosurus cristatus L. Common name: Crested dogstail

Holcus lanatus L. Common name: Yorkshire fog

Seed production is limited. Both are cross-fertilised and require adequate isolation.

Characteristics used for distinguishing between cultivars have been published only for crested dogstail (OECD 1995a, p. 136).

Poa species

There are several species of *Poa* which are used in agriculture and are also used for amenity purposes. However, only one of these is significant, namely *Poa pratensis* L. Six other species are listed in OECD (1995b, p. 20), all with very few or no cultivars:

Scientific name	Common name
Poa ampla Merr.	Big bluegrass
Poa annua L.	Annual meadowgrass
Poa compressa L.	Canada bluegrass
Poa nemoralis L.	Wood meadowgrass, Wood bluegrass
Poa palustris L.	Swamp meadowgrass, Fowl bluegrass
Poa trivialis L.	Rough-stalked meadowgrass, Rough bluegrass

All of these are perennial except the annual meadowgrass. Rough-stalked meadowgrass is the most frequently grown and there are nine cultivars listed in OECD (1995b, p. 21), one from the USA and the rest from Europe.

The meadowgrasses are very common, and can occur as weeds in seed crops of other grass species where they are very difficult to eliminate, particularly *P. annua* and *P. trivialis.*

Poa pratensis L. Common Names: Smooth-stalked meadowgrass, Kentucky bluegrass

Smooth-stalked meadowgrass is used in pastures, particularly for grazing by sheep; it is also a very important grass for amenity purposes.

Classification of Cultivars (based on UPOV 1990a)

Ligule pubescence: not pubescent, pubescent
Leaf sheath colour in young stage: green, red colouration
Upper leaf surface pubescence: absent, present

Distinguishing between Cultivars

Main features used for distinguishing cultivars in the field:

At beginning of flowering in second or subsequent year
Time of inflorescence emergence: early, late
Length of longest stem (including fully expanded inflorescence): short, long
Shape of rachis opposite lower side branches: straight, curved
Inflorescence colour: mainly white, mainly red
Angle of rachis branches: inclined upwards, horizontal, drooping

Other features used for distinguishing between cultivars:

Hairs on margin of leaf sheath: absent, present
Hair tuft on leaf sheath near top: absent, present
Leaf width in autumn of year of sowing: narrow, wide
Hair fringe at junction of leaf blade and sheath: absent, present
Hairs on lower leaf surface: absent, present
Collar of rachis at junction with lower branches: split, closed
Size of seed: small, large

Laboratory tests:

Electrophoresis: variation of the isoenzyme esterase using isoelectric focusing (Van Dreven et al. 1990).

Environmental Requirements for Seed Production

Poa pratensis requires fertile conditions with a well distributed rainfall for seed production. Its creeping rhizomes enable it to survive in less favourable conditions, but adequate fertilisers are needed to ensure a good seed crop.

Previous Cropping and Isolation

Grass weeds are a particular problem in this crop and previous crops should be planned to provide the best opportunity to eliminate them. At least two years are required for this, and when growing seed of early generations longer should be allowed.

Poa pratensis is normally regarded as apomictic, but cultivars differ in the extent to which this is so and may contain a proportion of plants which are cross-fertilised. Advice should therefore be sought from the breeder as to the isolation requirements for the cultivar to be grown. Most seed certification schemes assume apomixis and require only physical separation of the seed crop from other crops of *Poa* spp. However, attention should also be paid to the field boundaries to ensure that they do not harbour naturally occurring meadowgrass plants; it is good practice to cultivate a sterile strip about 1 m wide around the seed crop.

Smooth-stalked meadowgrass is somewhat susceptible to ergot (*Claviceps purpurea*) and it is advisable to avoid situations where ergot may spread into the crop from field borders. Some resistant cultivars are now available.

Seed Crop Establishment

It is useful practice to use a 'stale seed bed' technique; the field is cultivated to encourage the germination of weed seeds, particularly those of weed grasses, the seedlings are killed by the application of herbicides and the crop is then sown without disturbing the soil again by overall cultivation. Pre-emergence herbicides may also be used.

Seed crops are usually established in rows which are spaced 30 cm apart; inter-row spacing should not exceed 60 cm or reduced seed yield may result. A seed rate of 1 kg/ha is generally satisfactory.

Spring sowing without a cover crop is to be preferred. The seed is slow to germinate and growth is retarded by excessive shade. If it is necessary to use a cover crop it is advisable to choose a species which is harvested early to allow time for the *Poa* seedlings to establish strongly before the onset of winter. Autumn sowing is not generally successful.

The inclusion of seed of a quick-growing species to show the rows early (as described for *Festuca rubra*), so as to allow earlier inter-row cultivation or spraying, is often an advantage. The technique of sowing under a charcoal strip, as described for *Festuca rubra*, can also be used for meadowgrass.

Seed Crop Management

Some N can be applied to the seed bed, but it must be much reduced if a cover crop is used. A pH of about 6 is needed and both P and K status must be satisfactory. Further N is not generally necessary in the autumn of the sowing year unless the crop plants have been retarded by adverse weather or a heavy cover crop. All debris must be removed from the field as soon as a cover crop is harvested. Smooth-stalked meadowgrass is very sensitive to shading from excess herbage as inflorescences form in the spring, so any excess herbage in the autumn before a seeding year should be removed by grazing lightly. However, the best seed-bearing tillers will be those which are formed early in the autumn, so it is essential not to do anything which might discourage the development of strong tillers at this time.

N should be applied in spring of a harvest year as early as practicable. In areas with good rainfall and moderate summer temperatures 40–60 kg/ha is suitable, but in areas with less rain and higher summer temperatures up to 135 kg/ha may be required.

Poa pratensis seed crops may be harvested for up to four successive years, provided the crop does not become sod-bound, and some crops may last even longer. To maintain the crop it is essential to remove all herbage from the field immediately after harvest. This used to be done by burning, but this practice caused pollution problems and has now largely ceased. The most effective substitute is to chop the straw and stubble as close to the ground as possible with a forage harvester, which will blow all the cut material into a trailer for removal from the field; disease organisms and insects which would have been killed by burning then have to be dealt with by the application of chemicals. Subsequent management follows the same pattern as that in the autumn of the sowing year and following spring.

Seed production fields of smooth-stalked meadowgrass eventually become 'sod-bound': the plants fill in between the rows and form a dense mat. Fields which have reached this stage are not satisfactory for seed production as the seed yield is much reduced. Normally a sod-bound field is ploughed and put through a cycle of other crops; however, it is possible by cultivation (shallow ploughing and rolling) to rejuvenate sod-bound fields, but this practice is not advisable when cultivars are being grown to supply seed of high cultivar purity, because a proportion of the rejuvenated crop will be from shed seed.

Seed Crop Inspection and Roguing

Roguing of weed grasses is possible if the numbers per hectare are small. Similarly, markedly different plants (for example, plants which are showing inflorescence emergence much earlier than the rest of the crop) can be removed if they are not too numerous.

The best time for inspection is after emergence of the inflorescences but while it is still possible to distinguish those which have emerged early.

Harvesting

It is usual to harvest by swathing and picking up later with a combine. The crop is ready for swathing when the inflorescences have changed colour to yellow or brown; the seed is firm with a moisture content of about 28%.

The swath is picked up when the seed has dried to 14% or less. Picking up at higher moisture content is possible, but the seed must be dried immediately. The crop does not thresh easily and may require to be picked up twice to obtain a full yield; in the USA combines have been specially adapted to enable threshing to be more effective.

The seed is difficult to handle because it is enclosed in the paleae which are hairy at the base, and these hairs entwine so that the seeds cluster; these clusters are difficult to break up and so the seed does not flow freely.

Drying, Storage and Seed Cleaning

Seed threshed from the swath at 14% moisture content is dry enough to pre-clean and store for some time. Pre-cleaning should include breaking up the clusters of seed to enable it to be passed through the machines; this can be done in a de-awner, but care must be taken not to hit the seed too hard as this may cause loss of germination.

Seed which is above 14% moisture content must be dried immediately. This can be achieved either by spreading the seed thinly on a well ventilated floor where it can be turned frequently or on a flat-bed drier. The temperature of the drying air should not exceed 49°C if the moisture content is 30% or more and can be increased by 5°C as the seed dries. An air/screen cleaner is normally used to clean the seed.

Agrostis species

There are four species of *Agrostis* which are commonly grown for seed but they are used almost exclusively for amenity purposes for the creation of fine lawns. The four species are:

Scientific name	Common name
A. tenuis Sibth. Syn. *A. capillaris* L.	Brown top, Colonial bent
A. gigantea Roth. Syn. *A. alba* Auct.	Red top
A. stolonifera L. Syn. *A. palustris* Hudson	Creeping bent
A. canina L.	Velvet bent

A. tenuis and *A. gigantia* both have rhizomes, but the ligules of the former are broader than long whereas those of the latter are toothed and blunt but longer than broad. *A. stolonifera* and *A. canina* both have leafy runners which root at the nodes, but the former has purplish green panicles which are rather closed up whereas those of the latter are large and spreading.

The following information applies to all four species.

Classification of Cultivars

This is based on UPOV (1990b).

Ploidy: diploid, tetraploid, hexaploid
Rhizomes: absent, present
Stolons: absent, present

Distinguishing between Cultivars

Main features used for distinguishing cultivars in the field:

Time of inflorescence emergence: early, late
Flag leaf length: short, long
Flag leaf width: narrow, wide
Stem length: short, long
Inflorescence length: short, long

Other features used for distinguishing between cultivars:

In autumn of sowing year
Leaf colour: light green, dark green
Leaf width: narrow, wide
Plant growth habit: erect, prostrate
Number of flower heads formed: few, many

Growing the Seed Crop

The requirements for *Agrostis* spp. are similar to those described for *Festuca rubra* and other turf species of fescue. Stubble burning is not now generally practised because of the environmental damage caused by smoke in the atmosphere, but *A. tenuis* has given the most favourable response

to this practice while *A. stolonifera* and *A. canina* cannot tolerate it.

Koeleria cristata (L.) Pers. Syn. *K. macrantha* (Ledeb.) Schultes. Common Name: Crested hairgrass

This temperate grass species is native of chalk and limestone soils. It is included in OECD (1995b, p. 12) with one cultivar. OECD (1995a, p. 142) gives the following characteristics for distinguishing between cultivars:

At vegetative stage
Plant growth habit
Leaf colour, length, width and hairiness
At flowering
Plant height and growth habit
Time of inflorescence emergence
Flag leaf length and width
Inflorescence length

Phacelia tanacetifolia Benth. Common Name: California bluebell

This grass has recently been used as a quick-growing cover crop. OECD (1995a, p. 156) gives the following characteristics as useful for distinguishing between cultivars.

At time of bud emergence
Natural height of plant and plant growth habit
Green colour of leaf and anthocyanin colouration
Length of leaflet and division
At flowering
Time when 50% of plants begin to flower
Natural height of plant
Stem and inflorescence colour
At maturity
Length of stem including inflorescence; length of inflorescence
Stem thickness; pubescence of stem; number of internodes

Seed
 1000 seed weight
Seedling
 Ploidy number

References

Andrade, P.R., Grabe, D.F. and Ehrensing, D. (1994). Seed maturation and harvest timing in turf type tall fescue. *Journal of Applied Seed Production*, **12**, 34–45.

Booy, G., Van Dreven, F. and Steverink-Raben, A. (1993). Identification of cultivars (*Lolium* spp.) using allele frequencies of the PGI-2 and ACP-1 isozyme system. *Plant Varieties and Seeds*, **6**, 179–196.

Canode, P.P. (1980). Grass seed production in the intermountain Pacific Northwest, USA. In *Seed Production* (Ed. P.D. Hebblethwaite). Butterworths, London, 189–202.

Copeland, L.O. and McDonald, M.B. (1995). *Principles of Seed Science and Technology*. Chapman and Hall, New York, 224–227.

Hahn, H. and Schoberlein, W. (1995). Achieving variety identification of seed samples of *Festuca pratensis* Huds. *Journal of Applied Seed Production*, **13**, 49.

Halligan, E.A., Forde, M.B. and Warrington, I.J. (1991). Discrimination of ryegrasses by heading date under various combinations of vernalization and daylength: Westerwolds, Italian and hybrid ryegrass varieties. *Plant Varieties and Seeds*, **4**, 115–124.

Halligan, E.A., Forde, M.B. and Warrington, I.J. (1993). Discrimination of ryegrasses by heading date under various combinations of vernalization and daylength: perennial ryegrass varieties. *Plant Varieties and Seeds*, **6**, 151–160.

Hampton, J.G. (1989). The effect of row spacing, method and time of sowing on seed production of prairie grass (*Bromus willdenowii* Kunth.) cv. Grasslands Matua. *Plant Varieties and Seeds*, **2**, 171–178.

Hampton, J.G., Rolston, M.P. and Hare, M.D. (1989). Growth regulator effects on seed production of *Bromus willdenovii* Kunth. cv. Grasslands Matua. *Journal of Applied Seed Production*, **7**, 6–11.

Hare, M.D. (1994). New Zealand cultivars of Tall Fescue (*Festuca arundinacea* Schreb.). *New Zealand Journal of Agricultural Research*, **37**, 11–17.

Hare, M.D. and Rolston, M.P. (1990). Nitrogen effects on Tall fescue seed production. *Journal of Applied Seed Production*, **8**, 28–31.

Hides, D.H., Marshall, A.H. and Jones, M.H. (1996). Influence of spring defoliation on the potential seed

yield of tetraploid hybrid ryegrasses (*Lolium* × *boucheanum* Kunth.). *Plant Varieties and Seeds*, **9**, 77–86.

ISTA (1996). International Rules for Seed Testing. *Seed Science and Technology*, **13** (Supplement), 257–262; 268.

Kelly, A.F. (1988). *Seed Production of Agricultural Crops*. Longman, Harlow, 103–104.

Lallemand, J., Michaud, O. and Greneche, M. (1991). Electrophoretical description of ryegrass varieties: a catalogue. *Plant Varieties and Seeds*, **4**, 11–16.

McMichael, A.C. and Camlin, M.S. (1994). New methodology for the assessment of leaf colour in ryegrass (*Lolium* spp.). *Plant Varieties and Seeds*, **7**, 37–49.

OECD (1995a). *Control Plot and Field Inspection Manual*. Organisation for Economic Co-operation and Development, Paris.

OECD (1995b). *List of Cultivars Eligible for Certification, 1995*. Organisation for Economic Co-operation and Development, Paris.

Rose, F. (1989). *Colour Identification Guide to the Grasses, Sedges, Rushes and Ferns of the British Isles and North-western Europe*. Penguin, London, 112.

UPOV (1980a). *Guidelines for the Conduct of Tests for Distinctness, Homogeneity and Stability: Ryegrass*. Doc. TG/04/7, International Union for the Protection of New Varieties of Plants, Geneva.

UPOV (1980b). *Guidelines for the Conduct of Tests for Distinctness, Homogeneity and Stability: Sheep's fescue, Red fescue*. Doc. TG/67/4, International Union for the Protection of New Varieties of Plants, Geneva.

UPOV (1984a). *Guidelines for the Conduct of Tests for Distinctness, Homogeneity and Stability: Meadow fescue; Tall fescue*. Doc. TG/39/6, International Union for the Protection of New Varieties of Plants, Geneva.

UPOV (1984b). *Guidelines for the Conduct of Tests for Distinctness, Homogeneity and Stability: Cocksfoot*. Doc. TG/31/6, International Union for the Protection of New Varieties of Plants, Geneva.

UPOV (1984c). *Guidelines for the Conduct of Tests for Distinctness, Homogeneity and Stability: Timothy*. Doc. TG/34/6, International Union for the Protection of New Varieties of Plants, Geneva.

UPOV (1990a). *Guidelines for the Conduct of Tests for Distinctness, Homogeneity and Stability: Kentucky bluegrass, Smooth stalked meadowgrass*. Doc/ TG/33/ 6, International Union for the Protection of New Varieties of Plants, Geneva.

UPOV (1990b). *Guidelines for the Conduct of Tests for Distinctness, Homogeneity and Stability: Bent*. Doc. TG/30/6, International Union for the Protection of New Varieties of Plants, Geneva.

Van Dreven, F., Esselink, G. and Houuwing, A. (1990).

Esterase isoenzyme differences between cultivars of Kentucky bluegrass (*Poa pratensis* L.) from extracts using isoelectric focusing. *Plant Varieties and Seeds*, **3**, 89–97.

Young, W.C. III, Chilcote, D.O. and Youngberg, H.W. (1995). Seed yield response of Perennial ryegrass to spring applied N. at different rates of Paclobutrazol. *Journal of Applied Seed Production*, **13**, 10–15.

GRAMINEAE: PRAIRIE GRASSES

The prairie grasses are a varied group of species which are adapted to areas with relatively hot summers and very cold winters. The development of cultivars in these species has been relatively recent and for some of them specialist seed growing, as opposed to harvesting seed from old pastures, has only developed over the last 50 years. The main areas of production are in Canada and the USA in the mid-west, and in other continents with similar ecological conditions, for example, eastern Australia (Victoria, Queensland), South America (Argentina), eastern Europe (Hungary, Czech Republic) and Africa (South Africa). In Lower latitudes the tropical grasses take over and the boundaries between these and the prairie grasses are sometimes indistinct.

Agropyron species. Common Name: Wheatgrasses

There are 10 species of the wheatgrasses for each of which there are a few cultivars available. These are:

Scientific name	Common name
A. cristatum (L.) Gaertn.	Fairway crested wheatgrass
A. dasystachyum (Hook) Scribn	Northern or thickspike wheatgrass
A. desertorum (Fischer ex Link.) Schultes	Standard crested wheatgrass
A. elongatum (Host) P. Beauv	Tall wheatgrass
A. inerme (Scribn et J.G. Smith) Rydb.	Beardless wheatgrass
A. intermedium (Host) P. Beauve ex Baumg.	Intermediate wheatgrass
A. riparium Scribn et J.G. Smith	Streambank wheatgrass
A. smithii Rybd.	Western wheatgrass
A. trachycaulum (Link) Malte ex H. Lewis	Slender wheatgrass
A. trichophorum (Link) K. Richter	Pubescent wheatgrass

So far as seed production is concerned, the following points apply to all of them except where indicated.

Distinguishing between Cultivars

The following characteristics are given in OECD (1995a) and are the same for all 10 species.

> In the second year, four weeks after spring growth has started
> > Plant growth habit
> > Leaf colour
> At inflorescence emergence
> > Time of inflorescence emergence
> > Length of spike
> Just before maturity
> > Plant height
> > Stem pubescence
> > Leaf attitude
> > Leaf length and width
> > Leaf pubescence
> > Leaf glaucosity
> At maturity
> > Seed pubescence
> > Seed size
> > Seed shape
> > Awnedness

As secondary characteristics the following are listed:

> In autumn of the year of sowing
> > Plant growth habit
> > Plant vigour
> > Leafiness
> > Tendency to flower
> At flowering
> > Spikelet angle to rachis
> > Inflorescence pubescence
> At maturity: tendency for seed shattering
> Seedling: ploidy number

Environmental Requirements for Seed Production

These grasses are adapted to areas where there is a relatively short growing season. In the higher latitudes the winter is usually severe with snow and hard frost and the grass plants do not start growth until warmer weather is spring. At lower latitudes the winters may not be so severe but the summers may be drier. An adequate summer rainfall is needed for seed production; irrigation may be used in areas where rainfall is not adequate.

Previous Cropping and Isolation

There should normally be an interval of at least two years between seed crops, and longer when changing from one cultivar to another or where there is a history of weed grasses or ground-keepers from previous crops.

Nine of the 10 species listed above are cross-fertilised and require adequate isolation from sources of pollen outside the crop: this is 200 m for fields of 2 ha or less and 100 m for larger fields when the seed to be produced is for further multiplication, or half these distances when it is intended for sowing to produce a forage crop. The exception to this requirement is *A. trachycaulum* (slender wheatgrass) which is self-fertilised; for this species separation from other grass crops by a barrier or a space of 3 m is sufficient.

Seed Crop Establishment

Seed crops should be sown in rows spaced wide apart, at least 60 cm and wider when moisture may be limiting for subsequent growth. Copeland and McDonald (1995) state that tall and crested wheatgrasses are able to germinate even under conditions of moisture stress. Seed rates vary depending on the distance between the rows and species; for example, for rows 60 cm apart, 2–3 kg/ha seed of *A. intermedium* (intermediate wheatgrass) is suitable.

Seed beds must be clear of weed grasses, particularly other species of *Agropyron* (for example, *A. repens P.B.*) since their seeds are practically impossible to clean from the harvested crop seed.

Seed Crop Management

The wheatgrasses do not respond to inter-row cultivation except for a light cultivation to breakup any soil crust. Weed control is therefore best achieved by herbicides or by hand roguing of small numbers.

Seed crops may provide four or more harvests and it is essential to remove from the field any straw or plant debris each time after the seed has been harvested. Excess herbage should also be removed by grazing or cutting the field at this time. Apart from this, no further defoliation is necessary.

Wheatgrass responds to nitrogen in the spring of a harvest year. Up to 112 kg/ha have given economic returns in the USA.

Seed Crop Inspection and Roguing

Roguing is only possible when there are very few rogues or weed grasses, but is worth the effort if the seeds of the rogue plants are similar in size to those of the crop seed.

The usual time for inspection is about the time of inflorescence emergence but before anthesis; this allows time for any isolation defects to be corrected.

Harvesting

The seed ripens somewhat unevenly but also sheds easily, therefore harvest must not be delayed but should be timed to recover the majority of the first ripening seed. The species which are least liable to shedding are *A. dasystachyum, A. intermedium* and *A. trichophorum*. The crop is ready for swathing when the seed has reached the soft dough stage but is not quite hard. For direct combining the crop should be left until the seed is hard; at this stage it should be possible to knock out the seed by striking the inflorescence against the palm of the hand.

Crops which are grown in very widely spaced rows may be better if combined direct, as a swath which falls into the inter-row space will not dry satisfactorily in damp weather; picking up such crops can also be difficult.

When combining direct, the cutting table should be set as high as possible consistent with taking all the heads and leaving the bulk of the herbage for later removal.

It is also possible to harvest wheatgrasses with a stripper. This reduces the amount of material harvested with the seed.

The seed is chaffy and sometimes does not flow freely. It is generally necessary to reduce the volume of the air from the fan to avoid blowing seed out of the combine.

Drying and Storage

Seed which is threshed from the swath will normally be at 14% moisture content or below, and can safely be stored for a limited period before cleaning. However, seed which is harvested with more than 14% moisture, or which contains much green debris, must be dried immediately. Normally this can be achieved by spreading the seed thinly on a well ventilated floor and turning frequently until moisture content is reduced to a safe level.

Seed Cleaning

Any debris in the seed should be removed by precleaning. In most samples of harvested seed there will be unbroken spikelets, seed clusters or awns and it is necessary to reduce these in order to get the seed to flow freely. Pre-cleaned seed can normally be finished on an air/screen cleaner.

Andropogon species. Common Name: Bluestem

There are five species of *Andropogon* which are grown for seed on a limited scale; for each species a small number of cultivars is available but no descriptors have been published.

Scientific name	Common name
A. gayanus Kunth.	Gamba grass
A caucasius L.	Caucasian bluestem
A. geradii Vitman	Big bluestem
A hallii Hackel	Sand bluestem
A. scoparius Michaux	Little bluestem

The first of these is grown in Australia and Latin America and is a tropical grass. The others are grown mainly in the USA. Seed production of *Andropogon* follows the same lines as *Agropyron*. Caucasian bluestem is apomictic or self-fertilised; the other species are cross-fertilised and isolation should be adjusted accordingly. Adequate soil moisture is necessary during culm extension and seed filling and for adequate seed yield the temperature at this time should not be excessively high.

Andropogon grows vigorously once established and will soon fill in the inter-row spaces. However, in the early stages of growth some inter-row cultivation is desirable to keep weeds in check and prevent soil capping. Spring defoliation must be avoided. Seed crops respond well to nitrogen.

Seed maturity is usually late. At each joint on the raceme is a fertile, sessile spikelet and an infertile, stalked spikelet. The fertile spikelets should be checked after anthesis to make sure that seed is developing, as sometimes fertilisation is not effective. If there is no developing seed it is preferable to cut the crop for forage.

The harvested seed consists of the fertile spikelets together with the stalks of the infertile ones, and there may also be awns present. Consequently the seed is difficult to handle and will not flow freely. It is therefore necessary to break up the seed further in a de-awner before proceeding to other cleaning, which is usually achieved in an air/screen cleaner.

Elymus species. Common Name: Wild rye

There are two species of *Elymus* which are grown for seed to a limited extent in the northern states of the USA and in Canada. They are:

Elymus junceus Fischer. Common name: Russian wild rye
Elymus canadensis L. Common name: Canadian wild rye

Of the two, Russian wild rye is the more widely grown. OECD (1995b) gives the following descriptors for cultivars of this species:

In spring of harvest year four weeks after start
of spring growth
Plant growth habit
Leaf colour
At flowering
Time of inflorescence emergence
Leaf attitude
Spike length
At full flowering
Plant height
Leaf length and width
1000 seed weight

Secondary characteristics:

In autumn of sowing year
Plant growth habit
Tendency to flower
Leaf colour

Seed production methods are similar to those for *Agropyron*. Seed crops should be established in rows spaced wide apart and in very dry areas inter-row spaces of 2 m have been used. Russian wild rye is cross-fertilised and requires suitable isolation; Canadian wild rye is self-fertilised and requires separation from other crops by a barrier or cultivated gap of 3 m. Wild rye seed crops respond well to nitrogen.

Russian wild rye sheds seed very easily and if the crop is to be swathed it is advisable to start early when the seed is still in the dough stage; direct combining should be somewhat later but should not be unduly delayed. Canadian wild rye sheds seed less easily, allowing more latitude in harvesting.

Wild rye seed is awned and is usually de-awned during seed cleaning to enable the seed to flow more freely.

Bouteloua species. Common Name: Grama

There are several species of *Bouteloua* but only two are grown for seed to any extent. These are:

B. curtipendula (Michaux.) Torrey. Common name: Side-oats grama

B. oligostachya (Nutc.) Torrey ex A. Gray. Common name: Blue grama

Both are grown in North America in the mid-west. The former is more widely grown and is a more vigorous plant, growing to about twice the height of the latter. No descriptors for cultivars have been published.

Requirements for growing seed of grama are similar to those described for wheatgrass (*Agropyron* spp.). Both species are cross-fertilised, although some plants of *B. curtipendula* may be apomictic. Normally adequate isolation is needed: 200 m for fields of 2 ha or less and 100 m for larger fields when the seed to be produced is for further multiplication, otherwise 100 m and 50 m.

The seed matures irregularly on the spike and harvest must be timed carefully to obtain maximum seed yield. By sampling spikes from different parts of the field it is possible to identify the best time for harvest: by pinching the rachis mature seed will be forced out, and provided over 10% of the spikes contain mature seeds the field should be harvested at the first opportunity.

Direct combining is the usual method for harvest. However, the combined seed will contain much green material and it is essential to dry the seed immediately by spreading thinly on a well ventilated floor and turning it frequently.

The seed should be pre-cleaned as soon as possible. An air/screen cleaner will normally complete the cleaning.

Phalaris species

There are two species of *Phalaris* used as grass forage which are grown for seed:

Ph. arundinacea L. Common name: Reed canary grass
Ph. aquatica L. Common names: Harding grass, Phalaris

Harding grass is grown in Australia and South America, and reed canary grass in Europe and

North America. Descriptors for cultivars have been published by OECD (1995c) as follows.

Ph. arundinacesa
> In the year after the sowing year four weeks after start of spring growth
> > Plant growth habit
> > Plant height
> > Leaf colour
> At flowering
> > Time of ear emergence
> > Panicle length
> At full flowering
> > Plant height
> > Leaf length and width
> > Inflorescence shape
> > Tendency to seed shattering

Secondary characteristics in autumn of sowing year:

> Plant vigour
> Plant growth habit
> Leafiness of foliage
> Tendency to produce flower heads
> Leaf colour and attitude

Ph. aquatica
> Vegetative
> > Plant growth habit
> > Anthocyanin on leaf sheaths
> > Anthocyanin on rhizomes
> > Leaf colour
> At heading: time of inflorescence emergence

Requirements for seed production are similar to those for *Agropyron*. Reed canary grass is adapted to areas where flooding over winter may occur. It yields best if sown in rows 15–30 cm apart at a seed rate of about 2.4 kg/ha. The crop does not respond to spring defoliation and will not flower if cut too close or too late; defoliation should be avoided. The seed of *Phalaris* species shatters very easily, although this tendency has been reduced in some modern cultivars. Harvest therefore has to be timed to catch the best seed as it matures; seed will mature from the tip of the head downwards and the best time is usually when 40 to 50% of the seed has turned brown. Seed with the hulls removed gives the best germination.

Buchloe dactyloides (Nutt.) Engelm. Common Name: Buffalo grass

Buffalo grass is an important grazing plant over much of the Great Plains area of North America. No descriptors for cultivars have been published.

Seed production procedures are similar to those described for *Agropyron* spp. The stamens and styles are usually produced on different plants and so isolation as for a cross-fertilised species is required.

Buffalo grass is very vigorous and will soon fill in the inter-row spaces in a seed crop if not checked by cultivation; a solid stand will soon become sod-bound and seed yield will be reduced.

The seed is enclosed in the outer glumes which form a hard outer covering containing four to five seeds. These 'burs' are usually carried close to the rather short foliage so making direct combining difficult. However, the burs are very resistant to weathering and the seed may lie dormant inside for up to a year; it is therefore possible to retrieve burs which have dropped to the ground by use of suction harvesters.

Harvested burs will require treatment to break dormancy before the seed can be used. This is achieved by passing the burs through a hammer mill but with the mill running at a reduced speed consistent with reducing the burs without causing damage to the seed.

Eragrostis curvula (Schrader) Nees. Common Name: Weeping lovegrass

Weeping lovegrass is a native of South Africa which is now grown in Australia and the USA. Descriptors for cultivars have been published in OECD (1995d) as follows.

> At the beginning of flower emergence
> > Time of inflorescence emergence
> > Plant height
> > Plant growth habit

At full flowering
   Leaf length and width
   Leaf colour
   Culm length
   Inflorescence length
   Inflorescence density

Secondary characteristics at full flowering:

Foliage density
Leaf anthocyanin colouration
Leaf pubescence, upper and lower sides
Curling of leaf apex
Culm branching above ground level
Culm anthocyanin colouration at nodes
Culm length of uppermost internode
Inflorescence
   Length of second lowest primary branch
   Anthocyanin colouration on main axis at lowest verticil
   Number of primary branches
   Lowest primary branches arranged in a whorl
   Glands in or near axils of primary branches
   Pubescence of primary branches

Seed growing methods are similar to those for *Agropyron*. Weeping lovegrass is self-fertilised and so needs only separation from other crops by a barrier or cultivated gap of 3 m. The seed crop requires adequate moisture and does not yield well under dry conditions. Moderate dressing of nitrogen in spring of the harvest year will give a good response.

The seed ripens unevenly and it is necessary to time harvest to capture the best seed. Seed is ready for harvest when it has turned amber in colour.

Sorghastrum nutans (L.) Nash. Common Name: Indian grass

This prairie grass is tall growing and the seed matures relatively late. No cultivar descriptors have been published. Seed production methods are similar to those described for *Agropyron*.

Seed crops should be planted in widely spaced rows and the species responds well to nitrogen. Seed set depends on adequate moisture after anthesis and irrigation at this period is recommended. The species is cross-fertilised and adequate isolation is required.

The seed sheds very easily, particularly during periods of early frost. Seed ripening is irregular and it is necessary to allow some of the topmost spikelets to fall and to harvest when those in the middle are at the hard-dough stage.

Panicum virgatum L. Common Name: Switchgrass

Switchgrass grows up to 2 m tall. It occurs naturally in marshy situations and when grown for seed in widely spaced rows may require irrigation. No descriptors for cultivars have been published. Seed crops are handled in a similar manner to those of *Agropyron*.

The species is cross-fertilised and seed crops should be isolated accordingly.

The seed crop is suitable for combining direct but harvest should begin at the first signs of seed shedding. The cutting table should be set high to avoid taking too much of the foliage below the culms.

Stipa viridula Trin. Common Names: Green needlegrass, Feather bunchgrass

This species favours heavy soils and produces most seed when there is adequate moisture during seed development. No cultivar descriptors have been published. Seed production methods are similar to those described for *Agropyron*.

Green needlegrass is cross-fertilised and isolation should be provided accordingly.

The seed matures early. Seed crops are difficult to combine direct because the seed ripens from the top of the inflorescence downwards and the top may shatter before the lower seed is ripe. Swathing or using stripper headers is to be preferred.

The seed is awned and will not flow freely unless de-awned. There may be a high proportion of dormant seed after harvest, but this will reduce during storage.

Trisetum flavescens (L.) P. Beauv. Common Name: Golden oatgrass

Golden oatgrass is grown in eastern Europe; it has been introduced into the USA but is not cultivated there. The following cultivar descriptors are given in OECD (1995e).

> In summer of year of sowing
>> Plant growth habit
>> Natural height of plant
>> Leaf length and width
>> Leaf colour
> In year of sowing: tendency to form inflorescences
> In harvest year just before heading: leaf colour
> At heading when inflorescence is fully expanded
>> Time of inflorescence emergence
>> Plant growth habit
>> Length of longest culm
>> Flag leaf length and width
>> Inflorescence length

The species is cross-fertilised and requires adequate isolation. Seed growing methods are similar to those described for *Agropyron*.

Dichanthium annulatum Staph. Common Name: Marvel grass

Marvel grass is a native of northwest India form the arid and semi-arid region. Some improved cultivars have been selected and the species is included in the seed certification arrangements in India (Tunwar and Singh 1988). The isolation distances specified are 20 m for Foundation Seed and 10 m for Certified Seed production from other varieties of the same species. From fields of other *Dichanthium* species the distances are 200 m and 100 m.

References

Copeland, L.O. and McDonald, M.B. (1995). *Principles of Seed Science and Technology*. Chapman and Hall, New York, 62.

OECD (1995a). *Control Plot and Field Inspection Manual*. Organisation for Economic Co-operation and Development, Paris, 102–112.

OECD (1995b). *Control Plot and Field Inspection Manual*. Organisation for Economic Co-operation and Development, Paris, 137.

OECD (1995c). *Control Plot and Field Inspection Manual*. Organisation for Economic Co-operation and Development, Paris, 158.

OECD (1995d). *Control Plot and Field Inspection Manual*. Organisation for Economic Co-operation and Development, Paris, 138.

OECD (1995e). *Control Plot and Field Inspection Manual*. Organisation for Economic Co-operation and Development, Paris, 176.

Tunwar, N.S. and Singh, S.V. (1988). *Indian Minimum Seed Certification Standards*. The Central Seed Certification Board, New Delhi, 154–155.

GRAMINEAE: TROPICAL GRASSES

Tropical grasses are very varied. Those grown in the tropics and to some extent in the subtropics are adapted to short-day conditions. Some will not produce flowering heads if grown in areas with longer days. Some of the species are adapted to humid and others to dry conditions. As well as seed production, some species are propagated vegetatively, but this form of multiplication will not be considered here. Some prairie grasses can also be grown in the subtropics: *Andropogon gayanus* is one such which is grown in northern Australia and in parts of Latin America.

Tropical grasses produce seed irregularly and seed yields are generally low at any one harvest. Against this it is usually possible to harvest seed two or three times a year if conditions are suitable, since day length is not a factor. Also, seed is normally produced in the year of sowing unlike the perennial temperate grasses. The ripening period is extended because heading is prolonged both between and within plants, and flowering within each head is also extended. Harvest therefore has to concentrate on those heads which have a high proportion of ripe seed. In tropical areas where labour is available the heads can be harvested by hand, taking only those heads with ripe seed on each visit to the field. Where labour is not available harvesters have been developed which remove ripe seed by rotary brushes, leaving the immature heads untouched for a further visit. In some cases flowering can be made more concentrated by sowing in more closely spaced rows and the judicious use of nitrogen; these measures may establish higher early tiller density so that late tillers are suppressed.

Chloris gayana Kunth. Common Name: Rhodes grass

Rhodes grass is stoloniferous. There are several cultivars available, but descriptors have not been published. There are both diploid and tetraploid cultivars.

Environmental Requirements for Seed Production

C. gayana grows over a wide range of soil types. It is not particularly drought resistant and grows best in areas with the middle range of rainfall – neither too dry nor too wet. It is tolerant of saline soils. Seed production is best on fertile soils and the crop responds well to nitrogen.

Previous Cropping and Isolation

There should be an interval of at least two years free from Rhodes grass (whether seed, pasture or hay) before planting a seed crop. The field must be clean of other grasses, whether cultivated or weeds, and previous cropping should be planned to provide opportunity to eliminate any grass plants.

Rhodes grass is cross-fertilised and requires isolation: when the seed to be produced is for further multiplication 200 m for fields of less than 2 ha and 100 m for larger fields; for the production of other seed these distances can be halved. Diploid and tetraploid cultivars will not inter-cross but some infertility may result if pollen transfer occurs and this will reduce yield.

Seed Crop Establishment

Seed crops should be grown in rows. Rhodes grass is very tolerant of spacing between the rows and 25 cm has proved as satisfactory as 100 cm (Boonman 1972a). However, closer spacing helps to produce more uniform flowering so making harvest easier. Seed rates of 1 kg/ha of pure germinating seed are satisfactory. Any deficiencies of P or K should be corrected before sowing the crop and nitrogen should be applied to the seed bed if the field is deficient. The crop benefits from farmyard manure but it must be well rotted before application in order to kill any seeds which it might contain.

Seed Crop Management

Emerging seedlings are small and tend to be weak and therefore weed control in the early stages is very important. The first harvest can usually be taken about six months after sowing and by the end of the first year the inter-row spaces will have filled in. However, the crop responds to inter-row cultivation during the early establishment period sufficient to check weeds.

Seed crops should not be defoliated before the first seed harvest. After each harvest any crop residue must be removed by grazing, cutting and carting or (where this is permitted) burning. There is normally no need for further defoliation, although cutting not too severely can help to synchronise inflorescence development and provide a more uniformly ripening crop at harvest. However, care must be taken to avoid removing the apices of developing inflorescences as this can cause a delayed harvest and reduce seed yield.

Nitrogen should be applied after each harvest at the rate of about 100 kg/ha. Provided P and K are applied as necessary before sowing, there will usually be no need to apply more. However, a vigorous crop will remove P and K from the soil which will be taken from the field at defoliation. It is therefore advisable to be prepared to apply more in the later years of a crop; seed crops can normally continue production for five or more years if kept vigorous.

Seed Crop Inspection and Roguing

Roguing is not usually practicable, although it may be possible to hand pull small numbers of weed or volunteer grasses if these occur in the inter-row spaces during establishment.

Seed crop inspection should be after inflorescence emergence. At this time it is also worthwhile to check that the fertile florets are filling; usually only about half will be producing seed at any one harvest and if there are substantially fewer it could be advisable to take the crop for forage and seed from the next growth.

Harvesting

Entire spikelets are shed by abscission below the glumes and therefore ripe seed is not retained on the inflorescence. Seed crops may be cut when some of the early florets have dropped but the latest ones are still in the soft dough stage. For combining, the crop should be left longer and up to 10% of spikelets will have dropped. Normally this will be 20–30 days after peak anthesis.

Where labour is available the crop is suitable for hand harvesting as this allows ripe seed to be removed at more than one visit as the seed develops. The ripe seed is obtained by beating the panicles against the side of a container; this can be repeated seven to 10 days later but after this the crop should be prepared for the next harvest. Alternatively the ripest inflorescences can be cut from the plant, spread on a dry, well ventilated floor and threshed when dry. If the inflorescences are cut with a long enough stem, they may be tied into bundles and dried under cover.

Direct combining is preferable to swathing. Care must be taken to set the combine correctly so as to ensure that the maximum yield of ripe viable seed is obtained.

Strippers or brush harvesters are also used and can be set to take only the ripe seed, so permitting more than one visit, as with hand harvesting.

Drying and Storage

It is essential to dry the seed immediately after threshing as it will contain much green material. In humid areas a moisture content of under 10% is needed for storage; in dry areas 14% is satisfactory. Drying on a dry floor protected from rain is most usual. The seed does not flow freely and is therefore difficult to dry in continuous-flow driers.

Seed Cleaning

The threshed bulk will consist of fertile florets, the second florets (which are usually empty), a few rudimentary florets and debris (broken straw, leaf and dirt). The lemma is awned and usually hairy, the hairs being stiff. These factors cause difficulties in cleaning as the seeds will not flow freely and are resistant to airflow. Cleaning Rhodes grass seed is therefore a specialist operation requiring careful setting of the equipment on the basis of experience.

The cleaned bulk will contain much dormant seed but this is reduced after six to 12 months in store. Cleaned seed will normally contain about 30% fertile seed of which about 70–80% is viable.

Setaria sphacelata (Schum.) Staph et C.E. Hubb. Syn. *Setaria anceps* Staph et Massey. Common Names: Setaria, South African pigeon grass, Golden timothy grass

Setaria sphacelata is a tufted perennial, often with rhizomes, growing to about 1 m tall. Requirements for seed growing resemble those for *Chloris gayana*. No descriptors for cultivars have been published. There are both diploid and tetraploid cultivars.

Setaria is grown on a wide range of soils and in climates with medium to high rainfall in the tropics and subtropics. It flowers continuously, but flowering is accelerated by longer days and relatively cool temperatures (21°C).

The species is mostly cross-fertilised although weakly self-fertilised, consequently adequate isolation is needed. Inter-row spacing of 30–50 cm is satisfactory with a seed rate of up to 1 kg/ha (Boonman 1972b).

Anthesis extends over an even longer period than in *C. gayana* and judging the correct time for harvest is thus more difficult. However, there is a relatively constant rate of inflorescence production so that a period of relatively constant production of ripe seed should be reached (Jones and Roe 1976). Against this, weather may cause additional shedding at certain times.

The spikelets are subtended by bristles which are sterile branchlets of the rachis and remain attached to it when the seed sheds. The seed flows somewhat more freely than that of *C. gayana* and is thus rather more easy to clean.

Panicum coloratum L. Common Names: Coloured guinea grass, Small buffalo grass, Kleingrass

Coloured guinea grass is cross-fertilised and requires adequate isolation. It is grown for seed in a manner similar to that described for *S. sphacelata*. However, it sheds seed very easily and

needs close attention as harvest approaches; direct combining will usually recover more seed than swathing first. Peak seed production is reached 20–22 days after anthesis (Young 1993).

Sorghum X almun Parodi. Common Name: Columbus grass

Columbus grass should be grown for seed in a manner similar to that described for *C. gayana*. The species is cross-fertilised and should be isolated accordingly.

Sorghum sudanense (Piper) Staph. Common Name: Sudan grass

Sorghum bicolor X sudanense. Common Name: Sudan grass

The original Sudan grass came from the Sudan in Africa but has been much improved, particularly by crossing with *S. bicolor*. It is an annual and is grown extensively in the USA, and in parts of Latin America and Southern Africa. It can be grown in higher latitudes than many tropical grasses.

Sudan grass is annual and is grown for seed in a similar manner to a sorghum grain crop (*S. bicolor*). It is cross-fertilised and requires adequate isolation (normally 200 m for fields of 2 ha and under and 100 m for larger fields when the seed to be grown is intended for further multiplication; these distances are halved when producing seed for growing fodder crops). Johnson grass (*S. halepense*) is a perennial which can be a pernicious weed in Sudan grass. It was originally cultivated as a forage grass but soon became a weed as it is very difficult to eradicate. The seeds are slightly smaller than those of Sudan grass but are very difficult to remove during seed cleaning; a clean field is therefore essential for seed production.

Sudan grass is grown in rows spaced up to 100 cm apart with a seed rate of about 4 kg/ha. Sowing must be delayed in higher latitudes to ensure that the danger of frost is past and the soil has warmed. The best seed is taken from the first growth but it is sometimes possible to take an early fodder crop before laying up for seed.

The seed does not ripen evenly and it is necessary to time harvest to ensure that the seed in the main tillers is ripe. Swathing and picking up with the combine when the crop is cured is preferred to direct combining.

Panicum milaceum L. Common Names: Common millet, Proso millet

Common millet is an annual grass which is also grown as a grain crop. It is grown for seed in a similar manner to *Sorghum sudanense*. It is partially self- and partially cross-fertilised and therefore usually requires adequate isolation.

Seed crops are established in rows spaced about 20–25 cm apart and the seed rate is 10–14 kg/ha.

In some areas the crop may be harvested by hand. On a larger scale, swathing is preferred to direct combining.

Bracharia species

There are four species of *Bracharia* which are used on a limited scale for forage production. They are apomictic and generally rather shy seed producers; consequently they are often propagated vegetatively.

Scientific name	Common name
B. decumbens Staph.	Signal grass, Surinam grass
B. humidicola (Randle) Schweick.	Koronivia grass
B. ruziziensis Germ. et Evrard.	Signal grass, Ruzigrass
B. mutica (Forskal.) Staph.	Paragrass

B. decumbens is the most effective species of *Bracharia*. It is a native of tropical Africa and requires humid conditions. It is sown in rows spaced 30 cm apart or planted as divisions or stem cuttings spaced 30 cm apart in rows 30 cm apart.

Being apomictic the seed crops require only minimal isolation (a 3 m gap from other crops of similar seed size). Seed crops respond well to nitrogen and up to 150 kg/ha N should be applied on most soils; phosphate and potash status should also be corrected before sowing. Further top-dressing with N four to six weeks after establishment and after each seed harvest has been recommended in some areas.

During establishment it is important to keep the crop clean, especially from annual grass weeds, and there are some effective herbicides; advice should be sought as to the best ones to use under the local conditions. The plants will normally fill in the inter-row spaces by the end of the first year.

The seed sheds easily and the crop requires careful monitoring as it ripens. There is not much change in the colour of the crop as it ripens and it is necessary to check the hardness of the seed, which is tightly enclosed in the spikelet. This can be done by crushing the spikelet with the thumbnail or between the teeth; when a first flush of seed is found to be hard the crop can be harvested.

Signal grass is not particularly tall (between 25 and 70 cm) and is usually harvested by direct combining. The cutting table should be set low to pick up lodged heads. In some areas, suction harvesters have been used to advantage to pick up shed seed after combining.

Seed of *B. decumbens* is comparatively large and is therefore easier to handle and to clean than that of some other tropical grasses. It is very dormant after harvest and germinates best after storage for about a year. For this to be done safely the seed should be dried slowly to below 9% moisture.

Of the other *Bracharia, B. mutica* is a particularly shy seed producer and is usually propagated vegetatively. *B. humidicola* and *B. ruziziensis* are mostly grown for seed in northern Australia.

Cenchrus cilaris L. Common Name: Buffelgrass

Buffelgrass is adapted to drier areas of the tropics. It is a tall creeping perennial, growing to about 1 m high, although cultivars vary in height. The species is apomictic and requires minimum isolation. Hare and Waranyuwat (1980) recommend broadcasting in northern Thailand, but in the USA it is usually sown in rows up to 100 cm apart. The crop must not be allowed to become sod-bound, and grazing or cutting is recommended between each seed harvest. Buffelgrass responds well to burning of stubble and trash in areas where this is not prohibited. The crop responds well to nitrogen.

The crop flowers over an extended period and therefore there will be ripe and immature seed at any one time. Harvesting by hand is common to remove the ripe seed at intervals. For machine harvesting it is usual to use a beater or brush harvester which shakes off the ripe seed; several passes are made at intervals.

Buffelgrass seed is difficult to clean. The 'seed unit' is a cluster of spikelets surrounded by light fluffy bristle at the base; each unit contains up to five seeds. These units have to be broken up and the trash removed; this can be done in a hammer mill provided care is taken to set the mill to do as little damage to the seed as possible. Subsequent cleaning is by air/screen cleaner. The seed is dormant for six months or more after harvest.

Hyperrhenia rufa (Nees) Staph. Common Names: Jaragua grass, Thatching grass

H. rufa is less important than *C. cilaris*. It is a tall perennial which is apomictic. The seed has long twisted awns which make it awkward to handle.

Melinis minutiflora. P. Beauv. Common Name: Molasses grass

This species is adapted to low fertility soils but requires good drainage. It grows best in tropical and subtropical areas with medium to high rainfall and requires a photoperiod of less than 12 h to initiate flowering. It is apomictic and requires only 3 m isolation. Harvest ripeness is better synchronised than in some other species.

The seed units are small and awned, making the seed difficult to handle; de-awning before cleaning is necessary. Crops are less tolerant of stubble burning or excessive grazing than other species.

Paspalum plicatulum Michaux. Common Name: Plicatulum

This tall-growing perennial species is adapted to most soils in the medium to high rainfall tropics and subtropics. It is apomictic, requiring 3 m isolation. Plicatulum responds well to nitrogen. Harvest ripeness is reasonably well synchronised; some early florets will have shed and the seed will be hard in the ripe florets. Direct combining is suitable for this crop. The glumes are hairy but the seed presents fewer problems for cleaning than some other tropical grasses.

Urochloa mozambicensis (Hackel) Dandy. Common Name: Sabi grass

Sabi grass prefers well drained soils in the semi-humid subtropics. It is resistant to drought and to grazing. It is apomictic and requires only 3 m isolation. It responds well to nitrogen. Seed maturity is relatively well synchronised and harvest by direct combining is possible; however, seed shatters very easily and harvest must not be delayed if unfavourable conditions are forecast. Up to five seed harvests may be obtained per year. The seed is dormant when harvested and may be stored for up to a year before use.

Panicum maximum Jacq. Common Name: Guinea grass

Guinea grass is a perennial which grows from 1.5 to 3 m tall, depending on cultivar. The species is apomictic, but can cross-fertilise up to about 5%; isolation as for a cross-fertilised species may therefore be required and advice should be obtained from the breeder. The seed is small and should be sown on a well prepared fine seed bed. Established crops should be grazed or defoliated in preparation for each seed harvest. This tends to promote more uniform ripening. The crop responds well to nitrogen.

The entire spikelets containing the seed shed easily, making judging the best time to harvest difficult. When most of the inflorescences have emerged, the heads should shed seed easily if shaken; when this seed is hard the crop should be harvested without delay. The seed can best be harvested by hand; if the seed heads are harvested whole, care must be exercised to avoid taking too much leaf as this can cause problems through heating during drying. Germination is often poor.

Paspalum dilatatum Poiret. Common Names: Dallis grass, Paspalum

Dallis grass is native to tropical America. It is perennial, growing up to 1 m tall. It is grown in the humid subtropics and requires reasonably good fertility. It is predominantly apomictic, but there are some cultivars which cross-fertilise; advice on suitable isolation should be sought from the breeder. The seed heads may be attacked by ergot (*Claviceps paspali*) which can cause considerable loss. The earlier harvests in the year are usually less affected. The seed matures over a long period and it is difficult to judge the best time. Repeated hand harvesting to remove those that are mature is often practised. Alternatively, machines which shake or brush the seed heads can be used to remove the ripe seed. Direct combining is also possible but there may be much immature seed. The seeds are contained in the spikelets and complete cleaning usually requires a gravity separator. The seed is dormant and germination usually poor.

Paspalum notatum Flugge. Common Name: Bahia grass

Bahia grass is native to tropical America. It is more deeply rooted than dallis grass and quickly forms a sod. It is adapted to light-textured soils in humid areas. It is apomictic but some cross-fertilising cultivars are available; advice on suitable isolation should therefore be obtained from the breeder. Crops can be grazed to prepare an even sward which promotes more uniform flowering. The crop responds well to nitrogen. Harvest is usually by direct combining as the spikelets change colour. The seeds are tightly enclosed in the floret which develops a hard waxy coat; the seed is dormant after harvest.

Pennisetum clandestinum Hochst. ex Chiov. Common Name: Kikuyu grass

Kikuyu grass is rhizomatous and grows at high altitudes in high rainfall areas. It is very short and forms a dense sward with flowering stems only a few centimetres long so that the crop must be harvested no more than 1 cm above ground. Consequently seed bed preparation to provide a level, fine tilth is essential. It is a facultative apomict and may require isolation. Kikuyu grass can be grazed heavily before laying up for seed, and mowing is necessary to produce the right conditions for inflorescence production. It responds well to nitrogen. The seed can be collected by rotary mower set to about 1 cm above ground level; the cut material contains much leaf and stem and must be dried immediately by spreading thinly on a well ventilated floor. Swathing and picking up with the combine is sometimes possible if the stems are long enough. The harvested seed and attendant trash have to be broken up in a hammer mill after which it is relatively easy to clean in an air/screen cleaner. Dehulling before cleaning may be necessary.

Cymbopogon flexuosus (Nees ex Steud.) Wats. Common Name: Malabar grass

Confined to southern India, this grass is grown to produce an essential oil from the leaves. It is grown from seed but, once established, continues in production for about six years. Production is a village industry and no improved cultivars have been produced. There are, however, white- and red-stemmed types; only the latter is used for oil production.

Malabar grass is cross-pollinated by wind and seed crops should be isolated accordingly. Seed is usually harvested by hand.

References

Boonman, J.G. (1972a). The effect of nitrogen and planting density on *Chloris gayana* cv. Mbarra. *Netherlands Journal of Agricultural Sciences*, **20**, 218–224.

Boonman, J.G. (1972b). The effect of row width on seed crops of *Setaria sphacelata* cv. Nande. *Netherlands Journal of Agricultural Sciences*, **20**, 22–24.

Hare, M.D. and Waranyuwat, A. (1980). *A Manual for Tropical Pasture Seed Production in Northeast Thailand*. Northeast Livestock Development Project, Khon Kaen.

Jones, R.J. and Roe, R. (1976). Seed production, harvesting and storage. In *Tropical Pasture Research, Principles and Methods* (Eds N.H. Shaw and W.W. Bryan). Bulletin 51, Commonwealth Bureau of Pastures and Field Crops, Hurley, Chapter 16.

Young, B.A. (1993). Optimising seed production in Kleingrass *Panicum coloratum* L. *Journal of Applied Seed Production*, **11**, 13–19.

LEGUMINOSAE: AGRICULTURAL CROPS – TEMPERATE FORAGE LEGUMES

The legumes used for forage production in temperate climates are of various types. Some are annuals and are usually used to produce cut forage (e.g. hay or silage) with or without companion grasses. The biennials and perennials are mainly used in pastures with companion grasses but may also be used for cut forage production.

In some cases seed may be taken from a ley which is also used for other purposes; for example, *Trifolium repens* (white clover) seed may be taken from long-established pastures. However, when this is done in most cases management is designed to obtain the best possible seed yield and any forage production is regarded as a by-product.

Legumes grow in association with nitrogen-fixing bacteria (*Rhizobium* spp.). Generally the soils in temperate regions contain sufficient *Rhizobium* to supply the needs of a seed crop. However, if an area has a history of poor crop establishment this may be due to low numbers or the presence of poor strains of *Rhizobium*. In such cases it may be worthwhile to inoculate the seed before sowing. The inoculum can be obtained from specialist companies and is usually applied mixed with a finely ground peat, the mixture being wetted to produce a slurry which can be applied to the seed. This treatment must be applied just before the seed is sown as the seed will otherwise deteriorate rapidly. Mixing in a cement mixer for immediate transfer to the seed drill is satisfactory. Application in the hopper of the seed drill is sometimes used. However, using slurry in this way is inconvenient and various other methods have been tried, although none is as effective; the seed may first be wetted and then the peat/inoculum mix applied or the dry mix may be added to the seed without any water.

Trifolium repens L. Common Name: White clover

The species covers a wide range of cultivar types, from the relatively short-lived, large-leaved to the very persistent small-leaved. The former are usually classified as 'Dutch white' or 'Ladino white' and the latter as 'wild white' clover. To measure leaflet size it is recommended to take the central leaflet in the third or fourth leaf from the tip of a rapidly growing stolon within two weeks after the mean date of flowering.

Distinguishing Cultivars

Main features used for distinguishing cultivars in the field (based on UPOV 1985a):

Leaflet length: short, long
Leaflet width: narrow, wide
Frequency of white leaf marks: few many
Length of petiole: short, long
Thickness of petiole: thin, thick
Thickness of stolon: thin, thick
Time of flowering: early, late

Other features used for distinguishing between cultivars:

Leaf colour at the start of growth in the first seed harvest year.

Laboratory tests:

The picric acid test to determine the proportion of plants with cyanid glucoside in the leaves (Dayday 1955).

Environmental Requirements for Seed Production

White clover prefers soils well supplied with lime, phosphate and potash. Although some nitrogen is required during establishment, the symbiotic action of *Rhizobium* will supply sufficient during the life of the crop; nitrogen application to established white clover may restrict the amount of clover present in the sward. White clover does well on heavy soils in areas with a regular rainfall throughout the growing season, but dry weather is essential at harvest which is usually eight to 10 weeks after the field has been shut up for seed. Pasumarty et al. (1995) reported that weekly mean temperature seven weeks before anthesis influenced floret number/inflorescence and ovule number/floret (7–11°C being best) and that weekly mean temperature three to four weeks before anthesis influenced ovule fertility and pollen fertility (16–19°C); 95 h of sunshine during two weeks after anthesis and less than 51 mm rainfall during three weeks after anthesis were optimal for seed set.

Previous Cropping and Isolation

Clover seed is very persistent in the soil. Normally there should be a period of four years free from all white clover plants before sowing a seed crop. Cropping can be arranged so that herbicides can be used to eliminate any volunteer plants in preceding crops. Fields should be selected where the clover can remain for several years. Seed crops of the small-leaved type (wild white clover) can remain for 10 years and will produce seed each year. However, it is more usual to plan for four years as after this weeds can become a problem. The larger-leaved type is less persistent.

T. repens is cross-fertilised, the pollen being carried by insects, notably honey bees, and the usual isolation distances are required: 200 and 100 m for small fields under 2 ha and 100 and 50 m for larger fields, the longer distances being for crops to produce seed for further multiplication.

Seed Crop Establishment

Seed crops may be established either with or without a companion grass. A pure white clover stand will produce the greatest yield of seed but fodder production from the field will be relatively small. Sown with a companion grass which is not too aggressive, the seed yield will be somewhat less but there will be a greater amount of fodder. The choice will depend on the relative values of seed and fodder.

White clover seed is best sown in rows spaced 15 cm apart. The best seed yields are obtained from uniform but not dense stands so seed rate should not be too high and 1.5–3 kg/ha is usually adequate. If a companion grass is sown it too should be at a low seed rate; 2–3 kg/ha of meadow fescue has given good results in England. The clover and the grass can be drilled at right angles to each other to achieve a more uniform sward.

It is very important to ensure that the seed bed is fine, firm and level. White clover requires a firm base for establishment and the eventual crop to be harvested will be short, so a level surface to allow machines to work is necessary.

Seed crops are normally sown in spring, either on bare ground or under a cover crop. If a cover crop is used it must not be too aggressive as this would smother the developing clover plants. A stiff-strawed cereal from which nitrogen has been withheld can be used. The objective is a standing, not too heavy crop.

Seed crops may also be sown in late summer, but early enough to permit the clover seedlings to become established before the winter sets in. Cover crops should not be used at this time.

Seed of some local cultivars is produced from very old pastures which have been grazed consistently to produce a clovery sward. This provides a small-leaved, very persistent wild white. However, this method of production gives rather low seed yields and is now used mainly to maintain stocks for further multiplication in crops sown for seed production.

Seed Crop Management

There are several weeds which can cause problems in seed cleaning and these must be eliminated, preferably before the crop is sown. The most objectionable is dodder (*Cuscuta* spp.) and fields with a history of this parasitic weed must be avoided as it is very difficult to deal with in a crop. Other weeds which cannot be eliminated satisfactorily in a seed crop are trefoil (*Medicago lupulina*) and suckling clover (*Trifolium dubium*). Weed seeds of the following species will cause difficulty in seed cleaning and should be controlled by suitable herbicides: thistles (*Carduus* spp. and *Cirsium* spp.), mouse-eared chickweed (*Cerastium vulgatum*), fathen (*Chenopodium album*), cranesbill (*Geranium* spp.), campion (*Melandrium* spp.), forget-me-not (*Myosotis arvensis*), broad-leaved plantain (*Plantago major*), self heal (*Prunella vulgaris*), docks and sorrels (*Rumex* spp.), field madder (*Sherardia arvensis*) and chickweed (*Stellaria media*).

Seed crops sown in the spring will need attention in the autumn. Grazing when the seedlings have established will encourage the clover to spread into the gaps, but excessive defoliation at this time should be avoided; carefully timed grazing with sheep is ideal. If there are no livestock available or if the field has not been grazed uniformly, the field should be trimmed to remove any excess herbage; cut trimmings must be removed from the field to avoid smothering the developing plants. Working or grazing the field during wet periods should be avoided as this may cause poaching. If stones have come to the surface the field should be rolled.

The main factor in deciding seed yield is the number of flowers per unit area (Van Bockstaele and Rijckaert 1988). Marshall (1994) states that there is little difference between types of white clover in the numbers of flowers per unit area which they produce, and suggests that seed crop management is the most important aspect.

The flowers of white clover develop from the stolons. If the stolons are excessively shaded by leaf growth, either on the clover or on the companion grass if present, the floral primordia will be restricted in development. Therefore the crop has to be trimmed in the spring by cutting or grazing to allow light to penetrate to the stolons in the bottom of the sward. The decision when to lay up the field for seed is crucial: the first flowers to develop will provide the best seed and defoliation must therefore be timed to avoid destroying any of these. On the other hand, defoliating too early will allow further growth of leaf and the stolons will become shaded once more. The exact timing of defoliation will depend on several factors. Cultivars differ as to whether they start to develop floral primordia early or late; in areas or seasons which are wetter than usual this will be later than where conditions are dry.

Grazing by sheep is the best means of defoliation; the stocking rate should be heavy so

as to achieve quick results. Alternatively a light silage crop may be taken. After defoliation by stock, the droppings should be spread by light harrowing. The field may require rolling if stones have come to the surface. If the herbage is not required either for grazing or silage, it is best to trim the crop several times, keeping the herbage short until laying up for seed; the cut herbage can then be dispersed on the field.

After the first and subsequent seed harvests the field will again be available for grazing or other forage use, and treatment should follow a similar pattern to that described for the first year. There will usually be a greater growth of herbage so that defoliation can be a more time-consuming operation. After a few years of this treatment the sward may become dominated by grasses; if this happens, the grass can be checked by a light application of a contact herbicide (defoliant), taking care not to apply an excessive amount which could also reduce the clover.

If no seed crop is desired in any one year the field may be used for forage production. However, if forage is cut rather than grazed the vegetative growth must not be allowed to become too heavy as this can drastically reduce the clover in the sward.

Pollination by honey bees can be encouraged by placing hives in the crop. Most authorities suggest that adequate pollination can be achieved by placing one hive per hectare.

Seed Crop Inspection and Roguing

Roguing is not practicable in white clover seed crops as it is usually not possible to identify the extent of spread of individual plants. Hand pulling of low numbers of objectionable weeds may sometimes be worthwhile.

The start of flowering is the best time to check the identity of the cultivar and isolation. Special care is needed when assessing cultivar purity since the size of leaf of a particular plant may be influenced by the microclimate in which the plant is growing. Since it is usually difficult and time-consuming to assess the exact plant population in a seed crop because of the spreading growth of individual plants, standards for cultivar purity are often expressed as number of off-types per unit area rather than in percentage terms.

Harvesting

Flowering in white clover extends over a very long period and therefore seed ripening is not synchronised. The best seed is found in the earliest formed flowers and therefore harvest should be timed to recover as much of this as possible. However, harvesting may be delayed by poor weather conditions. As seed heads ripen they become brown and the seed hardens and changes to a light yellow colour. The best time to harvest is when 80% of the heads have reached this stage; however, if the weather is uncertain harvest at 70% seed head ripeness is advisable. There is a possibility that seed will germinate in the seed head in damp weather so harvesting should not be delayed; if such germination does occur it may be preferable to abandon the seed harvest and take the crop as forage.

The harvesting operation is difficult because the seed heads are not usually very far off the ground; the larger-leaved Ladino cultivars may be easier in this respect. Because of the shortness of the crop it is not often possible to combine direct and retrieve all the seed heads. Therefore the crop is usually windrowed and picked up later. It is sometimes advantageous to apply a desiccant to the windrow to hasten drying. Standing crops can also be desiccated but must be windrowed and threshed within 24 h as regrowth is rapid and this new growth can be more difficult to deal with than the old if it grows through the windrow. When crops are very short it is sometimes possible to cut them with a forage harvester or a lawn mower, collecting the cut material and threshing it immediately in a stationary combine or clover huller; material collected in this way will contain much green leaf and must be threshed immediately to avoid heating and loss of germination.

White clover seed is difficult to thresh as the seeds are tightly enclosed in the pods or 'hulls'. When using a combine, the cylinder clearance from the concave must be set close and the cylinder speed must be increased. However, great care must be taken to get these settings correct because if the hulls are hit too hard the seed can be damaged. If stationary threshers are used it is necessary to fit a second or 'hulling cylinder'; this normally works at a speed 15% faster than the threshing cylinder and the hulls must not be left too long in a hulling cylinder as this can damage germination. Hulling attachments can be obtained for most threshers.

Drying and Storage

Clover seed will heat very quickly if stored for too long at too high a moisture content. The threshed material will contain broken green leaf and other trash which can cause heating within a few hours even in a trailer. Pre-cleaning by aspiration to remove as much as possible of the green material is worthwhile as this will reduce drying time. Spreading on a well ventilated floor to a depth of not more than 10–12 cm and turning frequently will dry effectively. The seed is very small and if the floor has any cracks (for example, a wooden floor with gaps between the floor-boards) it is advisable to cover it with hessian or similar sheeting. Forced air can be used, although the bulk is often too small for this to work satisfactorily on a floor and drying in ventilated bins may be preferable. Air temperature should not exceed 10°C above ambient nor should it exceed 35°C. If continuous-flow driers are used, the air temperature should not exceed 35°C when the initial moisture content is high; this can be increased to 38°C as the seed dries.

When storing for a few months, a moisture content of 12% is satisfactory but for storage over a year 9% or less is needed. For very long-term storage the seed should be dried slowly to 5% moisture, sealed in moisture-proof containers and stored in air-conditioned rooms at 0–5°C.

Seed Cleaning

Cleaning white clover seed is a specialised operation requiring an experienced operator as machine settings are critical. Air/screen cleaners are used in the first instance, followed by specialised equipment such as a velvet belt or roll cleaner or a magnetic cleaner. Finally, a gravity separator is often used.

Trifolium pratense L. Common Name: Red clover

In the UK red clovers were traditionally divided into three groups according to the earliness of the start of growth in the spring and time of flowering. 'Broad red' or 'Double cut' clovers were the shortest lived and were normally no more than biennial, but they could be cut for fodder at least twice in the year after the sowing year. 'Single cut' might persist into a third year and could be cut twice a year, the second cut being late and smaller than the first. 'Late flowering' was the most persistent but could be cut only once a year. With modern plant breeding these groups have become less distinct and it is now usual to define two groups only, based on time of flowering. There are both diploid and tetraploid cultivars.

Classification of Cultivars

Spring growth and flowering: early, late
(Flowering: three heads per plant with open florets)
Ploidy: diploid, tetraploid

Distinguishing between Cultivars

Main features used for distinguishing cultivars in the field (based on UPOV 1985b and OECD 1995a, p. 76).

Number of stem internodes: few, many
(Longest stem, one to two weeks after flowering; lowest internode min. 0.5 cm)
Density of hairs on stem: sparse, dense
Flower colour: light red, dark red

Other features used for distinguishing between cultivars:

Two weeks after flowering
Length of longest stem including head: short, long
Stem thickness: thin, thick
Central leaflet of upper normally developed leaf below terminal flower
Length: short, long
Width: narrow, wide
White leaf marks: few, many
At maturity
Seed coat colour: yellow, bicolour, violet

Environmental Requirements for Seed Production

Red clover is a long-day plant and is very widely adapted in the temperate regions. However, late flowering cultivars which are adapted to the higher latitudes may not set seed if grown in lower latitudes. Any seed that is produced will be from the earlier flowering plants in the population and seed produced under these conditions may not be typical of the cultivar.

Red clover is reasonably tolerant of acidity (pH 5.8 to 6.2) but will not grow well on really acid soils. It does best on fertile soils which are well drained; it will not withstand waterlogging. Red clover requires moisture for growth but wet weather during flowering and at maturity can ruin a seed crop. Equally, an over-hot spell at flowering may prevent full seed development. It is reasonably frost hardy, but in areas with a late spring the earliest cultivars may not produce enough growth for a forage crop before the seed crop (see section on 'Seed Crop Management'). Red clover is pollinated by bees and an environment favourable to wild bees is preferable for seed growing, particularly of the late flowering types.

Previous Cropping and Isolation

Red clover seed is long-lived in the soil and the species is also prone to two soil-borne pathogens: *Sclerotinia* spp. (clover rot) and *Ditylenchus* spp. (eelworm). There should therefore be a long interval between seed crops and six years is normally recommended. However, if there is no history of seed-borne pathogens this period could be shortened, particularly if the seed to be produced is not intended for further multiplication. *Cuscuta* spp. (dodder, a parasitic weed) is a particular problem in red clover and its seed is very difficult to clean out of clover seed; areas with a history of dodder should not be used for seed production.

Red clover is cross-pollinated by insects, mostly bees. Seed crops therefore require isolation as for white clover (*T. repens*). Infertile florets may result if diploid and tetraploid cultivars cross-pollinate; at least 50 m separation is required to avoid seed yield reduction.

The length of the corolla tube and height of nectar in the tube will affect the effectiveness of different kinds of bees and these characteristics will differ in different cultivars. Generally, honey bees are able to pollinate the early flowering cultivars but do not have tongues long enough to reach the nectar in the long corolla tubes of late flowering cultivars. The latter are therefore largely dependent on wild bees such as *Bombus hortorum, B. ruderatus* and *B. subterraneus* and seed crops should be sited in areas with suitable nesting sites for the bees; generally these are rough areas extending from the field boundary. It is also necessary when relying on wild pollinators not to have too large a seed crop as this will place areas of the crop a long distance from the nesting sites; fields less than 5 ha are preferable. Some efforts have been made to domesticate wild bees for use as pollinators by importing queens to the field margins; however, this is not always successful. There is one wild species of bee, *B. terrestris*, which can rob nectar by cutting a hole at the base of the corolla tube and this avoids pollinating the floret; honey bees may also use these holes so that seed set is poor.

Seed Crop Establishment

Early flowering cultivars are usually sown without any companion. Late flowering cultivars give the best seed yields when sown alone, but in some areas where fodder for livestock is an important consideration, a companion grass may be added. It is an advantage if the companion grass will produce seed at the same time as the clover. Companion grasses must not be too aggressive; in the UK timothy (*Phleum pratense*) has proved satisfactory.

It is usual to sow red clover in the spring under a cover crop such as a cereal which must not be too heavy. The cover chosen should be early maturing so as to give opportunity for the young clover plants to establish well before winter.

In some areas, seed crops can be sown later in the season without a cover, but sowing must be early enough to allow time for the clover seedlings to become established before winter sets in.

It is also sometimes possible to sow without a cover in early spring and to take a seed crop in the same year. However, care must be taken to ensure that this system is used only for adapted cultivars as otherwise seed may be harvested from the earlier flowering plants in the population only.

Seed can be sown broadcast but it is generally preferable to sow in rows spaced 15 cm apart. For diploid cultivars a seed rate of 5 kg/ha is satisfactory, and for tetraploid 10 kg/ha. In some direr areas, wider spaced rows are used up to 60 cm apart provided lodging can be avoided, as lodged crops between widely spaced rows are difficult to harvest. In these wider spaced rows the seed rate can be reduced to 1 kg/ha or even less for diploids and double this for tetraploids.

Seed Crop Management

Seed crops of early flowering cultivars sown early in the season with the intention of taking seed in the same year should be allowed to grow unchecked until the seed is harvested.

The more usual practice for all classes of cultivars is to sow in the spring with the intention of taking the first seed harvest in the following year. Crops sown without a cover crop may be cut for forage or grazed, provided that the plants are allowed to establish strongly; grazing or cutting should not continue for too long to ensure strong but not rank growth before winter sets in. Crops sown under a cover crop should be allowed to grow for a while after the cover crop is harvested; when the plants are growing strongly they may be cut or grazed lightly to establish uniform growth before winter.

In the second year, early flowering cultivars are first cut for fodder or grazed as the first growth is normally too heavy to provide a good seed harvest. Late flowering cultivars, on the other hand, are taken for seed from the first growth, although a light, early fodder cut may be possible from the intermediate cultivars. After cutting or grazing, the early cultivars must be shut up for seed in time to allow the second growth to flower and set seed at a good time for harvesting (in England normal shutting up time is in late May or early June).

If late flowering cultivars are to be taken for seed a second time, the autumn and spring procedures described above are repeated. Seed yields are normally less from older stands.

Weeds regarded as objectionable in red clover seed crops are: *Carduus* spp. (thistles), *Cirsium* spp. (thistles), *Cuscuta* spp. (dodder), *Geranium* spp. (cranesbill), *Medicago lupulina* (trefoil), *Melandrium* spp. (campion), *Plantago lanceolata* (ribgrass) and *Rumex* spp. (docks and sorrels). Except for *Cuscuta* spp. and *Medicago lupulina*, weed control by herbicides is possible. For the two species mentioned it is essential to use only fields with no history of either weed.

Seed Crop Inspection and Roguing

Roguing is not generally practicable. Some very early flowering plants may be removed from late flowering cultivars and it is possible to hand pull small numbers of the objectionable weeds. However, roguing should not be relied upon to correct errors of earlier management.

Seed crop inspection should be at the start of flowering in a growth from which it is intended to take seed. Isolation can also be checked at this time.

Harvesting

Flowering may extend over several weeks and therefore seed ripening will also be over a long period. It is necessary to judge harvest when the maximum yield of good seed will be available. In general, the best seed will be from the early flowers and whenever the weather conditions make it possible, it is preferable to concentrate on harvesting this seed rather than to attempt to gain yield by waiting. At this stage the proportion of heads which appear to be ripe will be between 60 and 90%, depending on the cultivar and the local conditions. A seed head is judged to be ripe when it has changed colour to brown and the seed can be rubbed out and is hard and dark coloured.

Whenever possible the crop should be direct combined. In some cases it is advisable to desiccate the crop two or three days before combining, but only when a few days of dry weather can be expected; if a desiccated crop stands too long in showery weather it will start to grow again and this regrowth can be more troublesome to harvest than the original.

The alternative to direct combining is windrowing and picking up later; the correct time to windrow is a few days before the stage for direct combining as the seed will complete ripening during the time in the windrow. However, it is essential not to leave the windrows too long as the seed may germinate in the head in damp weather.

Red clover is easier to thresh than white clover but careful setting of the combine is required. Cylinder clearance and speed can be somewhat greater for red than for white clover but care must be exercised to avoid damage to the seed. If the crop is to be windrowed and then taken to a stationary thresher, it is advisable to use a clover hulling attachment.

The small seed of red clover flows very easily and any gaps in the casing of combines and other equipment should be sealed to prevent seed loss.

Drying and Storage

The suggestions for white clover (*T. repens*) apply also to red clover. However, the seed is somewhat larger and the quantities to be handled are also larger. Bulk drying facilities are therefore easier to use.

Wang and Hampton (1991), working with cultivar 'Grasslands Pawera', recommend a moisture content of 10% or less for storage of up to a year at ambient temperature, but only if the seed vigour is satisfactory; they found that low vigour lots had to be held at 5°C to maintain germination.

Seed Cleaning

The specialised cleaning referred to under white clover also applies to red clover.

Trifolium hybridum L. Common Name: Alsike

This short-lived perennial is grown for seed in a manner similar to that described for red clover (*T. pratense*) except for the following points.

Distinguishing between Cultivars

Two additional characteristics are listed in OECD (1995a, p. 171): anthocyanin colouration of the stem, and flower head size.

Previous Cropping and Isolation

Pollination is mainly by honey bees. One hive per 3 ha is normally sufficient. In Canada, leaf-cutting bees (*Megachile* spp.) have been effective (Fairey and Lefkovitch 1993).

Seed Crop Management

Seed is normally harvested from the first growth in spring without previous defoliation. Some autumn grazing is beneficial and excess growth should be removed from the field before winter sets in.

Trifolium fragiferum L. Common Name: Strawberry clover

Strawberry clover is perennial and is grown for seed in a similar manner to that described for white clover (*T. repens*), but differs in the following points.

Distinguishing between Cultivars

The following features are listed in OECD (1995a, p. 170).

Plant growth habit
Tendency to perennial flowering habit at beginning of flowering
Time of flowering (three heads per plant)
Colour of corolla
Length and width of central leaflet (as in white clover)
Colour of crescent and associated anthocyanin marking of leaf
Pubescence of underside of leaf
Glossiness of lateral leaves
Thickness of stolon (as in white clover)
Testa colour
Ornamentation on seed coat
1000 seed weight
Plant colour at start of growth in second year

Environmental Requirements for Seed Production

Strawberry clover is tolerant of extremes of temperature and will grow in very high summer temperatures followed by severe winters. The crop is reasonably tolerant of saline soils, although normal soils of good fertility are best for seed production.

Previous Cropping and Isolation

Several seed crops can be taken from an established crop, but strawberry clover will not set seed before the first winter after sowing. There are self-pollinating cultivars but the crop is generally regarded as cross-pollinating. Advice should be sought from the breeder as to the isolation required.

Seed Crop Establishment

Dense crops are said to be best for seed production.

Seed Crop Management

Strawberry clover withstands hard grazing, which helps to promote a dense sward. However, late autumn grazing should not be too severe as a very short crop is susceptible to winter kill.

Harvesting

As the crop ripens the calyx around each pod becomes inflated and of a light brown or grey colour. These capsules break off from the seed head and when the seed is hard and the capsules have changed colour to light brown the crop can be harvested. Harvesting too soon may result in a high proportion of shrivelled seed; too late will result in much lost seed through shedding. Handling the crop when it is slightly damp from morning dew may prevent excessive shedding.

Trifolium alexandrinum L. Common Names: Berseem clover, Egyptian clover

Berseem clover is an annual and is grown for seed in a similar manner to that described for early flowering red clover (*T. pratense*), except in the following particulars.

Distinguishing between Cultivars

The following characteristics are listed in OECD (1995a, p. 169):

    At beginning of flowering in second year
        Time when 50% of plants begin to flower
        Plant growth habit
        Plant height
        Basal branching
        Stem branching
        Stem length and thickness (as in red clover)
        Shape of central terminal leaflet
        Length and width of central terminal leaflet
            (as in red clover)
        Leaf pubescence
        Leaf colour
    At flowering
        Number of flowers
        Length of petals
        Ramifications of flower
        Length of bracts in relation to calyx
        Flower colour
        Presence or absence of peduncle
    Seed
        Shape
        Colour of seed coat (as in red clover)
        Ornamentation of seed coat
        1000 seed weight
    In year of sowing and during first winter
        Tendency to flower
        Winter hardiness

Environmental Requirements for Seed Production

Berseem clover is more tolerant of wet conditions than red clover. Although not generally winter hardy, in some areas it can be sown in autumn and taken for seed in the following year. Otherwise it is treated as an annual.

Previous Cropping and Isolation

Berseem clover is generally self-pollinated but must be visited by insects to 'trip' the flower for pollination to be effective. Thus cross-pollination is possible and isolation as for a cross-pollinating species is usually required.

Seed Crop Establishment

Seed crops can be established in rows 15 to 20 cm apart and the seed rate is higher than for red clover.

Seed Crop Management

Seed is best harvested from the first growth in spring. Crops sown in spring are harvested from the first growth. Overwintered crops may be trimmed or lightly grazed before winter but not in the following spring.

Harvesting

The crop can usually be combined direct.

Trifolium incarnatum L. Common Name: Crimson clover

Crimson clover has an elongated flower head which distinguishes it from other *Trifolium* species with globular or ovoid flower heads. It is an annual but can survive the winter.

It is grown for seed in a similar manner to the early flowering cultivars of red clover (*T. pratense*) except in the following particulars.

Distinguishing between Cultivars

OECD (1995a, p. 172) lists the following characteristics:

In the second year at budding
  Plant growth habit
  Leaf colour and hairiness
  Length and width of middle leaflet (see also red clover)
At the beginning of flowering
  Plant height
  Time when 50% of plants have flowers
At full flowering
  Stem length and thickness
  Stem hairiness
  Flower colour
In year of sowing
  Plant growth habit
  Tendency to flower
Time of maturity
Seed size
Hard seededness

Environmental Requirements for Seed Production

Crimson clover requires cool, humid weather and winters which are not too severe. It has virtually no seed dormancy and the seed will germinate when mature if there is any light rain, although some modern cultivars have been selected to incorporate some hard seededness. Soils of medium fertility are preferred as very fertile soils tend to produce too much leaf.

Previous Cropping and Isolation

The species is self-fertile but requires a visit from insects to 'trip' the flowers. Isolation as for a cross-fertilised species is therefore required.

Seed Crop Establishment

Seed crops are usually established by autumn and taken for seed in the following year. The seed is about twice the size of that of red clover and consequently seed rates must be higher.

Harvesting

Seed is taken from the first growth in spring. The plants can grow up to 1 m tall. The seed is easy to thresh but, because of the amount of green material on the plants when the crop is ripe, it is not easy to harvest by direct combine. The crop is therefore usually windrowed when 75% of the seeds are becoming hard. Windrows should be kept small to allow the green material to dry and to ensure an even feed to the combine when picking up. Although easy to thresh, the seed is difficult to hull and careful setting of the combine is required.

Seed Cleaning

Because of the difficulty of hulling, the threshed seed usually has to be hulled a second time before cleaning.

Trifolium resupinatum L. Common Name: Persian clover

This native of Mediterranean countries is not particularly frost hardy but in areas with mild winters it will produce good fodder crops well into the winter.

It is grown for seed in a similar manner to that described for early flowering cultivars of red clover (*T. pratense*), except for the following points.

Distinguishing between Cultivars (based on OECD (1995a, p. 173).

In the second year when plants begin to flower
  Plant growth habit
  Time of flowering (50% of plants with open flowers)
  Number of internodes
  Pubescence of stem
  Flower shape and colour
  Leaf colour
  Length and width of middle leaflet

In the year of sowing: tendency to flower
One to two weeks after flowering
    Stem length and thickness
        Length of central leaflet of leaf below ter-
          minal flower
        Frequency of plants with leaf marks
At maturity: pod shape
Seed coat colour

Environmental Requirements for Seed Production

Persian clover is not winter hardy but is a winter
annual, i.e. it is sown in the autumn and taken for
seed in the following year. It favours heavy, moist
soils.

Previous Cropping and Isolation

Although self-pollinated, Persian clover is visited
by insects seeking pollen and nectar, and placing
hives of bees in the crop is said to increase seed
yield. Isolation may therefore be required and
advice should be sought from the breeder of the
cultivar to be grown.

Seed Crop Establishment

Seed crops are normally sown in autumn when the
proportion of hard seeds is reduced in seed stored
from the previous harvest. The plants will not
produce a lot of growth but will grow slowly
throughout the winter. Some grazing may be
possible.

Seed Crop Management

In spring after the sowing year the crop will grow
rapidly and it is advisable to remove the first
growth by grazing or cutting. The crop should be
shut up for seed at least a month before flowering.
Defoliation before this reduces the amount of
foliage to be harvested with the seed and makes
for more uniform ripening.

Harvesting

The seed is ready for harvest when most of the
capsules have turned light brown. The capsules

swell when ripe and are easily dispersed by the
wind. Harvest must therefore be timed to secure
the best seed before it shatters. It is normally
preferable to windrow the crop and pick up later
as the seed can be protected within the windrow
and so prevented from shedding.

Other Clovers Grown for Seed in Limited Quantities

Scientific name	Common name
Trifolium balansae Boiss.	Balansa clover
Trifolium semipilosum Fresn.	Kenya clover
Trifolium versiculosum Savi.	Arrow-leaf clover
Hedysarum coronarium L.	Sulla

Balansa clover is grown in Australia, and one
cultivar is listed in OECD (1995b, p. 36).

Kenya clover is a perennial similar to white
clover (*T. repens*). The *Rhizobium* requirement is
very specific and seed usually has to be inoculated.

Arrow-leaf clover is an annual, sown in the
autumn for seed harvest in the following summer.
It is cross-pollinated, mainly by bees. Cultivar
descriptors are published in OECD (1995a, p.
174).

Sulla is adapted to Mediterranean climates, but
will tolerate some frost. It may be either perennial
or biennial. Sulla prefers deep, calcarious soils.
Cultivar descriptors are published in OECD
(1995a, p. 140).

Medicago sativa L. and *Medicago X varia* T. Martyn. Common Names: Lucerne, Alfalfa

Lucerne is the name used in most of Europe,
South Africa and Oceania; alfalfa is used in
Spanish- and Portuguese-speaking countries,
Canada and the USA. Lucerne is the most
widely grown forage crop. It is perennial and,
depending on the situation, may be cut up to 12
times each year for fodder. Various crosses have
been used in the production of interspecific

hybrids (*Medicago X varia*), the main one being with *M. falcata*. Apart from the usual objectives of plant breeding – improved yield and quality of forage – major developments from plant breeding have been improved winter hardiness and resistance to bacterial wilt.

Cytoplasmic male sterility has been found in lucerne and efforts have been made to produce hybrid cultivars. However, the hybrids which have been produced have been poor seed yielders and this has prevented their widespread use. Most modern cultivars are synthetic in the sense that they result from the cross-pollination of a selected number of parent plants.

Distinguishing between Cultivars

The following is based on UPOV (1988a) and OECD (1995a, p. 156).

Cultivars are classified as early, medium or late depending on the time when growth starts in the spring, which influences the time of flowering. Early cultivars are usually more erect in growth habit than late cultivars. Late cultivars start growth about four weeks later than the early ones, are later to flower and consequently are later to harvest. There are a few cultivars which have a more of less creeping habit with rhizomes.

Main features used for distinguishing cultivars in the field:

Flower colour
  Frequency of plants with very dark blue/ violet flowers
  Frequency of plants with variegated flowers

Other features used for distinguishing between cultivars:

At early flower bud: width and length of central leaflet
At full flowering: length of longest stem, including head

Environmental Requirements for Seed Production

A dry climate in an area where there is adequate irrigation water available is the best for seed production. Lucerne is deep rooting and requires a deep, well drained, fertile soil for best results; it will not do well on wet soils or soils which have been badly compacted. It requires frequent light rain or irrigation to give uniform development (little and frequent water application). Production of seed in general requires rather less water than production of fodder but it is important to ensure that growth is not restricted by stress at any time. Temperature above 20°C during flowering and seed ripening is preferable. Hacquet (1990) reported that temperature between 25 and 30°C increased flower fertility, pollinator activity and seed set, while excessive water or drought stress reduced seed yield. Soil should be well provided with major nutrients (Ca, P and K).

Previous Cropping and Isolation

Lucerne seed is long lived and can survive in the soil for several years. A period of four years free from lucerne should be maintained between seed crops. For earlier generations in the multiplication cycle, this period should be increased to at least six years. During these interim periods the cropping should be arranged so that it is possible to suppress any lucerne plants which establish. If authentic seed of the same cultivar is used to sow preceding lucerne crops, these periods can be shortened. Dodder (*Cuscuta* spp.) is parasitic on lucerne and infected fields must be avoided for seed production as dodder seed is most difficult to remove from lucerne seed.

Lucerne is mainly cross-pollinated (see, for example, Brown and Bingham 1984) and for successful pollination the flowers have to be tripped by insects. For small fields of 2 ha or less the isolation requirements are 200 and 100 m, and for larger fields 100 and 50 m, the greater distance in each case being for the production of seed intended for further multiplication.

Pollination is by bees. Honey bees are widely used as pollinators but are not very efficient; when collecting pollen they usually trip the flowers to effect pollination but when collecting nectar they may manage to avoid tripping. Pedersen et al. (1975) recommended a density of one bee per square yard (0.84 m^2) when collecting pollen but 10 times this density when collecting nectar.

Wild bees have been used as pollinators in some areas by encouraging their presence by providing artificial nest sites to attract them. In some cases it has been possible to transfer a wild population to a prepared site.

In the more temperate climates, *Bombus terrestris* (the humble bee) is a useful pollinator. This species favours rough grass at the edges of fields for nesting. Generally *B. terrestris* does not fly great distances and hence is more effective in smaller fields. Copeland and McDonald (1995) mention also *Melissodes* spp. but state that they too are not very efficient pollinators.

In warmer areas there are two other species which have been domesticated in some degree for use as pollinators. *Megachile* spp. (leaf-cutting bees) nest in small holes in wood (e.g. trees) either using existing holes or boring them themselves. Artificial nesting sites can be provided by erecting shelters containing pieces of wood in which holes are drilled to a depth of 15 cm; alternatively plastic drinking straws may be used, placed horizontally in bundles, but plastic is not suitable in humid conditions. In humid conditions the holes should not be as deep, so as to ensure adequate ventilation. The bees create a nest in these holes by cutting pieces of leaf which are used to line the hole; eggs are then laid, provided with a supply of nectar and pollen, and the hole sealed by more leaf pieces. Several layers of eggs may be laid and sealed in this way if the hole is deep enough. The bees remain in the nest when the temperature is below 20°C and are most active when the temperature rises above 30°C; they are therefore only suited to the warmer areas. In the USA, colonies of leaf-cutters can be purchased and are relatively easy to establish in artificial nests sited in seed production fields. A density of 1235 female bees/ha is regarded as satisfactory (the females do most of the foraging). These colonies can be stored over winter in suitable conditions by removing the cells from the holes. There are some predators (beetles and chalcid wasps) which can reduce colonies and these can be removed during the storage period. In spring the

cells are taken to the field and encouraged to reoccupy the nest site by placing them in trays next to the prepared holes; they should be prevented from drying out by covering them (e.g. with sawdust).

The other bee is the alkali bee (*Nomia melanderi*) which is more difficult to manage than the leaf-cutter but is a very effective pollinator. This species nests in moist loams with a high salt content on a bare, level site. Artificial sites have been prepared and cores containing bees transferred from an established site. However, these efforts have met with only limited success. Alkali bees do not like rain and only nest in areas with warm, dry weather.

Seed Crop Establishment

The best seed is produced from vigorous plants which are not too large; the aim should be to provide plants which are regularly spaced so as to allow maximum development without stress at any time. Generally this is best achieved by sowing in rows which are separated by a space suitable for the available moisture in the soil during the growing season; the drier the conditions the wider should be the inter-row spaces. In temperate areas with well distributed rainfall or well supplied irrigation water, inter-row spacing can be between 35 and 60 cm, but in drier areas up to 150 cm or more may be required.

For seed production it is preferable to sow without a companion grass. The aim should be to establish plants spaced about 10 cm apart in the row. Normally this can be achieved with a seed rate of 1 kg/ha in rows 90 cm apart, increasing to 3 kg/ha for rows 35 cm apart. Higher seed rates may be required when the conditions are less than optimum.

In temperate areas such as Europe, lucerne seed crops are usually sown in spring under a light cereal crop or in the warmer areas under a sunflower crop (*Helianthus annuus* L.). In more favourable conditions, as in western parts of the USA, it is possible to sow in autumn and achieve a satisfactory seed yield in the following year. In the latter area it is also possible to sow in spring without a cover crop and to obtain a seed yield from the first growth in the same year.

Seed Crop Management

Lucerne is deep rooted and soils which are supplied with nutrients to a good depth are required. Any deficiency of calcium, phosphate or potash should be corrected before sowing. A seed crop will normally continue in production for five years or longer; therefore any loss of P or K should be made good in later years after establishment.

Seed is best taken from the first growth in spring. In some areas it is possible to take a forage cut in spring before taking the second growth for seed; however, this can result in a seed harvest in late autumn or even early winter and can only be practised where conditions are likely to be favourable at that time.

The best seed is produced from strong plants which have formed a rosette, thus defoliation must avoid damage to the crowns of the plants. Crops sown in the spring should be trimmed to a uniform height just above the level of the rosette in autumn, and this will normally provide enough foliage for fodder. At other times, when a forage cut precedes the growth for seed production, the same rule applies: cutting must not be so low as to damage the plant. When seed is taken from the first growth in spring, the harvest will normally be early enough to allow a forage cut to be taken in the autumn or early winter.

Weed control is essential as lucerne will not compete well in the seedling establishment stage. Once established, the crop will compete well so the aim must be to achieve early weed control. Measures appropriate for forage crops are equally appropriate for seed crops. The most objectionable weeds in seed crops are: *Cuscuta* spp. (dodder), *Amaranthus* spp. (pigweed), *Brassica* spp. (charlock, mustard, rape), *Cenchrus* spp. (sandbur), *Convolvulus arvensis* (bindweed), *Galium aparene* (cleavers), *Melilotus* spp. (sweet clover), *Rumex* spp. (docks and sorrels), *Sida* spp. (mallow) and *Sorghum halepense* (Johnson grass).

There are several insects which attack lucerne and they should be controlled by the application of suitable insecticides using the treatments recommended for forage crops. However, there are also some which attack the flowers and seeds and these may require control measures additional to those for a forage crop. Examples are *Lygus* bugs, chalcids and crickets. For these latter insects control is usually necessary during the period that the flowers are being pollinated by bees and it is essential to avoid spraying insecticides at times when the bees are active; for example, spraying at night may be needed. Using insecticides during the flowering period should be avoided whenever possible; in some areas it is possible to delay flowering to a period when insects are less active by taking appropriate forage cuts.

Recent work suggests that the use of growth regulators may increase seed yield (see, for example, Askarian et al. 1994).

Seed Crop Inspection and Roguing

Roguing to improve cultivar purity is not usually possible unless the rogue plants have a distinctly different flower colour or are markedly taller than the crop. Removal of weeds is sometimes possible provided they are not too numerous. Patches of dodder should be pulled up and burnt on the spot, together with the infected lucerne plants; it is essential to prevent the dodder from seeding.

Seed crop inspection for quality control should take place when the crop is in full flower, when obvious impurities can be seen and isolation can be checked.

Harvesting

The plants in a seed crop will flower over an extended period and so the seed will not ripen all at the same time. The seed is contained in small, spirally twisted pods. Harvest has to be timed so as to secure the maximum amount of good seed. Inevitably some seed will shed and there will be some unripe seed still on the plants; harvest can begin when three-quarters of the pods have changed colour to brown or dark brown and before too much shedding has occurred.

Because ripening is not uniform, it is not usually possible to combine the crop direct without some pre-treatment. The most usual is to windrow and pick up after the crop has dried. The more uneven the flowering the more likely it is that this will be the preferred option because some seeds may ripen while the crop is in the windrow. The alternative is to apply a desiccant and then to combine the desiccated crop direct when it has dried. Using this method, the standing crop can be allowed to ripen for a longer period before treatment but it is essential to combine as soon as the desiccated crop is dry (usually after two to three days) otherwise shedding may take place, or in damp weather secondary growth may occur.

In either method the combine can be used as soon as the leaves are dry (below 20% moisture) even though the stems may still be somewhat green and have a higher moisture content than the leaves.

Combines require careful setting for harvesting lucerne seed. They are able to thresh the seed well but care must be taken to avoid damage to the seed by threshing too hard. The seed is small and flows very freely and may leak from the combine in places which are designed for the larger seeds of cereal crops, for example; such places should therefore be carefully sealed before using the combine for lucerne seed.

After combining, the straw and chaff must be removed from the field. If there are patches where harmful weeds or insects have affected the crop and these are not too numerous, they should preferably be burnt *in situ* provided that there are no local restrictions which would prevent this. Burning at this stage will not harm the lucerne plants.

Drying and Storage

Seed must be dried immediately after combining otherwise germination may be reduced. For storage up to one year, 12% moisture is satisfactory but for longer periods moisture should be reduced to below 8%. When drying seed with an initial moisture content above 20% the tempera-ture of the drying air should not exceed 30°C; as moisture content falls below 20% the drying air temperature may be increased to a maximum of 38°C.

Seed Cleaning

Cleaning lucerne seed is a specialised task. An air/screen cleaner followed by a gravity separator is usual; a roll separator and magnetic separator are also required to remove weed seeds. The roll separator will remove dodder but may not do so entirely, and also gives rise to much loss of good seed during cleaning; it is preferable to ensure that there is no dodder in the field.

Inoculation of the seed with *Rhizobium* is often required. Seed which has been infected by stem eelworm (*Ditylenchus* spp.) should be fumigated; this is a highly specialised operation and must be done by a specialist service.

Onobrychis viciifolia Scop. Common Name: Sainfoin

Sainfoin is in many ways similar to red clover (*T. pratense*) but is much more open in growth and so less well able to compete with weeds, especially in the early establishment stage. Simmonds (1976) states that the species is native to Europe and western Asia and is nutritious and very palatable; in England sainfoin hay was considered a very good fodder for race horses. Sainfoin has been introduced to North America. However, in recent times sainfoin has not been able to compete with the modern cultivars of red clover or lucerne.

There are two types of sainfoin: giant sainfoin is a biennial which will not survive more than three years; single-cut sainfoin is perennial.

Distinguishing between Cultivars

The following are based on OECD (1995a, p. 135).

Main features used for distinguishing between cultivars in the field:

At flowering
   Plant height
   Time of flowering
   Flower colour

Other features used for distinguishing between cultivars:

At the vegetative stage
   Plant growth habit
   Seedling height at 12 weeks
   Plant canopy
   Size of secondary leaf
   Leaf colour
   Leaf hair type

OECD also suggests seed tannin content as a further characteristic.

Martiniello (1992) found height and smoothness of shoot, standard petal profile and seed colour as the most effective characteristics.

Environmental Requirements for Seed Production

Sainfoin will not do well on wet soils; good drainage is essential. Soils high in calcium are preferred and in England it has been mostly grown on the chalk soils. It resists drought well after it has established beyond the seedling stage.

Previous Cropping and Isolation

Fields should have been free of sainfoin for at least four years before establishing a seed crop. Previous crops should be planned so as to give opportunity to eliminate weeds as far as possible. Where seed to be produced is intended for further multiplication, an interval of six years is advised.

Sainfoin is self-incompatible and therefore cross-fertilised. Isolation as recommended for red clover (*Trifolium pratense*) is also suitable for sainfoin. Pollination is mostly effected by honey bees and from two to 10 hives per hectare have been recommended. To achieve a good set of seed it is estimated that each flower should be visited by a bee on at least three occasions.

Seed Crop Establishment

The seed is usually under-sown in a cereal crop in spring. The cereal crop must not be too heavy as the sainfoin seedlings are slow to establish. Autumn sowing is not suitable.

Crops are sown in rows 15–20 cm apart and the seed rate is about 50 kg/ha of hulled seed or 80 kg/ha if the seed has not been hulled. Hulled seed will normally germinate better than unhulled.

Seed Crop Management

In the autumn of the sowing year the crop should not normally be defoliated (either by cutting or grazing), although if there is rapid development of foliage after the nurse crop has been removed and the growth is excessive it may be trimmed or lightly grazed; any cut foliage should be removed from the field.

The biennial type is cut for hay in the spring of the year after the sowing year, and seed is taken from the second growth. The perennial type is taken for seed from the first growth in spring; the best seed yields for this type are normally obtained in the second or third harvest years and in the first year only hay is taken or the crop grazed. The perennial type can continue to produce good yields for seven or more years and may be taken for hay one year and returned to seed production thereafter if desired. However, care should be exercised to ensure that there is not excessive seed shedding during the hay harvest.

Harvesting, Drying, Storage and Seed Cleaning

These operations follow a similar pattern to that described for red clover (*Trifolium pratense*).

Ornithopus species

The most commonly grown species in Europe is *O. sativus. O compressus* L. is used to a limited extent in Australia; it can be difficult to establish because of the very high percentage of hard seed, although once established this characteristic may be an advantage in providing continuity. Fu et al. (1994) describe 25 accessions of different species.

Ornithopus sativus Brot. Common Name: Seradella

Seradella is an annual which is native to south-west Europe and northwest Africa and is grown for fodder mainly in southern and central Europe. In recent years it has been introduced to New Zealand where at least one cultivar has been bred. Seradella is an annual fodder crop.

Distinguishing between Cultivars

OECD (1995a, p. 153) lists the following for use at the beginning of flowering:

Plant growth habit
Time of flowering
Density of hairs on stem
Flower colour
Leaflet: shape, density of hairs, colour

Pod length is also listed as a useful characteristic at maturity.

Environmental Requirements for Seed Production

Seradella requires a Mediterranean climate but not too hot in summer. It does best on sandy soils which are adequately supplied with moisture.

Previous Cropping and Isolation

An interval of three years free from seradella is recommended before establishing a seed crop.

Seradella is mostly self-fertilised and therefore requires only a clear gap or physical barrier from other crops.

Seed Crop Establishment

When mature, the pods break into short segments, each containing one seed; seed for sowing is usually still enclosed in these pod segments but they may have been removed by hulling. Hulling can, however, damage germination. The seed rate for unhulled seed is between 20 and 40 kg/ha, depending on germination of the lot and on the conditions at time of sowing.

Since seradella is very weak-strawed it is sometimes grown for seed with a companion crop of rye to provide support at harvest.

Inoculation with *Rhizobium* will be needed in situations where Seradella is not a regular crop.

Seed Crop Inspection and Roguing

Roguing is not of great practical value unless the off-type plants are very different in some obvious feature from the plants of the cultivar. The most usual time for effective roguing is when flower colour can be seen.

Field inspection is best conducted at the same stage of growth as that described for roguing.

Harvesting

Seed sheds easily and harvest therefore has to be judged to secure the maximum yield of mature seed. Because of this it is usually best to swath the crop so as to retain shed seed within the swath and to allow time for the completion of maturation while in swath. The crop can be picked up by combine after a few days in swath.

Drying, Storage and Seed Cleaning

Seed is usually dry enough at harvest to store for some time. However, if moisture content is above 10% the seed should be dried to this level for storage to the next sowing season. The seed is normally left in the pod segments and cleaning in an air/screen cleaner, possibly followed by a gravity separator, is generally satisfactory.

Seed treatment against *Colletotrichum trifolii* is sometimes necessary.

Vicia species used as Fodder Crops

There are three species of *Vicia* which are used for fodder production:

Scientific name	Common name
Vicia pannonica Crantz.	Hungarian vetch
Vicia sativa L.	Common vetch, Tare
Vicia villosa Roth.	Hairy vetch

Of these, the most widely grown is *V. sativa*. All are annuals, although sometimes sown in the autumn in favourable areas.

The description of methods for *V. sativa* (common vetch) applies also to the other two species except where stated.

Distinguishing between Cultivars

The following are listed in UPOV (1988b) and OECD (1995a, p. 50) for *V. sativa*:

Seedling: length/width ratio of leaflet of second primary leaf

During vegetative growth: anthocyanin colouration of leaf axil and of stipule nectaries

At time of flowering
    Hairiness of upper stem internodes
    Shape of tip of leaflet: concave, convex
    Colour of standard petal: white, pink, violet
    Time of 30% open flowers

After flowering: hairiness of pods, seed size

At maturity, characteristics of the seed
    Size and shape
    Ground colour of testa: white, grey-brown, grey-green, blue-black
    Brown ornamentation: presence and extent
    Blue-black ornamentation: presence and extent
    Colour of cotyledons: coffee, pink-violet, orange

Martiniello (1992) lists the following characteristics: height, colour and growth habit; attitude of the shoot; axial length; surface and leaflets of the leaf; inflorescence number; standard petal profile and calyx sepals shape; length and shape of pod; smoothness and segmentation of the marginal leaf; seed colour; and colour of the hilum.

For the other two species, which are given in OECD (1995a, pp. 178–179) only, the list is somewhat different:

Stage of growth	Vicia villosa	Vicia pannonica
Seedling (three/four weeks old)	Stipule shape, colour marks	
10% plants flowering	Time	
	Height of main stem	
	Leaflet tip shape	
	Leaflet hairiness	
	Leaflet shape, length, width	
At flowering	Colour of standard	Colour of flower
	Colour of inner keel	Main stem height
	No. Flowers/inflorescence	Stem hairiness
	Peduncle length v. leaf	Leaflet length/width
		Leaflet shape
		Leaflet tip shape
		Leaflet hairiness
At beginning of ripening: pod	Shape	
	Pubescence	
	Number of seeds/pod	
	Tip shape	
	Constrictions between seeds	
seeds	Shape, size	
	Colour	
	Colour of hilum	
	Colour of cotyledons	

Environmental Requirements for Seed Production

V. sativa is an annual which can be sown in autumn in the warmer areas. *V. villosa* is more winter hardy and is more adapted to autumn sowing. Stress during growth can cause failure to set seed, so a uniform distribution of rain or irrigation is necessary. Best results are obtained on soils rich in lime and well supplied with phosphate.

Previous Cropping and Isolation

Vetch seed can remain viable in the soil for long periods and an interval free from vetch of at least three years is recommended for the final generation, and double this for the earlier generations. Tares (common vetch) can be an objectionable weed of arable land. Previous crops should be planned to make possible the eradication of any ground-keepers from previous crops.

V. sativa and *V. pannonica* are almost entirely self-fertilised and therefore only require separation from other crops by a gap of 3 m or a physical barrier. *V. villosa* is cross-fertilised and requires adequate spatial isolation; for fields of 2 ha or less, 200 m and 100 m are needed for crops of earlier generations and Certified Seed production respectively; for fields of more than 2 ha the distances are 100 m and 50 m.

Bees are advantageous to aid pollination and two hives per hectare should be provided, especially for *V. villosa*.

Seed Crop Establishment

Seed crops are normally drilled without a companion crop in rows 15–20 cm apart. Some cultivars have weak stems and it is sometimes advantageous to sow with a companion crop such as beans (*V. faba* L.) to provide support; the large bean seeds are easily separated after harvest. However, it is necessary to ensure that the two species ripen at the same time, which can be difficult, so this practice is not recommended unless absolutely necessary.

Seed rate is between 20 and 40 kg/ha depending on seed bed conditions.

Seed Crop Management

Management of a seed crop does not differ from that for a fodder crop except that the first growth in spring must be left to set seed.

Seed Crop Inspection and Roguing

Roguing when the crop is in flower to remove plants with different flower colour may be worthwhile in the earlier generations. However, crops are not easy to walk through as the vetches become entwined and it is difficult to distinguish individual plants. Roguing should not therefore be relied upon.

Seed crop inspection should also take place when the crop is in flower. Because of the difficulty of identifying individual plants, cultivar purity standards are often set as number of off-types per unit area rather than in percentage terms. OECD (1995b, p. 141) gives one/30 m^2 for Basic Seed and one/10 m^2 for Certified Seed.

Harvesting

Crops should be examined carefully prior to harvest to ensure that the pods are well filled. If growing conditions have been unfavourable pollination may not have been effective and the crop should be cut earlier for fodder. The lower pods ripen first and it is generally best to concentrate on these by timing harvest to secure this seed in good condition. *V. sativa* and *V. villosa* shed seed very easily, so harvest of these species should not be delayed. *V. pannonica*, on the other hand, does not shed seed so readily and harvest may be delayed until the majority of seeds are ripe.

Seed crops are normally harvested by direct combining. For *V. sativa* and *V. villosa*, combining can begin when the seed in the lower pods has hardened and has changed colour; seed in the upper pods will still be soft and the pods green. It is necessary to complete combining in three to four days, otherwise seed will be lost through shedding. For *V. pannonica*, 90% of the seed should be ripe and there is less urgency to complete combining.

Crops can be windrowed and picked up by combine after about a week, but this method is not normally as effective as direct combining and much seed may be lost while the crop is in the windrow.

It is important to set the combine carefully to avoid damage to the seed, especially for *V. pannonica* which is harder to thresh.

Drying, Storage and Seed Cleaning

Seed from the combine will usually contain much green material and should be dried as soon as possible; pre-cleaning to remove this green material before drying is advantageous. The drying procedures outlined for white clover (*Trifolium repens*) can also be followed for vetch.

Vetch seed is larger than that of the clovers and is therefore more easily handled, otherwise the procedures are the same. Seed may be infected with vetch weevil (*Bruchus branchialis*) and such seed can sometimes be cleaned out as it is generally lighter in weight; severe infections can only be dealt with by fumigation, for which specialist operators should be employed.

References

Askarian, M., Hampton, J.G. and Hill, M.J. (1994). Effect of Paclobatrazol on seed yield of lucerne (*Medicago sativa* L.). *Journal of Applied Seed production*, **12**, 9–13.

Brown, D.E. and Bingham, E.T. (1994). Selfing in an alfalfa seed production field. *Crop Science*, **34**, 1110–1112.

Copeland, L.O. and McDonald, M.B. (1995). *Principles of Seed Science and Technology*. Chapman and Hall, New York, 229–236.

Dayday, H. (1955). Cyanogenesis in strains of white clover. *Journal of the British Grassland Society*, **10**, 266–274.

Fairey, D.T. and Lefkovitch, L.P. (1993). Pollination of *Trifolium hybridum* by *Megachile rotundata*. *Journal of Applied Seed Production*, **11**, 34–38.

Fu, S.M., Hampton, J.G. and Hill, M.J. (1994). An investigation of seed characteristics in serradella (*Ornithopus* L.) accessions. *Plant Varieties and Seeds*, **7**, 127–133.

Hacquet, J. (1990). Genetic variability and climate factors affecting lucerne seed production. *Journal of Applied Seed Production*, **8**, 59–67.

Marshall, A.H. (1994). Seasonal variation in the seed yield components of white clover (*Trifolium repens*). *Plant Varieties and Seeds*, **7**, 97–106.

Martiniello, P. (1992). Morphological aspects involved in varietal registration and the production of certified seed of typical Mediterranean leguminous forage crops. *Plant Varieties and Seeds*, **5**, 71–82.

OECD (1995a). *Control Plot and Field Inspection Manual*. Organisation for Economic Co-operation and Development, Paris.

OECD (1995b). *List of Cultivars Eligible for Certification*. Organisation for Economic Co-operation and Development, Paris.

Pasumarty, S.V., Higuchi, S. and Murata, T. (1995). Environmental influences on seed yield components of white clover. *Journal of Applied Seed Production* **13**, 25–31.

Pedersen, M.W., Bohart, G.E., Marble, V.L. and Klostermeyer, E.C. (1975). Seed production practices. In *Alfalfa Science and Technology* (Ed. C.H. Hansen). American Society of Agronomy, Madison, Wisconsin.

Simmonds, N.W. (1976). *Evolution of Crop Plants*. Longman, London, 313–314.

UPOV (1985a). *Guidelines for the Conduct of Tests for Distinctness, Homogeneity and Stability: White clover*. Doc. TG/38/6, International Union for the Protection of New Varieties of Plants, Geneva.

UPOV (1985b). *Guidelines for the Conduct of Tests for Distinctness, Homogeneity and Stability: Red Clover*. Doc. TG/01/4, International Union for the Protection of New Varieties of Plants, Geneva.

UPOV (1988a). *Guidelines for the Conduct of Tests for Distinctness, Homogeneity and Stability: Lucerne*. Doc. TG/06/04, International Union for the Protection of New Varieties of Plants, Geneva.

UPOV (1988b). *Guidelines for the Conduct of Tests for Distinctness, Homogeneity and Stability: Common Vetch*. Doc. TG/32/6, International Union for the Protection of New Varieties of Plants, Geneva.

Van Bockstaele, E.J. and Rijckaert, G. (1988). Potential and actual seed yield of white clover varieties. *Plant Varieties and Seeds*, **1**, 159–170.

Wang, Y.R. and Hampton, J.G. (1991). Seed vigour and storage in 'Grasslands Pawera' red clover. *Plant Varieties and Seeds*, **4**, 61–66.

LEGUMINOSAE: FORAGE LEGUMES FOR WARM CLIMATES

The following species are adapted to regions with temperate climates but with warm summers (typically a Mediterranean climate). Some are adapted to particular areas which may experience cold winters. Some of the species used in North America are used mainly for soil conservation.

Lotus species

There are three species of *Lotus* which are used in agriculture:

Scientific name	Common name
L. corniculatus L.	Bird's-foot trefoil
L. tenuis Waldst. et Kit ex Willd.	Slender bird's-foot trefoil
L. uliginosus Schk.	Greater bird's-foot trefoil

Of these, *L. corniculatus* is the most widely grown. All are perennial. *L. corniculatus* and *L. tenuis* are not stoloniferous and produce a crown but the latter has many more slender stems; *L. uliginosus* produces spreading stems which can root at intervals.

The comments which follow apply to all three species except where stated.

Distinguishing between Cultivars

The following characteristics are listed in OECD (1995a, pp. 146–9) as applicable to all three species.

   Commencement of flowering in second year (three heads with at least one flower open).
   At flowering
      Whole plant: growth habit, height, horizontal spread
      Stem: colour, hairiness, solidity
      Leaflet: length and width of terminal, distance of widest part below tip, shape of terminal, colour, pubescence, veins on lower surface
      Flower: average number per head, length from base of calyx to keel tip, basal colour, colour of tingeing, intensity of flowering
      Calyx: length and attitude of teeth, pubescence, shape of upper sinus
   At maturity
      Pod: number per head, length, style
      Seed: 1000 seed weight, shape, colour
   Seedling
      Ploidy number

Laboratory tests:

   Gardiner and Forde (1988) describe differences in the electrophoretic pattern of bulk seeds which distinguish between the species, but report less clear differences between cultivars.

Environmental Requirements for Seed Production

L. corniculatus is the most drought resistant of the three species and is less tolerant of poor drainage. *L. uliginosus* is more tolerant of wet soils and *L. tenuis* of soil salinity. The latter also tolerates higher summer temperatures than the other two.

Previous Cropping and Isolation

Trefoil seed can remain viable in the soil for several years and an adequate interval free from all trefoil crops is required. When producing seed for further multiplication this interval should be at least five years, but can be reduced to three when the seed to be produced is intended for sowing forage crops.

The trefoils are cross-fertilised, generally by bees, and the provision of two hives per hectare is recommended. Isolation distances for crops to produce seed for further multiplication should be 200 m for fields of 2 ha or less and 100 m for larger fields. When the seed to be produced is intended for the sowing of forage crops these distances can be halved.

Seed Crop Establishment

Establishment is slow and therefore weed competition in the early stages of growth in particular should be reduced to a minimum. Some effective herbicides are available against such weeds as convolvulus (*Convolvulus arvensis*) and docks and sorrels (*Rumex* spp.). In a seed crop it is also necessary to avoid fields which contain plants of other forage legumes such as clovers and lucerne, which have seeds which are difficult to separate from trefoil seed. A clean seed bed is therefore essential.

The aim is to establish a uniform stand. Drilling is preferable to broadcasting and the rows may be 15–60 cm apart depending on expected moisture availability. However, closer spacing gives plants which have a shorter range of flowering time and hence improves evenness of ripening. Seed rates vary but generally 3–5 kg/ha is adequate.

Autumn sowing is not recommended and spring sowing is usual. A cover crop may be used, but must not be too heavy as the seedlings take some time to establish. Nowak et al. (1993) state that it may be possible to improve seedling vigour by plant breeding. Sowing with a companion grass is possible but will generally result in a reduced seed yield although improving the value of any forage harvest.

Hard seed content is generally high and scarification before sowing is sometimes worthwhile.

Inoculation with *Rhizobium* may be needed. A different strain is required for *L. ulginosus* from that for the other two species.

Seed Crop Management

The crop should be trimmed in autumn of the sowing year to provide a uniform stand going into the winter for seed harvest in the following year, and grazing at this time will enhance seed yield. However, when trimming the crop, grazing or taking a forage cut, care must be exercised not to defoliate the plants too close to the ground as this can damage the crowns of the plants which will provide the flowering stems next spring.

Seed harvest is usually taken from the first growth. However, in some areas it may be desirable to provide for harvesting at a later, more favourable period by taking an early forage cut or by grazing; the time from shutting up the field to seed harvest is about two months, but as flowering may also be controlled by day length, shutting up should not be unduly delayed.

Trefoils are perennial and five or more seed harvests are usually taken from an established crop. Management in subsequent years is similar to that described for the first year.

High fertility is not desirable as this may produce weak, rank growth which does not produce good seed yields and is more difficult to harvest. However, adequate phosphorus and potash are needed. *L. uliginosus* will tolerate higher acidity and lower phosphorus levels.

Seed Crop Inspection and Roguing

Flowering time is recommended as the best for roguing and crop inspection. At this time isolation can also be checked.

Harvesting

Trefoil is one of the more difficult legumes to harvest for seed because flowering, and therefore ripening, takes place over a long period, although some cultivars have been selected for more uniform ripening. The pods dehisce quickly and much seed can be lost through shedding. Generally the first ripening seed will be of the best quality and the timing of harvest should concentrate on this seed. The plants remain green as the seed ripens and must be watched closely; the pods change colour first, becoming brown or black, and the seed hardens changing to yellow-brown or yellow-green.

Windrowing is preferred to direct combining as this will allow some more seed to ripen in the windrow. Defoliants can be used and will reduce the amount of green material to be dealt with. However, regrowth can be rapid and harvest should proceed within a week of defoliant application. Windrowing and picking up when the crop is slightly damp (e.g. from dew) can help to reduce shedding losses.

Drying and Storage

Requirements are similar to those described for *Trifolium repens* (white clover). The stems and leaves will still be green during harvest and therefore the combined seed will contain much broken green material; pre-cleaning before drying to remove as much as possible of this material is usually worthwhile. However, drying must not be delayed as the seed will heat very quickly.

Seed Cleaning

The requirements are similar to those described for white clover (*Trifolium repens*).

Lespedeza species

According to Langer and Hill (1991) there are three species of *Lespedeza* used in the USA. The perennial *L. cunenata* (Dum.) G. Don. (sericcea lespedeza) is cross-fertilised and is adapted to eroded soils of low fertility. The other two are annuals and are mainly self-fertilised: *L. stipulacea* Maxim. (Korean lespedeza) is shown in OECD (1995b, p. 28) with one cultivar, but *L. striata* (Thunb.) Hook et Arn. (striata lespedeza) is not mentioned. The following comments are therefore confined to *L. stipulacea*.

Lespedeza stipulacea Maxim. Common Name: Korean lespedeza

No descriptors have been published internationally for cultivars of Korean lespedeza.

Environmental Requirements for Seed Production

Lespedeza is adapted to areas with hot summers and cold winters and is grown mainly in the south of the USA, particularly in the east. It grows well on poor soils of low fertility.

Previous Cropping and Isolation

AOSCA (1971) requires an interval of five years free from lespedeza for early generations, decreas-ing to two years for the final generation of production. Dodder (*Cuscuta* spp.) is very objectionable in lespedeza seed crops and fields which are free from this weed should be chosen.

Lespedeza is mainly self-fertilised and AOSCA (1971) requires only 2 m isolation. However, as well as the flowers which do not open and are therefore wholly self-pollinated, there are also some flowers which do open and are visited by bees; provision of better isolation may therefore be desirable for the earlier generations. Setting two hives of honey bees per hectare in a seed crop is said to improve seed yield.

Seed Crop Establishment

The crop is usually sown in spring and is established in rows. The seed rate for unhulled seed is about 25 kg/ha.

Seed Crop Management

The first growth after sowing can be taken for seed and if there are any signs of stress (loss of lower leaves) this course is desirable. However, if the first growth is very heavy, a forage cut may be taken before the seed harvest. Crops should not be cut too low as some live buds must be left below the cutting point to allow regrowth. Beuselinck and McGraw (1994) advise that in Missouri, cutting for fodder after mid-June will reduce seed yield.

The crop does not compete well with weeds and control is essential. Johnson grass (*Sorghum halapense*) and ragweed (*Ambrosia* spp.) are particularly objectionable.

Seed Crop Inspection and Roguing

Flowering time is the best time for both crop inspection and roguing.

Harvesting

Lespedeza seed sheds very easily as the crop ripens. The crop is usually combined directly and harvest should not be delayed as the first ripening seed is likely to be the best. If weed control has not been successful, combining will be more difficult and the cylinder speed will have to be increased. However, care must be taken to ensure that the seed is not damaged.

Drying and Storage

The seed is usually harvested with a moisture content low enough for storage. However, if there is much green material in the bulk, it is essential to pre-clean and dry immediately. The seed is left unhulled. The dormancy period is short; the seed does not store well and will lose germination after one year.

Seed Cleaning

An air/screen cleaner is usually all that is required. If dodder is present a dodder mill will be needed (velvet roll mill).

Melilotus species

There are three species of *Melilotus* which are used in agriculture:

Scientific name	Common name
M alba Medikus	White sweet clover
M. officinalis (L) Palles	Yellow sweet clover
M. indica (L.) All	Indian sweet clover, Senji

The first two are listed in OECD (1995b, p. 32). They are grown in the USA and Canada, whereas the third is grown mainly in India, but it is bitter and therefore less palatable to livestock. The following comments apply to all three species except where stated.

Distinguishing between Cultivars

The following characteristics are given in OECD (1995a, pp. 151, 152) as applicable to white and yellow sweet clover (not for Indian):

In sowing year: plant growth habit, leaf colour

In second year at beginning of growth: plant growth habit and vigour, leaf size and colour

In second year at flowering: plant height, stem thickness, time of flowering, flower colour and size

Environmental Requirements for Seed Production

White and yellow sweet clovers are widely adapted to temperate climates. There are winter hardy cultivars of white sweet clover but it is less drought resistant and requires irrigation in dry areas. Yellow sweet clover is more tolerant of dry conditions. Indian sweet clover is grown under irrigation in dry areas of India on neutral or slightly alkaline soils.

Previous Cropping and Isolation

Generally, five years free from sweet clover are required before sowing a crop to produce seed of the earlier generations, but an interval of two years is sufficient for the production of seed for sowing a fodder crop.

Both white and yellow sweet clover are cross-fertilised, although there can be a significant degree of self-fertilisation in white sweet clover. Indian sweet clover is mostly self-fertilised. For white and yellow sweet clover, isolation distances should be 200 m for fields of 2 ha or less and 100 m for larger fields when producing seed of earlier generations; for the later generations these distances can be halved. For Indian sweet clover, isolation distances of 50 m for earlier generations and 25 m for the later generation are considered satisfactory (Tunwar and Singh 1988).

Pollination is by insects, mostly bees, and from two to 25 hives per hectare have been reported to improve seed yields.

Seed Crop Establishment

Crops should not be too thick since rank, tall growth will inhibit pollination and will be much more difficult to harvest. Drilling in rows 15–100 cm apart is usual: the drier the situation, the wider the spacing. At the closer spacing a seed rate of 15 kg/ha is required, while in rows spaced further apart 3–4 kg/ha will be sufficient. Companion grasses may be used and will improve the value of any fodder harvested from the crop; however, seed yield may be reduced.

Seed crops are usually sown in spring for harvest the following year, but there are some cultivars which are not winter hardy and are harvested in the sowing year. The best seed yields are obtained from crops which have been grazed in the sowing year, but stock should be removed to leave 15 cm of uniform growth as winter approaches.

Seed Crop Management

Sweet clover is very vigorous and will compete well with weeds provided that it is sown into a clean seed bed. The best seed yields are obtained from the first growth in spring. When seed is harvested in the sowing year this presents no difficulty but when seed is taken in the year after the sowing year this growth can be very vigorous and may grow to 120 cm high. This vigorous growth can be difficult to harvest and it may be advisable to graze lightly or take an early forage cut to reduce the amount of vegetation. However, this must not be done too late or seed yield may be very much reduced.

Seed Crop Inspection and Roguing

Inspection and roguing are most effective when the crop is in full flower. However, crops can be very dense at this stage and may be difficult to walk through. Leaving access gaps at sowing time may be worthwhile in the earlier generations of production.

Harvesting

There is a wide range in time to maturity and there can be as much as two months difference between cultivars.

Seed crops are usually windrowed and picked up by combine when dry. The seed sheds very easily and every effort should be made to retain the shed seed in the windrow. The crop is ready for windrowing when 60% of the pods have changed colour to brown or black. At this stage the stems are still sappy and few of the leaves will have withered and fallen. In very dry areas crops may be harvested at a slightly earlier stage.

Direct combining is possible after chemical desiccation but it is essential to combine within three days of applying the chemical, otherwise secondary growth will probably occur and this will be more difficult to deal with than the crop in its original state.

Seed shedding is least when the crop is damp, therefore it may be best to concentrate harvest in the early morning, avoiding the hottest part of the day.

Drying, Storage and Seed Cleaning

The combined seed will usually contain much green debris which will heat very quickly if not dealt with. Pre-cleaning to remove this trash is worthwhile as this reduces drying time. The comments on drying made for white clover also apply to sweet clover.

There will be a high proportion of unhulled seed as it comes from the combine and this can be reduced by scarification. This will reduce hard seed and improve germination provided it is done carefully.

Seed storage and cleaning are as described for white clover (*Trifolium repens*).

Medicago lupulina L. Common Name: Black medic trefoil

Black medic trefoil is of minor importance. It is adapted to calcareous soils in temperate areas. The species is biennial and is sown one year for producing seed the next. OECD (1995a, p. 150) lists the following characteristics for distinguishing between cultivars; two cultivars only are shown in OECD (1995b, p. 29):

In year of sowing: plant growth habit, tendency to flower, leaf colour

At full flowering: plant growth habit, height of main bulk of foliage, leaf colour, shape of middle leaflet, flower colour and size

Excess herbage should be removed before winter in the sowing year. Grazing in the spring of the following year will promote stem production (and hence increase potential seed yield). The plant is generally low growing and normally does not exceed 40 cm in height, but in dry conditions may grow to only 8–10 cm. The species is self-fertilised and seed crops need isolation by a gap of only 2 m. The first flowers occur on the lower part of the stems and these seeds may shed before the pods produced later are ready for harvest. Ripe pods are black; the seed is small and difficult to handle. The harvest period is short and the crop must be watched carefully during the ripening period which lasts only about a week. It is usual to windrow and pick up within two days so as to reduce seed loss. The seed is closely enclosed in the hulls which should be removed in a scarifier before cleaning.

Aestragalus cicer L. Common Names: Chick-pea milkvetch, Cicer milkvetch

Cicer milkvetch is of minor importance and no descriptors for cultivars have been published internationally. A firm fine seed bed is needed as the seedlings are slow to establish. The seed rate is 3–6 kg/ha in rows 60–90 cm apart. The species is cross-fertilised and therefore adequate isolation is needed. Pollination is best achieved by humble bees, although honey bees are also effective. Seed shedding is not a problem but the seed is difficult to thresh; the crop is normally harvested by windrowing and picking up by combine, but the windrow must be allowed to dry thoroughly and the combine will need careful

adjustment with a high cylinder speed. It may be economic to pick up and thresh the swath a second time.

Coronilla varia L. Common Name: Crown vetch

OECD (1995a, p. 134) lists the following characteristics for distinguishing between cultivars:

Twelve weeks after seedling emergence: plant growth habit, seedling height, canopy size, leaf colour and hair type, size of secondary leaf

At flowering: plant height and growth habit, time of flowering, flower colour

Seed: tannin content

Crown vetch is of minor importance and, being perennial, is used mainly for purposes of soil conservation in difficult situations. It is grown in North America and in eastern Europe. The species is cross-fertilised and requires adequate isolation; the best pollinating agents are humble bees, but honey bees are also effective. Adequate isolation should be provided. The flowering stems grow to a height of about 90 cm with a dense mat of foliage below reaching 30 cm high. The seed sheds easily and the crop is normally combined direct, with the cutting table set to avoid as much as possible of the foliage consistent with taking as much of the seed as possible. The harvested seed requires dehulling and may also require to be scarified to reduce hard seed content. Both operations must be done with care as the seed coat is easily damaged.

References

AOSCA (1971). *AOSCA Certification Handbook*. Pub. No. 23. Association of Official Seed Certifying Agencies. Raleigh, North Carolina, 53.

Beuselinck, P.R. and McGraw, R.L. (1994). Management of annual lespedeza for seed and herbage. *Journal of Production Agriculture* **7**, 230–232.

Gardiner, S.E. and Forde, M.B. (1988). Identification of cultivars and species of pasture legumes by sodium dodecylsulphate polyacrymlamide gel electrophoresis of seed proteins. *Plant Varieties and Seeds*, **1**, 13–26.

Langer, R.H.M. and Hill, G.D. (1991). *Agricultural Plants*. Press syndicate of the University of Cambridge, Cambridge, 247.

Nowak, J., Pillay, V.K. and Papadopoulos, Y.A. (1993). *In vitro* assessment of seedling vigour in birdsfoot trefoil, *Lotus corniculatus* L. *Plant Varieties and Seeds*, **6**, 161–168.

OECD (1995a). *Control Plot and Field Inspection Manual*. Organisation for Economic Co-operation and Development, Paris.

OECD (1995b). *List of cultivars Eligible for Certification*. Organisation for Economic Co-operation and Development, Paris.

Tunwar, N.S. and Singh, S.V. (1988). *Indian Minimum Seed Certification Standards*. The Central Seed Certification Board, New Delhi, 151–152.

LEGUMINOSAE: FORAGE LEGUMES FOR VERY DRY AREAS

These species have a unique method of survival. Although annuals and not particularly drought resistant as plants, they are able to survive over many years by reseeding themselves. The crop is established in the autumn and will survive over winter requiring a cold period to induce flowering, although not frost hardy. A reasonable winter rainfall is necessary (about 500 mm). As summer drought approaches the plants flower and set seed in inflorescences which also comprise sterile upper florets. The calyces of these sterile florets are reflexed and come together to form a burr with a series of barbs on the outside enclosing the seed pods. The burrs are borne on long hairy stems which spread out from the plant, and as the seed matures the stems bend over to deposit the burr on the soil surface, or in some cases they may force the burrs into the surface of the soil. In this state the species is able to survive without moisture until the next time that conditions are favourable for germination. The seeds are provided with varying periods of dormancy so that they do not germinate all at the same time, and thus survival will be ensured for several years.

This unique method of survival by reseeding presents particular problems for the seed grower, since once a seed crop is established the field will continue indefinitely to produce further crops. It is not possible to use generation control as a means of preserving genetic purity as the plants in crops after the first will be of differing generations depending on the dormancy period of the seed. For this reason, OECD has agreed a special scheme for these species: the OECD Scheme for the Varietal Certification of Seed of Subterranean Clover and Similar Species Moving in International Trade (OECD (1995a). This Scheme differs from other OECD Schemes: it is confined to self-pollinating species; Basic Seed must be produced on the basis of generation control from the parental material; Certified Seed of the first generation is so designated in the first year after crop establishment but subsequent harvests must be designated as 'Mixed Generations'. However, some cultivars, particularly of subterranean clover

(*Trifolium subterraneum*), have specific marker characters and in the case of Certified Seed do not have to be of direct descent from either Basic or Certified Seed, provided it is produced from a crop which did not contain more than 5% of plants not showing the specific marker characters.

Trifolium subterraneum L. Common Name: Subterranean clover

Distinguishing between Cultivars

The following characteristics are listed in OECD (1995b):

> At cotyledon stage: presence/absence of pale green mark, anthocyanin flecking
> At vegetative stage
>   Leaf: pattern of pale central mark, pattern of associated white bands, hairiness of upper surface
>   Petiole: hairiness
>   Stipule: pigmentation
>   Runner: hairiness
>   At flowering: time of flowering, calyx pigmentation
>   At maturity: colour of seed coat, isozyme analysis

Environmental Requirements for Seed Production

Subterranean clover is grown extensively in South and Western Australia where most of the development work on this species has been carried out. It is also grown in South America and, to a more limited extent, in southern Europe. The climatic conditions required for satisfactory growth are outlined above.

A fertile soil with an adequate supply of calcium, phosphorus and potash is best for seed production.

Previous Cropping and Isolation

Wherever possible, new cultivars should be sown on land which has not previously carried any subterranean clover. However, as the plant occurs extensively in areas where it is grown, it is usual to allow six years free from crops of the species, during which time a rigorous regime of control of volunteer plants is exercised. There are herbicides which can be used for this purpose.

The crop is self-pollinating so that a clear gap of 3 m is all that is required.

Seed Crop Establishment

Seed is sown in the autumn for harvest in the following year. A pure stand is best for seed production; companion grasses may be sown but may suppress the clover if not carefully controlled, although they will improve the grazing value. The seed rate is 6–10 kg/ha in rows spaced close together. It is essential that the soil be carefully prepared to provide as flat a surface as possible.

Seed Crop Management

Some grazing may be possible in early winter, especially if a companion grass has been sown. In spring, the first growth is normally taken for seed unless there is excessive vegetation which may be trimmed early. However, dense stands provide the best possibility for successful harvest as the seed will be kept further from the soil surface. Management must be directed towards maintaining a flat, firm soil surface. Stones should either be collected and removed or rolled well into the soil; moles or other underground animals should be discouraged. The aim is to keep as much seed as possible above ground so that it can be recovered at harvest.

Grazing by sheep to control herbage growth except when laid up for seed is satisfactory. However, during drought after seed harvest, the sheep may dig up and eat the burrs which are buried in the soil surface and they must be removed before they have reduced the seed to a level which is insufficient to establish the next crop. When they are moved it should be remembered that seed may be excreted for some time afterwards and they must not be transferred to a field where a different cultivar is grown.

Subterranean clover seed is generally larger than other clover seed so that rogue plants of other similar species are not so objectionable as in other situations because the seed can be cleaned with reasonable accuracy. The weeds listed as objectionable in a red clover (*T. pratense*) seed crop should be controlled in subterranean clover seed crops.

Seed Crop Inspection and Roguing

Roguing of small areas is possible when the cultivar has specific markers. Plants of other species, particularly medics (*Medicago* spp.), can also be removed. Markers are usually associated with leaf characteristics so that the best time for this operation is in the spring before flowering; however, a second visit when the crop is in flower is often needed.

Field inspection should take place at the same growth stages as recommended for roguing.

Harvesting

Because the seed heads or burrs are borne so close to the ground, specialised equipment is required to recover a maximum yield. Dense stands provide the best possibility for a good seed yield. With a thin stand more of the soil surface is exposed and cracked and therefore more seed becomes buried. A dense stand with a firm, level soil surface will ensure that most of the seed is on or close to the soil surface.

The seed is ready for harvest when the plants are dead with dry vegetation and the seed is hard. Because the seed will be very close to the ground it is not practicable to combine direct; the crop should be mown as close to the ground as possible and the swath can then be picked up by combine. However, this will leave a large amount of seed lying on the soil surface which can be recovered with a suction harvester.

Drying and Storage

Seed is usually very dry when harvested. However, because of the way in which it is necessary to recover the seed, the harvested material will contain much debris, including soil particles and stones. These should be removed by pre-cleaning by aspiration before the seed is stored for any length of time.

Seed Cleaning

Cleaning as described for other clovers is usually all that is required. Treatment with chemical seed dressings is not required but inoculation with *Rhizobium* may be necessary in some areas.

Other Species of Similar Type

There are six species of *Medicago* and one of *Centrosema* included in OECD (1995c). This is, however, a developing field of research and it is possible that other species will be added.

Medicago species

The six species currently listed in OECD (1995c) are:

Scientific name	Common name
M. littoralis Rohde ex Lois.	Strand medic, Harbinger's medic
M. polymorpha L.	Burr medic
M. rugosa Desr.	Gama medic
M. scutellata (L.) Miller	Snail medic
M. tornata (L.) Miller	Disc medic
M. truncatula Gaertn.	Barrel medic

Distinguishing between Cultivars

The following characteristics are listed in OECD (1995b) and apply to all six species:

At flowering: time of flowering, corolla pigmentation
After flowering: pod shape and size, direction of pod spiral, pod spininess
Vegetative: leaf pattern of marking, shape and colour
At maturity: seed shape, 1000 seed weight

Growing the Seed Crop

The comments made for subterranean clover (*Trifolium subterraneum*) apply also to the medics.

Centrosema pascuorum Mart ex Benth. Common Name: Centurion

Distinguishing between Cultivars

The following characteristics are listed in OECD (1995b):

Vegetative: hairiness of petioles and petiolules, hairiness of stems and internodes, size of fully expanded leaves and leaflets
At flowering: time of flowering, flower colour
At full maturity: 1000 seed weight

Growing the Seed Crop

The comments made for subterranean clover (*Trifolium subterraneum*) apply also to centurion.

Additional Species

There is one other species from which some cultivars have been developed, but which is not listed in OECD (1995c). This is *Trifolium hirtum* All., known as rose clover. It is essentially similar to subterranean clover (*Trifolium subterraneum*).

References

OECD (1995a). *OECD Scheme for the Varietal Certification of Seed of Subterranean Clover and Similar Species Moving in International Trade*. Up-to-date Version of the Seed Schemes as of 15 June 1995. Organisation for Economic Co-operation and Development, Paris, 119–141.
OECD (1995b). *Control Plot and Field Inspection Manual*. Organisation for Economic Co-operation and Development, Paris, 86, 90–96.
OECD (1995c). *List of Cultivars Eligible for Certification 1995*. Organisation for Economic Co-operation and Development, Paris, 77.

LEGUMINOSAE: FORAGE LEGUMES FOR THE TROPICS AND SUBTROPICS

The development of forage legumes suitable for growing in the tropics and subtropics is comparatively recent. Seed production is specialised and some species are best harvested by hand if labour is available. There is a wide range of plant types, some being shrub-like and others climbing in habit. The climbing species may be grown on trellises. Seed shedding is often a difficulty and seed may be recovered from the ground by suction harvesters or swept up by hand. Most of the species are 'short-day', requiring 12 h or less to induce flowering. Seed crops are best grown on the lighter soils but generally require good fertility, responding particularly to phosphate. Sowing is generally in the wet season, timed to provide harvest in the dry season.

Centrosema pubescens Benth. Common Name: Centro

C. pubescens is perennial and a vigorous climber. The trifoliate leaves are dark coloured and may be hairy on the under-surface. Flowers are large and lilac in colour. Pods are brown and up to 15 cm long containing about 20 black, mottled seeds. No descriptors for cultivars have been published internationally.

Environmental Requirements for Seed Production

Centro responds to fertile conditions and requires more moisture than some other species during the growing season. Soils which dry out quickly at the start of the dry season should be avoided. The species is not frost resistant.

Previous Cropping and Isolation

Centro produces a high proportion of hard seeds which can remain viable in the soil for long periods. An interval of five to six years free from centro should be allowed before planting a seed crop. During this time every effort should be made to prevent any volunteer plants from flowering and setting seed.

The species is self-fertilising and requires only sufficient isolation to prevent mixture during harvest.

Seed Crop Establishment

When seed is to be harvested by hand, the crop should be established in rows spaced 1 m apart which will allow trellises or support fences to be erected. When machine harvesting is intended, the rows should be closer together. Seed rates are between 4 and 8 kg/ha.

Soaking the seed in warm water before sowing may help to reduce the proportion of hard seed and so provide more uniform germination. Rapid establishment is encouraged by inoculating with an appropriate strain of *Rhizobium*, although the species will normally nodulate with the strain of *Rhizobium* present in the soil.

Seed Crop Management

Additional fertiliser is not usually required if the soil is provided with an adequate supply of the major plant nutrients before sowing. In some instances, however, additional application of superphosphate has been beneficial.

Best seed yields are obtained from the first growth each season. Moisture, if necessary from irrigation, will stimulate flowering and should be followed by a period of stress to encourage pod formation.

If there is excessive vegetative growth it is preferable to reduce this by grazing or cutting, otherwise there will be too much green material to deal with at harvest. It is also possible to adjust harvest to a suitable dry period by judicious defoliation. However, cultivars differ in their response to management and it is advisable to seek local advice.

Seed Crop Inspection and Roguing

Although no descriptors have been published, there is some variation in flower colour between cultivars and it is likely that the best time for both roguing and inspection will be when the crop is in full flower. Roguing is unlikely to be successful unless the crop is grown on supports, which allows individual plants to be identified.

Harvesting

Ripening is comparatively well synchronised. Seed is ready for harvest when the first ripening pods open and start to shed seed. Harvesting by hand will normally require two visits.

There is usually dense foliage surrounding the pods and it is unlikely that direct combining will be possible. For machine harvesting the crop should be windrowed for about a week to allow this material to dry before picking up with a combine.

Drying and Storage

Harvested material will contain much broken leaf and stem which will heat very quickly if not dried immediately. Some seed will be contained in unbroken pods and these will split open if dried, so making subsequent cleaning easier.

Drying on a tarpaulin or concrete base is possible in the less humid areas. In humid conditions, forced-air drying will be needed but the temperature of the drying air should not exceed 35°C.

Seed in store is vulnerable to insect attack and it may be beneficial to apply insecticides in the store; specialist advice should be sought as to the safe and most effective material to use in the area where the seed is stored.

Seed Cleaning

To avoid having to dry twice, it is essential to clean the harvested material without delay as leaf and broken stem may pick up moisture in humid climates. An air/screen cleaner is usually the only equipment required. Hand harvested seed may be cleaned by winnowing.

Seed may require treatment to reduce hard seed content or inoculation with *Rhizobium* as outlined above. Mechanical scarification is also possible

but must be done with care to avoid damage to the seed; scarified seed deteriorates rapidly and is not suitable for long storage.

Leucaena leucocephala (Lam.) de Wit. Common Names: Jumbie bean, White popinac

L. leucocephala is an erect, shrub-like perennial plant. The inflorescences consist of flowers held in small ball-like clusters. The flowers are white and develop long, flat pods each containing up to 25 bright brown seeds. There are no cultivar descriptors published internationally.

Environmental Requirements for Seed Production

Jumbie bean requires an alkaline soil which is well drained. It is adapted to the higher rainfall areas.

Previous Cropping and Isolation

The hard seed content of shed seed is high and therefore an interval of up to six years may be required to allow for the eradication of volunteer plants.

The species is self-fertilised and requires isolation sufficient to avoid mixture at harvest.

Seed Crop Establishment

The seedlings are slow to establish and may be raised in a nursery for later transplanting. For transplanted crops the rows may be up to 2 m apart with 20 cm between plants. For direct sowing, rows may be 60–120 cm apart and the seed rate is 15–40 kg/ha.

Hard seed may be reduced by scarification or by a hot water soak: 80°C for 2 min followed by immersion in cold water for up to three days; the seed must be dried immediately after treatment.

The species has a very specific *Rhizobium* requirement and inoculation is usually necessary, using the adapted strain.

Seedlings are very susceptible to shading and will not tolerate weed competition. Seed beds must be clean and weed control must be effective.

Seed Crop Management

The shrubs will grow tall unless controlled. It is best to cut them back at the beginning of the wet season. If the crop is grown in wide rows it may be interplanted with a suitable grass which can be cut for fodder.

Seed Crop Inspection and Roguing

Cultivars vary in plant height and in the amount of basal branching. There are also some differences in the size of the flowers. Roguing and inspection should be timed when these characters can be distinguished.

Harvesting

The seed is harvested by hand. The pods change from green to brown when they are ready to harvest. Several visits will be needed.

Drying, Storage and Seed Cleaning

The pods must be threshed after harvesting and the resultant material will contain broken pods and other green debris which must be dried immediately.

After drying, the seed should be cleaned by winnowing or on an air/screen cleaner. Subsequently it may be stored satisfactorily until the next growing season.

Macroptilium atropurpureum (DC) Urban.
Common Name: Siratro

Siratro is perennial and has a climbing habit with stoloniferous stems which can root at the nodes. The leaves are trifoliate, slightly glaucose below and hairy on the upper, dark green surface. The flowers are dark purple/deep red and the pods straight. The seeds are brown/black when ripe. No cultivar descriptors have been published internationally.

Environmental Requirements for Seed Production

The species is adapted to the warm tropics but will survive a light frost. It will withstand a moderate drought but does not do well on free-draining soils which dry quickly. It responds well to irrigation but some moisture stress during the early flowering period promotes flowering and seed set. However, it will not tolerate flooding and on heavy, damp soils may produce excessive vegetative growth with few flowering stems. Siratro is more tolerant of saline conditions than some other species.

Previous Cropping and Isolation

Although siratro establishes quickly, it requires a weed-free environment to develop. The seed may exhibit high dormancy and thus will survive in the soil for several years. Six years free from siratro before establishing a seed crop is therefore desirable.

Siratro is self-fertilised and requires only sufficient isolation to prevent mixture at harvest.

Seed Crop Establishment

If the intention is to harvest by hand, the crop should be sown in rows spaced about 1 m apart to permit the erection of support trellises. For machine harvesting it is best to establish the crop in narrowly spaced rows. The seed rate is 2–3 kg/ha. The seed should be covered to a depth of about 2 cm. A companion grass will improve the grazing and may offer some support to the seed crop, but will usually reduce seed yield.

Germination can be improved by breaking dormancy either by hot water treatment or by scarification. Inoculation with the cowpea strain of *Rhizobium* is usually required.

Seed Crop Management

The crop grows very quickly during the wet season and, if it is intended to harvest seed only once in the dry season, it should be grazed during this time. However, grazing should not be too hard and it is essential to preserve the crowns of the plants; generally this requires that about 15 cm of growth should always be left on the plants.

Siratro is not demanding in day length requirement and in many situations will produce flowers at any time of the year, so permitting two seed harvests if desired. If this is the intention the crop should not be grazed, although there may be a large amount of foliage to deal with. However, at the end of each growth period the plants should be trimmed to a height of about 15 cm to permit new, vigorous growth to develop.

Phosphate is the main nutrient required and the phosphate status of the soil should be maintained throughout the life of the crop.

Aphids and other insects may damage the plants and can be controlled by a suitable programme of insecticide application.

Seed Crop Inspection and Roguing

Crops grown on supports may be rogued, usually when both leaf and flower colour can be seen. It is less easy to rogue crops grown as swards.

Seed crops may be inspected when in full flower.

Harvesting

The seed is ripe when the pods have become dark brown in colour. Because flowering is continuous there will be pods in all stages of maturity. The seed sheds easily and the decision when to harvest is therefore critical.

The best seed yields are obtained by hand harvesting with two or more pickings. For machine harvesting it is possible to allow the first ripening seed to shed into the dense mass of herbage below and then to recover it by combining the crop when more seed has matured. However, this involves cutting and threshing a large mass of herbage and some growers prefer to go over the crop twice, the first time with the combine cutter set high.

Following both hand and machine harvest there will be much shed seed on the ground when the herbage is cleared. This can be recovered by sweeping (hand harvest) or by suction harvester.

Drying, Storage and Seed Cleaning

Drying is usually necessary to reduce seed moisture content to about 8%. The seed is vulnerable to insect attack in store and should be fumigated if stored for long periods.

There will usually be much green material with machine harvested seed and pre-cleaning to remove it will be worthwhile. Subsequent cleaning may be by winnowing or air/screen cleaner.

Other Species of *Macroptilium*

There are two other species which are grown on a minor scale: *M. phaseoloides* and *M. axillare*. Both are perennial and self-fertilised.

Stylosanthes species

There are four species of *Stylosanthes* which are listed in OECD (1995):

Scientific name	Common name
S. guianensis (Aublet) Sw.	Stylo
S. hamata (L.) Taubert.	Caribbean stylo
S. humilis H.B.K.	Townsville stylo
S. scabra J. Vogel	Shrubby stylo

Of these, *S. humilis* is annual and the rest perennial. *S. guianensis* has hairy, more or less erect stems growing to about 1 m; the flowers are borne in spikes and vary in colour from yellow to orange; the seed pods are hairy and contain only one seed each; seeds are yellow/brown and enclosed in a brown hull.

S. hamata has semi-erect stems which have short, white hairs down one side. The flowers are borne on spikes on a long stem; each flower produces two articulations, one without hairs with a coiled beak or hook and the other hairy and hookless.

S. humilis has small yellow flowers in short spikes; the grey/brown pods are hooked and one-seeded.

S. scabra grows to about 1.5 m in height with hairy stems; flowers are small and yellow. As in *S. hamata*, each flower produces two articulations, one with and one without a hook but both are pubescent. The seeds are yellow/light brown.

There are no international publications with cultivar descriptors.

The following comments apply to each of the species except where noted.

Environmental Requirements for Seed Production

The species are generally adapted to the medium to high rainfall areas. *S. hamata* is day-neutral but the other species generally require short days to induce flowering, although exact requirements may differ for different cultivars. Soils which do not waterlog are needed, and generally free-draining soils but with good moisture retention are to be preferred. *S. hamata* and *S. scabra* are more tolerant of dry conditions than the other two. None of the species is tolerant of saline conditions, but *S. guianensis* is more tolerant of low fertility than the others.

Previous Cropping and Isolation

Seeds may lie dormant in the soil for long periods and therefore an interval of up to six years should be allowed when none of the species is planted. Fields intended for seed production must be weed-free and previous cropping should be such that measures can be taken to eradicate weeds.

Stylosanthes species are usually treated as self-fertilising although some cross-fertilisation may take place. Generally a physical barrier or clear gap of 3 m is regarded as sufficient isolation. In India, however, isolation distances of 50 m are specified for Foundation Seed production and 25 m for Certified Seed (Tunwar and Singh 1988).

Seed Crop Establishment

Seed crops are generally best established in rows up to 50 cm apart. Hare and Waranyuwat (1980) suggest that in Thailand *S. hamata* should be broadcast. In Cuba, Perez (1994) found that rows 100 cm apart gave the best results for *S. guianensis*.

Seed rates are generally low, at 1–3 kg/ha.

Seed Crop Management

The status of soils in relation to the major plant nutrients should be satisfactory. The crops should be kept free from weeds; Hare and Waranyuwat (1980) mention *Hyptis* spp., *A. vaginalis*, *Sida* spp. and *Chromolaena* spp. as being particularly objectionable in Thailand.

In the first year the first growth can be taken for seed. Defoliating the crop before shutting up for seed production should be done with caution. For *S. guianensis*, defoliation one month before flower initiation will help to give a more uniform canopy for harvest, but delay in shutting up for seed may severely reduce seed yield. Defoliation should not be too severe, and plants must not be cut or grazed to ground level. Generally, at least 20 cm growth should be left on the plants.

In subsequent years the perennial species are treated in the same way as the first year crop.

Seed Crop Inspection and Roguing

Inspection during flowering is generally preferred. Roguing is not always practicable in very dense crops.

Harvesting

Hand harvesting is generally possible for all four species but they shed seed easily. For *S. guianensis* and *S. scabra* it is possible to place plastic sheets beneath the plants and, by shaking the plants a few times each week, to collect the seed as it ripens; when seed has been collected the plants can be cut and the crop prepared for the next harvest. With *S. hamata* and *S. humilis* the plants are cut and allowed to dry; the dry material can be swept up and in doing so most of the seed will fall to the ground and can then be recovered by sweeping the soil surface and removing the seed and debris for cleaning. It is usually also worthwhile to sweep up shed seed from the other two species.

Machine harvesting by combine is possible but should usually be followed by suction harvesting or collecting the material from the soil surface by rotary brushing to recover shed seed. Combines work best in conditions of low humidity. Various types of flail or brush harvesters are also used to harvest seed as it ripens.

The seed is generally difficult to handle. *S. guyanensis* produces a sticky exudate on the leaves and the hooked seed of *S. humilis* tends to stick together; consequently the seed does not flow freely.

Drying, Storage and Cleaning

Seed brought in from the field will usually be accompanied by much soil and plant debris. This must be removed by hand sieving or on a pre-cleaner. Following this pre-cleaning, final cleaning is by hand winnowing or on an air/screen cleaner.

Seed does not normally require drying except when it is to be stored for longer than six months.

Other Legume Species Grown for Fodder in the Tropics

Aeschynomene americana L. (common name: Joint vetch) originates in tropical America. It is also grown as a green manure.

Desmodium intortum (Miller) urban (common name: Greenleaf) is a perennial which can be grown in the wetter areas. It is cross-pollinated, but with some self-pollination. It is hardier and more persistent than silverleaf.

Desmodium uncinatum (common name: Silver-leaf, Tick clover) is similar to greenleaf but less tolerant of frost.

Glycine wightii (Wight and Arn.) Verde is a self-fertilised perennial. It can be grown from sea level up to 2000 m.

Lotononis bainesii Baker is a self-pollinating, slender perennial adapted to sandy soils in areas with an annual rainfall of 1000 mm in the sub-tropics. It requires high temperatures for active growth but also shows some frost resistance. It tolerates acid conditions. *Rhizobium* is a specialised form.

Pueraria phaseoloides (Roxb.) Benth. is a self-fertilised perennial which is adapted to frost-free areas of the tropics. It is intolerant of shade but tolerates wet conditions. It can be grown from sea level to 1000 m.

References

Hare, M.D. and Waranyuwat, A. (1980). *A Manual for Tropical Pasture Seed Production in Northeast Thailand.* Northeast Livestock Development Project, Khon Kaen, Thailand, 33–44.

OECD (1995). *List of Cultivars Eligible for Certification.* Organisation for Economic Co-operation and Development, Paris, 35.

Perez, A. (1994). Effect of sowing density and spacing on seed production in *Stylothances guianensis. Pastos y Forrages,* **17**, 307–315.

Tunwar, N.S. and Singh, S.V. (1988). *Indian Minimum Seed Certification Standards.* The Indian Seed Certification Board, New Delhi, 164–165.

LEGUMINOSAE: AGRICULTURAL CROP FOR FIBRE AND FORAGE

Crotalaria juncea L. Common Name: Sunn hemp

Sunn hemp is an important fibre crop and is also grown for forage and for green manure. It is a tall, shrubby, cross-fertilised short-day annual. It is fast growing and competes well with weeds. For seed production the seed rate should not be too heavy, about 12–15 kg/ha. It is preferred on well drained soils which retain moisture; it is tolerant of drought but not of salt or frost. Seed matures in four to five months after sowing.

LEGUMINOSAE: AGRICULTURAL AND VEGETABLE CROPS

Pisum sativum L. *sensu lato*. Common Names: Pea, Garden pea

This species has a wide range of uses spanning both agriculture and horticulture. Peas are widely grown in temperate areas and also in the cool seasons of the tropics. There are some cultivars which are winter hardy. The cultivated crops can be divided into four main groups, and some cultivars may be classified in more than one of these groups.

1. Vining peas: mechanically harvested while the seeds are still immature for immediate freezing or canning.
2. Picking or Garden peas: hand picked while seeds are still immature, either for shelling or consuming the complete pod. The edible podded cultivars are usually referred to as 'mange tout' or 'snap peas'.
3. Combining peas: cultivated for production of dry seeds which are either harvested by machine or by hand. These may be used for either human consumption or as animal feed.
4. Forage peas: produced for grazing, ensiling or hay-making.

Classification of Cultivars

The following is the cultivar classification outline on which the UPOV (1988) system is based:

> Seed: shape of starch grains in dry seed
> Seed: colour of cotyledons in dry seed
> Plant growth habit
> Stem: number of nodes, up to and including the first fertile node, including scales
> Pod: parchment, absent or present
> Pod: character of apex
> Seed: colour at green shell stage

Environmental Requirements for Seed Production

The pea crop requires a soil pH of 6.5. The fertiliser requirement will depend on residues from the previous crop and the amount of leaching which has occurred since. Generally, an NPK ratio of 1:1:1 is used for large-scale agricultural crops, while 2:1:1 is nearer the norm for small-scale horticultural crops. Soils deficient in manganese should either be avoided or supplementary dressings of manganese sulphate provided.

Previous Cropping and Isolation

The usual minimum duration since a previous pea crop is two years. However, important soil-borne pathogens and nematodes include *Fusarium oxysporum* Schlecht. ex Fr. f.sp. *pisi* (van Hall) Snyd, and Hans (wilt), *Fusarium solani* f. *pisi* (Jones) Snyder and Hansen (foot rot), *Ascochyta pisi* Lib. (leaf and pod spot), *Mycosphaerella pinodes* (Berk. and Blox.) Vesterg. (foot rot, leaf spot, black spot), *Phoma medicaginis* Malbr. and Roum. var. *pinodella* Jones (foot rot, collar rot) and *Heterodera gottingiana* (pea cyst nematode). Where these are known to be present the duration since previous cropping should be increased to at least four years.

The pea crop is self-pollinated and the minimum isolation distance is 3 m, although greater distances should be allowed for in seed crop planning if air-borne pathogens are known to be prevalent.

Seed Crop Establishment

Winter hardy cultivars are sown in the autumn, otherwise the crop is sown in the spring with the objective of obtaining a plant population of approximately 75/m^2 for most cultivars or 100/m^2 for cultivars with modified foliage ('leafless'). Rows are sown 12 cm apart. When a high degree of selection is required or it is not possible to use herbicides for weed control, the inter-row distance may be increased to 20 cm.

Seed Crop Management

Weed control is essential in order to ensure the physical purity of the subsequent seed lot harvested.

The pea crop is sensitive to water stress from the start of anthesis through to petal fall.

Major pests include *Contarinia pisi* Winn. (pea midge) and *Cydia nigricana* Steph. (pea moth): adequate chemical controls should be applied when these are known to be a potential problem.

Seed Crop Roguing

Seed crops are rogued when anthesis commences. At this stage the individual plants which display off-type flowers or flower colour are removed; in addition, off-types with excessive height or small leaflets ('tare leaved' or 'rabbit eared') are removed.

Harvesting

When the crop starts to senesce at the base and there is no further growth at the tops of plants, the seed pods start to turn a pale colour and to appear parchment-like. Pea seeds are very susceptible to mechanical damage when their moisture content is either too high or too low. Biddle and King (1977) demonstrated that the subsequent quality is reduced if they are harvested with a moisture content above 30–35%.

In situations where machine harvesting is not possible, the seed crop is cut by hand and further dried on racks either outside or under cover, depending on weather conditions. Large-scale areas are either combined direct or cut and left in windrows. The mature seed is prone to mechanical damage when the seed moisture content is 12% or less. Crops which remain green, are slow to dry off and if relatively weedy can be treated with a desiccant prior to harvesting.

Drying

In arid areas further sun drying is normally very successful, but in most temperate areas further artificial drying must be done to ensure that the quality of the seed lot is retained. If the initial moisture content is above 25% the material is dried in two stages, with the bin or drying floor's ventilating air temperature not exceeding 5°C above ambient. The drying temperature for continuous-flow driers should not exceed 38°C when the moisture content is 25% or more, or 43°C when below this moisture content.

Seed Cleaning

The seed lot can be further upgraded by air/screen cleaners. Seeds which are discoloured can be removed by passing the seed lot through an electronic colour separator. Any which are cracked or holed by pea moth larvae can be removed by a needle drum separator.

Seed Storage

The pea seed moisture content for short-term storage should not exceed approximately 14%, but for storage periods in excess of six months it should be reduced to less than this. For storage in vapour-proof containers the seed moisture content is reduced to 7%.

Trigonella foenum-graecum L. Common Name: Fenugreek

The seeds of fenugreek are used as a spice and the leaves are used as a vegetable. One cultivar from Greece is listed in OECD (1995a) and it is also included in the certification arrangements in India (Tunwar and Singh 1988). Cultivar descriptors have been published in OECD (1995b) as follows:

At beginning of flowering: time of flowering
At full flowering: plant height, pod shape and pilosity
Seed: shape, size, basic colour, ornamentation, colour of hilar ring.

Fenugreek is largely self-fertilised and isolation of 5–10 m is considered adequate. The seeds turn brown when ripe and the pods are indehiscent.

Vicia faba L. Common Names: Broad bean, Field bean, Faba bean

This species is widely cultivated in northern Europe and West Asia, either as an overwintering crop in the areas with milder winters, or as a spring sown crop where winters are more severe; there are some cultivars which are more winter hardy than others, which assists the possibility of overwintering young plants. Although the vegetative plant will usually survive under high temperatures, it will not set a successful market or seed crop. The immature seed and the dried seed are used as a vegetable while the whole plant and/or the seed are used for stock feed.

Classification of Cultivars

Earlier classifications which divided the species into three subspecies (namely, tick bean, horse bean and broad bean) are no longer valid because of the large degree of overlap of the different types.

The main uses of the range of cultivars currently maintained provides the most appropriate classification.

1. Broad beans: this is the cultivar group regarded as a vegetable; the seeds are either harvested while still immature for use as a fresh, frozen or canned commodity, or allowed to produce mature seed which can be stored and cooked as a vegetable after soaking.
2. Field beans: this includes those cultivars which are harvested as mature seed and used for stock feed. The group can be subdivided according to seed size into those with small seeds (tick beans) and large seeds (horse beans). Plants of the large-seeded field beans can also be harvested as whole plants for stock feed in the form of hay or silage.

Cultivars are generally identified according to the following outline (a more detailed scheme is provided by UPOV 1984):

Season, specific use (or uses), winter hardiness and maturity period
Presence or absence of melanin spot on wing petal
Seed: tannin content of testa and colour immediately following harvest, seed size (expressed as 1000 grain weight), shape
Plant: height, pod-bearing tillers, time of 50% flowering
Flower: presence or absence of standard petal anthocyanin
Pod: length, breadth, number of seeds per pod

Environmental Requirements for Seed Production

The broad bean crop succeeds on a water-retentive soil with a pH of 6.5. The NPK ratio applied during seed bed preparation is 1:1:1, although for autumn sowing the nitrogen is usually reduced; due consideration should be given to nitrogen residues from the previous crop.

Previous Cropping and Isolation

A minimum of two years should elapse between a *Vicia faba* seed crop and any preceding related crop. Some fungal pathogens can survive on debris from previous crops. *Ascochyta fabae* Speg. (leaf and pod spot) and *Fusarium* spp. are important pathogens noted for their survival between crops.

Broad beans are largely cross-pollinated, although self-fertilisation also occurs. Bees are the main pollinating agent. Some authorities require a greater isolation distance for the cultivars which are regarded as vegetables than for those to be grown for agricultural purposes. This is because a higher degree of cultivar purity is required for horticultural crops produced for market outlets such as the processing industry. The minimum isolation distance for vegetable cultivars is 1000 m, while for the agricultural cultivars produced in fields exceeding 2 ha the recommendation is 200 m and for fields of less than 2 ha it is 100 m for production of seed for further multiplication; for Certified Seed production these distances are halved.

Seed Crop Establishment

There are three factors which directly influence sowing rate: seed size; whether sowing in the autumn or spring; and whether weed control will be predominantly by herbicides or cultivations (hand or mechanical).

For autumn sowing crops are generally sown in rows from 12 to 60 cm apart, with 30–35 cm as the most frequently used spacing.

Spring sown crops are generally sown in rows from 12 to 50 cm apart, with 20 cm most frequently used.

When inter-row cultivations are used for weed control the wider spacings are adopted. The sowing rate is approximately 250 kg/ha for the larger seeded cultivars or 175 kg/ha for those which are relatively small seeded. In some areas, especially the Middle East and Asia, double rows 25 cm apart with 70 cm between the double rows is the normal practice, with a sowing rate of 150kg/ha.

It is important that the seed is sown sufficiently deep; approximately 8 cm is generally accepted as the optimum, although the cultivars with large seeds are sown approximately 10 cm deep. It is particularly important to sow relatively deep in relation to the seed size when pre-emergence herbicides are planned to be used.

Seed Crop Management

Cultivations to control weeds will be necessary, especially in the early stages of crop establishment if appropriate and timely herbicides have not been applied. Broad beans are responsive to irrigation before and during anthesis.

If the wild bee population is sparse, supplementary hives placed in the crop will assist cross-pollination within the plant population; one hive per hectare is the usual recommendation.

Harvesting and Threshing

As the seeds mature, the pods turn black and lose their characteristic sponginess; this is associated with blackening and loss of foliage, especially lower down the plant. In relatively dry areas, such as the Middle East, the crop can be left to stand until it is cut and threshed. Small areas can be cut by hand. The optimum seed moisture content at harvesting is approximately 16–20%. In temperate areas the crop is either combined, in which case it is left standing until this moisture content is reached, or the ripening crop is cut when the moisture content is 30% or less and left in windrows. Care must be taken to ensure that threshing drum speeds and clearance settings are such that seed is not mechanically damaged.

Drying and Cleaning

These are as described for *Pisum sativum*.

Phaseolus vulgaris L. Common Names: French bean, Common bean, Haricot bean, Kidney bean, Snap bean

This crop is cultivated either for consumption of the immature green pods or for its mature seeds, which are used as a dehydrated vegetable in the household and as the main vegetable ingredient in the preparation of canned beans, especially 'navy' or 'baked' beans. The immature pods are used as a fresh vegetable, canned or frozen.

Phaseolus vulgaris is cultivated in the temperate, subtropical and tropical areas of the world, but is not frost hardy.

Classification of Cultivars

The following is based on UPOV (1982).

Market use: suitability for specific market outlets, fresh, processing or dried
Seed: length, colour/pattern of testa
Plant habit: dwarf (determinate and semi-determinate), climbing (indeterminate)
Leaf: shape, texture, colour and relative size
Flower: colour of individual parts
Pod: length, shape, colour, curvature, characters of beak, presence or absence of string
Resistance to specific pathogens
    Colletotrichum lindemuthianum (Sacc. and Magn.) Bri. and Cav. (anthracnose)
    Bean common mosaic virus
    Blackroot virus

Environmental Requirements for Seed Production

Phaseolus vulgaris will succeed on soils with a pH range from 5.5 to 6.5. An NPK ratio of 1:2:2

applied during seed bed preparation will provide these elements in satisfactory proportions, although residual nutrients from previous crops should be taken into account. This species is not frost hardy. The cultivars grown as a market crop in temperate areas are day-length neutral although for other regions there are short-day cultivars. Seed production is most successful if the crop matures under relatively dry conditions.

Previous Cropping and Isolation

There should be an interval of at least one year between crops intended for seed or previous crops grown for pulse production. There should be a break of at least four years when major pathogens such as *Colletotrichum lindemuthianum* (Sacc. and Magn.) Bri. and Cav. (anthracnose) and *Ascochyta* spp. (*Ascochyta* leaf spot) have been prevalent.

The flowers are mainly self-pollinated although some cross-pollination occurs. A minimum isolation distance of 3 m between a seed crop and any other *Phaseolus vulgaris* crop is required, although some authorities may stipulate greater distances than this, especially for Basic Seed production.

Seed Crop Establishment

Dwarf cultivars are sown in rows 40–90 cm apart depending on the cultivar's vigour, while the distance between rows for climbing ('pole') cultivars is 90–120 cm; the sowing rates are approximately 100 kg/ha and 50 kg/ha respectively. The climbing cultivars require supports, which in tropical areas are usually made from local cane, otherwise poles, strings or nets supported by stakes can be used.

Seed Crop Management

Weed control, especially during early plant growth, is essential, especially for the dwarf cultivars, in order to reduce competition and minimise contaminating weed seeds in the harvested seed lot.

In arid areas, irrigation is usually necessary on a regular basis throughout the season until the seed crop starts to mature, but in temperate areas the best response to irrigation is during anthesis and pod development.

Seed Crop Roguing

The first roguing is normally done during the vegetative stage to check that foliage, vigour and general plant habit are true to type. The main roguing is at the start of anthesis; at this stage flower colour and plant form are confirmed as true to the cultivar's description. A third roguing is done when the morphological characters of the young pods can be assessed. Individual plants showing virus symptoms are removed from the crop as soon as observed. Bean common mosaic (syn. bean mosaic virus) and cherry leaf roll virus are both seed transmitted.

Harvesting

The optimum seed moisture content at harvest time is 20–25 %; the visual signs are that the pods have become pale brown in colour with a parchment-like texture. The maturing pods of climbing cultivars grown on supports are hand picked over a period as they reach this stage. Large areas grown without supports can be combined direct when the entire crop is mature; this is especially the case in areas with satisfactory climatic conditions and when risk of seed loss from shattering is minimal. Alternatively, the plants are cut and further dried in windrows prior to threshing.

Threshing and Cleaning

These are as described for *Pisum sativum*. Colour sorters can be used to remove off-type seeds with testa colours which do not conform to the cultivar in addition to separating stained seed.

Drying

In arid and tropical areas, open air drying can be used, although direct sunlight must be avoided to minimise the risk of seed damage. If warm air drying is necessary, the air temperature should not exceed 38°C.

Storage

The safe moisture content for vapour proof storage is 7.0%.

Phaseolus coccineus L. Common Name: Runner bean

Although not frost hardy, this species grows better in cooler conditions than *Phaseolus vulgaris* and is not usually cultivated in the tropics. The runner bean is a popular garden crop in northern Europe where it is also known as 'scarlet runner bean'. Although there are both bush and climbing cultivars available, the former are mainly cultivated by commercial growers, while the climbing cultivars are popular for private garden production. The climbing cultivars are also grown as a commercial market crop, but the plants are often maintained as bushes by pinching out the leaders when about 30 cm high.

Classification of Cultivars

Use: fresh market, processing, private garden
Growth habit: dwarf or climbing
Flower colour: white, red or bicolour
Pod: length, 'string' (present or absent), texture
Seed colour when mature and dry

See also UPOV (1983).

Environmental Requirements for Seed Production

The crop responds to bulky organic manures incorporated during soil preparation. The resulting improvement in the soil's water-holding capacity increases the available moisture which can significantly increase the seed set. This is further enhanced by maintaining available soil water during anthesis. The crop is cross- and self-pollinated by bees which 'trip' the stigma, i.e. the stigmatic surface is slightly ruptured which assists pollen germination.

Seed Crop Production and Processing

The production and harvesting methods are the same as for *Phaseolus vulgaris*.

Pachyrrhizus erosus (L.) Urban. Common Name: Yam bean

Pachyrrhizus tuberosus (Lam.) Spreng. Common Names: Yam bean, Potato bean

These two species are cultivated for their edible tubers, and *P. erosus* is also cultivated for its immature pods which are eaten as a green vegetable. The pods of *P. tuberosus* have hairs which are an irritant. Both species are grown in the tropics of Southeast Asia and South America. The yam beans are climbing or trailing herbaceous perennials, although when cultivated as a seed crop are grown on supports. There are no reported cultivars of either species.

The crop is sown at a rate of 30 kg/ha. The mature seeds are hand harvested and dried at ambient temperatures.

Voandzeia subterranea (L.) Thou. ex DC. Common Name: Bambara groundnut

This is cultivated in central and West Africa, and also to a lesser extent in Southeast Asia. It is of particular value as a food crop in relatively dry areas with prevailing high temperatures and poor soil conditions. It is often intercropped with maize or sorghum by subsistence farmers, although this practice is not recommended for seed production.

The species is an annual; after fertilisation the flower stems push into the soil and subsequently produce fruits similar to groundnuts.

There are local selections and land races, although there are no listed cultivars in the literature.

Previous Cropping and Isolation

Ideally a period of at least two years should elapse since a previous bambara groundnut crop in order to eradicate ground-keepers. This is especially important when working with selected material.

According to Purseglove (1984), the species is self-pollinated, and some self-pollination occurs in unopened flowers (cleistogamy), therefore there is no need for isolation. However, mosaic has been reported as a seed-transmitted virus of this species (Richardson 1979) and it would be prudent to produce the seed crop in isolation from commercial or garden crops.

Seed Crop Establishment

The species is most productive when following a crop for which bulky organic materials were incorporated, and harvesting is easier on light, friable soils.

Shelled seed is sown at the rate of approximately 50 kg/ha 45 cm apart within rows, with 50–60 cm between the rows.

Seed Crop Management

The crop should be cultivated in its early stages to minimise competition from weeds. The crop is earthed up once anthesis commences.

Seed Crop Roguing

The plants are rogued just before the start of anthesis. Foliage characters and general plant vigour are checked when there is a description of a specific selection to follow. The crop is rogued at this stage to remove any plants showing virus symptoms.

Harvesting

When the subterranean crop is getting near to maturity, the aerial foliage tends to become yellow and the state of the pods should be confirmed. The plants are either ploughed out of the ground or hoed out, depending on the scale of operation.

The plants remain in windrows for two to three days before the pods are hand gathered.

Threshing and Cleaning

The pods are either hand threshed or the seed extracted with a portable groundnut thresher. The seed is cleaned as described for *Arachis hypogea*.

Drying and Storage

Shelled seed is stored at 13% moisture content. The moisture content is reduced to 11% for storage in vapour-proof containers.

References

Biddle, A.J. and King, J.M. (1977). Effect of harvesting on pea seed quality, *Acta Horticulturae*, **83**, 77–81.

OECD (1995a). *List of Cultivars Eligible for Certification*. Organisation for Economic Co-operation and Development, Paris, 38.

OECD (1995b). *Control Plot and Field Inspection Manual*. Organisation for Economic Co-operation and Development, Paris, 175.

Purseglove, J.W. (1984). *Tropical Crops: Dicotyledens*. Longman Group Ltd, Harlow, pp. 229–332.

Richardson, M.J. (1979). *ISTA Seed Health Testing Handbook, Section 1.1* International Seed Testing Association, Zurich, 254.

Tunwar, N.S. and Singh, S.V. (1988). *Indian Minimum Seed Certification Standards*. The Central Seed Certification Board, New Delhi, 279–280.

UPOV (1973). *Guidelines for the Conduct of Tests for Distinctness, Homogeneity and Stability, Runner beans (Phaseolus coccineus L.)*. Doc. TG/9/1, International Union for the Protection of New Varieties of Plants, Geneva.

UPOV (1982). *Guidelines for the Conduct of Tests for Distinctness, Homogeneity and Stability, French bean (Phaseolus vulgaris L.)*. Doc. TG/12/4, International Union for the Protection of New Varieties of Plants, Geneva.

UPOV (1984). *Guidelines for the Conduct of Tests for Distinctness, Homogeneity and Stability, Broad bean, Field bean*. Doc. TG/8/4, International Union for the Protection of New Varieties of Plants, Geneva.

UPOV (1988). *Guidelines for the Conduct of Tests for Distinctness, Homogeneity and Stability, Peas*. Doc. TG/7/4, International Union for the Protection of New Varieties of Plants, Geneva.

LEGUMINOSAE: AGRICULTURAL AND VEGETABLE CROPS – PULSES

Vigna unguiculata (L.) Walp. Common Name: Cowpea

This species is cultivated in the tropics and subtropics especially in parts of Africa, Asia and southern USA. It is cultivated for its long green pods which are used as a fresh vegetable, and in some areas these are allowed to develop fully for the production of the seeds which are consumed after rehydration and cooking. The young leaves are also cooked fresh or first dried, when they form a useful contribution to the diet in the dry seasons. The plant is also grown for fodder which is either grazed *in situ* or used to produce hay and silage.

Classification of Cultivars

Some of the seed stocks used in Africa and Asia are local selections. There are two main cultivated types:

(i)   dwarf, which are determinate or semi-determinate;
(ii) climbing, which are indeterminate

The main characters used to distinguish cultivars are:

Shape of terminal leaflet
Plant hairiness
Pod: angle of attachment, pigmentation, curvature, length
Seed: shape, external texture, pigment pattern associated with hilum ('eye')

Environmental Requirements for Seed Production

The optimum soil pH for the crop is 5.6–6.5. Nitrogen is usually omitted from base-dressings; the usual NPK ratio applied is 0:1.5:1. If growth is subsequently weak, a top-dressing of a nitrogenous fertiliser is applied.

The species will tolerate high temperatures, and thrives between 20 and 35°C, but not lower than 15°C. It is not frost hardy.

Previous Cropping and Isolation

An interval of at least one year since a previous crop is recommended in order to minimise the risk of ground-keepers. However, cowpea is usually rotated with an unrelated crop which, depending on local conditions, may be maize, mustard, sorghum or rice. The main soil-borne pathogens include *Fusarium oxysporum* Schlecht. (wilt) and *Xanthomonas vignicola* Burkholder (bacterial blight); where either of these or other known pathogens have been observed in previous crops, longer rotations should be implemented.

It is generally believed that the crop is self-pollinated; however, according to Purseglove (1984, p. 321) the degree of cross-pollination varies between different areas, therefore seed crops should be produced on sites which are at least 500 m apart. The use of a surrounding barrier crop, such as maize or sorghum, may also reduce the risk of pollen contamination from outside the intended crop. Satisfactory isolation is especially important for crops intended to produce early generations, including Basic Seed stocks.

Seed Crop Establishment

Dwarf and semi-erect cultivars are sown in rows 30–60 cm apart, the distance between rows depending on cultivar vigour. The optimum distance between plants within the rows is 15 cm. Depending on seed size of the cultivar, a sowing rate of approximately 35 kg/ha will achieve this.

The climbing cultivars require supports approximately 2 m high; these are usually made from local materials. The seed is sown in rows 75 cm apart, at approximately 15 cm apart within the rows.

Seed Crop Management

Weeds should be controlled in the early stages of the crop. The dwarf cultivars are usually 'stopped' by pinching out the growing points when they have reached approximately 15 cm high; this assists the formation of a bushy plant.

Seed Crop Roguing

Any plants which display significant off-type characters are removed during routine cultivations, as are any plants showing virus symptoms. Important viruses which are seed-borne include cowpea mosaic virus, cowpea ringspot virus and tobacco mosaic virus.

The optimum stage for the main roguing is when the seed pods are nearing maturity; at this time the external pod characters can be confirmed. As the pods become more mature it is possible to check samples for seed characters. It is important to ensure that the whole of each off-type plant is removed when a rogue is identified. Removal of plants showing virus infection is also continued at this roguing stage.

Harvesting

The climbing cultivars are hand picked as they reach maturity; this is confirmed by their yellow-brown or brown pod external colour. The same criterion is used to judge the harvesting stage of the bush types. Those cultivars in which the whole plant's pods mature relatively evenly can be cut and left in windrows for further drying; although this can also be done for cultivars with a wider maturity span, it is more productive to hand harvest as the pods ripen.

Threshing, Drying, Cleaning and Storage

Crops which have been left in windrows can either be picked up with a combine or taken to a stationary thresher. Care must be taken during threshing to adjust the drum speed in order to minimise mechanical damage to the seed crop.

The seed usually requires further drying as the maximum moisture content for storage is 13%. This can be done on a drying floor; the air temperature must not exceed 35°C when artificially drying.

Seed cleaning methods are as described for *Pisum sativum*. The seed should be reduced to 11.5% for storage in vapour-proof containers.

Phaseolus radiatus L. Syns. *Vigna radiata* (L.) R. Wilczek and *Phaseolus aureus* Roxb. Common Names: Green gram, Mung bean

This is an important crop in parts of Asia, where its main uses are as a pulse and for the production of bean sprouts. The green immature pods are used as a fresh vegetable in some areas. This species is sometimes used as a cover crop for production of hay and the haulm remaining from the pulse crop is used for fodder.

Classification of Cultivars

Two main criteria are used for classifying cultivars, namely seed colour (yellow, green, through to brown) and plant type (indeterminate or determinate).

The main cultivar distinguishing characters are:

Seed: shape, lustre, mottling, character of hilum
Hypocotyl: colour
Leaf: shape and length of terminal leaflet, pubescence, colour, length of petiole
Inflorescence: length of peduncle, position of raceme, colour of calyx, corolla and ventral suture
Pod: colour and shape at maturity, angle of attachment to peduncle at maturity, pubescence, constrictions, curvature

Environmental Requirements for Seed Production

This species tolerates high temperatures and succeeds in temperatures of 30–35°C. It is drought tolerant; yields tend to be reduced by excessive rainfall. There are day-neutral, short- and long-day types.

The crop tolerates a soil pH of 5.5–6.5. The NPK ratio of 1:8:5 is applied during seed bed preparation, although this can be modified according to the soil's nutrient status. If the young plant growth is weak, a nitrogenous top-dressing is applied when they have established, but no further nitrogen is applied once anthesis has started.

Previous Cropping and Isolation

A minimum of two years should elapse if there have been preceding soil-borne pathogens, but normally a break of one year is sufficient unless otherwise stipulated by the seed legislation authority.

The species is considered to be self-pollinated and the only isolation necessary is a distance of at least 3 m between crops in order to avoid admixture.

Seed Crop Establishment

The crop is sown in rows 20–45 cm apart, at a rate to provide a plant at approximately every 8 cm, although the overall plant density may be reduced for the taller cultivars. The seed size varies between cultivars but a sowing rate of up to 10 kg/ha will achieve the above plant density.

Seed Crop Management and Subsequent Operations

These are as described for *Vigna unguiculata*.

Phaseolus mungo L. Syn. Vigna mungo (L.) Hepper. Common Name: Black gram

This is an important pulse crop in the Indian subcontinent, where it is also cultivated for fodder.

No detailed descriptors have been published, but according to Purseglove (1984, p. 301), cultivars are grouped according to their seed characters.

The seed crop requirements and subsequent management are as described for *Phaseolus radiatus*. The pods are black at maturity and are much less prone to shattering than mung bean.

Phaseolus aconitifolius Jacq. Common Names: Mat bean, Moth bean

This species is mainly cultivated in the Indian subcontinent as a green manure, fodder crop and as a vegetable; both the immature pods and ripe seeds are cooked. It is also used as a pasture and fodder crop in parts of North America.

Classification of Cultivars

There are two distinct types: one has leaves which have five relatively long narrow leaflet divisions; the other has leaves with three broader divisions. The main classification is based on background colour of seed, and plant growth as either indeterminate or determinate.

The main cultivar distinguishing characters are:

Plant habit: erect, intermediate or prostrate, tendency to twine

Leaf: terminal leaf base shape, lobing and tip of terminal leaflet, pubescence, petiole pubescence

Inflorescence: days to 50% flowering, position of raceme

Pod: angle of attachment to peduncle, pubescence, curvature, colour at maturity

Seed: relative size, shape of hilum

Seed Crop Production

The seed crop requirements and subsequent management are as described for *Phaseolus radiatus*, except that as the species produces a plant with a more trailing habit distances between rows are increased to 1 m and the sowing rate is reduced to 3–4 kg/ha.

Phaseolus angularis (Willd.) W Wight. Common Name: Adzuki bean

This species is mainly cultivated in Japan and China, and to a lesser extent in the USA, South America and parts of West Africa as a vegetable. It is also used as a fodder crop in some areas.

There are no detailed cultivar descriptors published. The species forms a relatively erect bushy annual plant, although there are some climbing or prostrate forms. It is largely cross-pollinated. Seed production is as described for *Phaseolus radiatus*.

Psophocarpus tetragonobolus (L.) DC. Common Names: Winged bean, Goa bean

Although cultivation was largely confined to Southeast Asia there has been considerable interest in adopting it as a crop in other tropical areas. This species is largely regarded as a vegetable crop especially suited for inclusion in cropping programmes for home gardens. The semi-mature pods and the leaves are cooked as a green vegetable. The seeds are also eaten after special preparation. Some of the cultivars form a tuberous root which is also edible. Plants can also be used for forage, some types being leafier than others.

Classification of Cultivars

> Photoperiodic response
> General vigour of vegetative growth, formation of tuber, pod yield
> Stem characters, pigmentation
> Leaf and leaflet: size, shape
> Inflorescence: number of days to start of anthesis, detail of flower colour (or colours)
> Pod: relative length, shape in transverse section, texture, pigmentation, character of wings, including wing colour
> Seed: external characters when mature and dry, colour (including any mottling), shape, surface texture, detail of hilum

Environmental Requirements for Seed Production

Winged bean is a perennial but is usually grown as an annual in the tropics; however, subject to remaining free of major diseases, it can be retained beyond the first year.

The crop responds to the incorporation of bulky organic manures during soil preparations. There is little detailed information on the crop's nutrient requirements; it responds to a low nitrogen to phosphorus ratio, therefore an NPK ratio of 1:6:4 would seem appropriate, provided that nutrient residues and supplements from organic manures are taken into account.

Previous Cropping and Isolation

Plots for seed production should not have grown winged bean for at least one year in order to ensure that there is no carryover of vegetative parts from a previous crop. If soil-borne pathogens or nematodes have infected the site in the past, then rotations should be longer. Special attention should be given to the eradication of perennial weeds, especially if it is planned to retain the plants for seed cropping beyond the first year.

Winged bean is largely self-pollinated although some cross-pollination occurs, therefore seed crops should be isolated from all possible sources of contaminating pollen by at least 500 m, or the minimum distance stipulated by prevailing seed regulations.

Seed Crop Establishment

The crop requires vertical supports to an approximate height of 2 m, which can be constructed from local materials such as canes. The support system needs to be more permanent when it is planned to continue seed production beyond the first year.

The required distance between rows will depend on the vigour of the cultivar; it is normal practice for rows to be at least 1 m apart. The distance between plants within the rows is approximately 50 cm. Seed is sown direct at the final plant stations allowing for the potential germination of the seed stock.

Seed Crop Management

The young plants are thinned as necessary before they start to climb the supports. The crop should be hoed to control weeds by the same stage.

Crops which will be retained for a further year's seed production should have the vines pruned back when seed crop harvesting has been completed.

Seed Crop Roguing

Any plants displaying off-type characters are removed during the routine cultural operations.

The main roguing is done at the start of anthesis when inflorescence characters are confirmed. Plants with virus infection are removed from the crop.

Seed crops continued beyond their first year will normally only need roguing to remove diseased plants.

Harvesting, Threshing, Drying and Storage

Seed pods are hand harvested when they become fibrous; this can be up to six weeks from flowering, depending on the cultivar. The pods are usually further dried before hand threshing; care must be taken to ensure that the material is protected from rain during the final drying. The seed is further dried to 13% moisture for open storage or 11.5% for storage in vapour-proof containers.

Dolichos lablab L. Syn. *Lablab purpureus* (L.) Sweet. Common Names: Lablab bean, Hyacinth bean

The lablab bean is mainly cultivated in the Indian subcontinent for use as a vegetable, pulse and for feeding livestock. It is grown in some other tropical areas including Africa, where it is also used as a green manure or cover crop.

Classification of Cultivars

According to Purseglove (1984, p. 273), there are two botanical varieties:

(i) var. *lablab*: a short-lived twining perennial usually treated as an annual, with pods which are longer and more tapering with the long axis of the seeds parallel to the suture;

(ii) var. *lignosus* (L.) Prain (syn. *Dolichos lignosus* L.): a semi-erect bushy perennial having shorter, more abruptly truncated pods with the seed's long axis at right angles to the suture.

Both varieties are reputed to cross-pollinate freely. There are no reports of distinct cultivars. Separate seed stocks which show some distinct characters should be isolated by a minimum of 500 m from all other sources of pollen contamination. Seed is sown at a rate of 30–60 kg/ha in rows up to 1 m apart. The crop should be rogued to remove individual plants which display characters outside the accepted norm for the stock. Plants for a seed crop are usually hand pulled and left in windrows before threshing. If grown on supports the seed pods are harvested by hand as they reach maturity.

Lens culinaris Medikuc. (Syn. *L. esculenta* Moench.). Common Name: Lentil

Although most of the cultivars which are eligible for the OECD Herbage and Oil Seed Scheme originate in Greece and Canada, the countries with the highest annual production are India and Turkey (OECD 1995a, p. 28; FAO 1995).

Distinguishing between Cultivars

Cultivars can be divided into two groups, based mainly on seed size. According to Langer and Hill (1991, p. 260), the large seeded cultivars have large flat pods with flattened seeds 6–9 mm in diameter, with yellow to orange cotyledons and large white or rarely blue flowers; small seeded cultivars have seeds 3–6 mm in diameter with small convex pods, convex cotyledons and small violet-blue or pink flowers.

The following list of characteristics is based on OECD (1995b, pp. 144, 145) and IBPGR (1985a).

Main features used for distinguishing cultivars in the field:

At the beginning of flowering
    Leaflet size and shape
    Number of leaflets per leaf
    Shape of leaflet tip
    Leaflet colour

Dentation of leaflet margin
Leaflet: glossiness of upper side and/or pubescence of lower side
At flowering
Time of flowering
Plant growth habit
Height of first flowering node
Flower size
Ground colour of standard petal and colour of petal veins
Location of flowers and number per inflorescence
Size of bracts
Anthocyanin colouration of peduncle
Sepal length, width and shape of tip
Presence/absence (rudimentary) of tendrils

Other features used for distinguishing between cultivars:

During vegetative growth
Plant growth habit
Stem thickness and degree of branching
Anthocyanin colouration at base and/or tip of stem
At maturity
Time of maturity
Plant height
Number of pods per inflorescence and number of seeds per pod
Pod pigmentation (absent/present)
Pod shape and curvature
Seed
Shape and size
Colour of testa, hilum and cotyledon
Surface texture
Ground colour of testa, pattern of markings and colour of markings
Type: whole yellow or split red

Laboratory tests

Cooke and Marchylo (1992) used sodium dodecylsulphate polyacrylamide gel electrophoresis to detect seed of a lentil-like vetch in commercial samples of French dark speckled lentils.

Environmental Requirements for Seed Production

Lentil is a cool season crop and is grown as a winter annual in temperate regions. It is not suitable for the tropics but may sometimes be grown in the dry season at higher altitudes. It is generally grown between latitudes 15° and 40°.

Lentil is intolerant of wet conditions but otherwise can be grown on a wide range of soil types. Phosphorus is the main requirement in fertiliser; when there is good nodulation additional nitrogen may reduce yield.

The species generally requires vernalisation, although there is variation between cultivars which may be day-neutral or long-day.

Previous Cropping and Isolation

Lentils are usually grown in a rotation with other crops. An interval of two seasons free from lentil is advised before sowing a seed crop. Areas which contain *Orobanche* or ground-keepers of other legume species such as *Lathyrus* or *Vicia*, should be avoided.

The species is considered to be self-fertilised, but some cross-fertilisation may occur. The isolation distance specified in seed certification schemes is usually 5–10 m.

Seed Crop Establishment and Management

When grown for seed, lentils are best sown in rows. The distance between rows varies between regions. When there is adequate moisture, spacing at 15–30 cm is appropriate, but in very dry conditions wider spacing is used. The seed rate depends on seed size of the cultivar and spacing between rows, and can vary between 15 to 70 kg/ha.

Inoculation with a suitable *Rhizobium* may be required in areas which have not carried lentil before, but is not usually necessary.

Lentils are not demanding of fertiliser but respond well to maintenance dressings of phosphorus.

Seed crops do not require management different from that given to a food crop. However, in areas where it is customary to grow lentils in a mixed crop for food, it is advisable to grow seed as a pure crop.

Irrigation should be used sparingly, but gives best response when applied at flowering or during pod formation.

Fungus diseases which affect lentils are *Fusarium* spp. (wilt), *Sclerotinium* spp. (collar rot, stem rot), *Rhizoctonia* spp. (brassicol) and *Uromyces* spp. (rust).

Seed Crop Inspection and Roguing

Crop inspection and roguing will be most effective when flower colour can be observed.

Harvesting, Drying, Storage and Seed Cleaning

The crop is suitable for combining direct. Smaller areas may be pulled by hand and threshed after drying. When ripe, the stems and leaves turn yellow and the pods are brown.

Small quantities of seed can be spread on a well ventilated floor to dry. Continuous-flow or batch driers can be used; the temperature of the drying air should not exceed 35°C.

The main seed cleaning equipment is an air/screen cleaner. Optical sorters can be used to remove stained seeds. Cracked or damaged seeds can be removed on a pin drum. The seed is easily damaged and should be handled with care.

Cajanus cajan (L.) Millsp. Common Name: Pigeon pea

Pigeon peas are grown mainly in India – according to Langer and Hill (1991, pp. 253–5), approximately 93%. Botanically it is a short-lived perennial tree or shrub, but can also be grown as an annual. Seed yields are highest from the first harvest, although for some small hand harvested areas it may be satisfactory to maintain the crop for two to four years.

Distinguishing between Cultivars

The following characteristics are based on IBPGR (1981a) and Purseglove (1984, pp. 236–41).

In India, two groups of cultivars are recognised, and these may be given the status of botanical varieties.

(i) Tur group, also known as var. *flavus* D.C. These mature early and are short stemmed; flowers are yellow; pods glabrous, green, light green when ripe, with (usually) three seeds.

(ii) Arhar group, also known as var. *bicolor* D.C. These are later maturing, bushy perennials; standards striped or splotched with red or purple; pods pubescent, dark coloured with four to five seeds.

Main features used for distinguishing cultivars in the field:

Time of flowering: early, medium, late
Growth habit: erect (compact), semi-spreading, spreading, trailing
Main flower colour: ivory, yellow, orange, red, purple
Pod colour: green, purple, mixed or streaked

Other features used for distinguishing between cultivars:

Stem colour: green, sun-red, purple
Secondary flower colour: none, red, purple
Streaking of second colour: absent, few, many
Flowering pattern: determinate, semi-determinate, indeterminate
Seed colour pattern: plain, mottled, speckled, mottled and speckled, ringed
Seed main colour: white, cream, orange, brown, grey, purple, black
Seed secondary colour: as for main colours
Eye colour (round hilum): as for main colours
Eye size: none, narrow, medium, wide
Seed shape: ovoid, globular, cubic, elliptical

Environmental Requirements for Seed Production

Pigeon peas are tolerant of drought and are less well suited to the humid tropics. They do not tolerate frost. Pigeon peas are adapted to a wide range of soils, provided they are not calcium deficient or subject to waterlogging. Most cultivars are short-day.

Previous Cropping and Isolation

Fields which have grown legumes in the last two years should be avoided.

Pigeon peas are mainly cross-fertilised and isolation should be provided accordingly. In India the relevant distances are 200 m for earlier generations and 100 m for Certified Seed production (Tunwar and Singh 1988, pp. 91–2).

Seed Crop Establishment

Early sowing usually gives higher yields. For seed production, pigeon pea should be treated as an annual and sown as a pure stand. There is considerable variation in the lay-out of crops; in general, earlier maturing cultivars can be grown at closer spacing than those maturing later. Plant spacing varies between 50 × 20 cm and 120 × 60 cm. Seed rates vary between 8–10 kg/ha and 12–15 kg/ha.

The crop is very susceptible to weed competition in the early stages; hand weeding of smaller areas or the use of herbicides on larger areas is necessary.

Seed Crop Management

Seed crops are managed in a similar manner to food crops in pure stand. The crop will respond to moderate application of phosphate fertiliser.

Seed Crop Inspection and Roguing

It is important to rogue out any ground-keepers of other legume species. Plants with flower or pod characteristics different from those of the cultivar should also be removed.

While it is advisable to inspect the crop twice (during flowering and after pod formation), the most important time is at the start of flowering when isolation can also be checked.

Harvesting, Drying, Storage and Seed Cleaning

Small areas are usually harvested by hand pulling the plants and allowing them to dry in the field before threshing. For larger areas it is best to windrow the crop and allow the pods to dry before picking up with a combine or carting to a stationary thresher.

Seed harvested above 14% moisture should be dried immediately, either on a well ventilated floor or on a continuous-flow or batch drier. When using forced air, the temperature should not exceed 35°C.

An air/screen cleaner will normally clean pigeon pea seed satisfactorily.

Cicer arietinum L. Common Names: Chickpea, Bengal gram

Cicer arietinum is more tolerant of cool weather than *Cajanus cajan* and is therefore grown more widely, extending into countries with a Mediterranean climate. The main production area is in India, but it is also grown in southern Europe and North Africa, Australia and Central/South America. Seed production requirements are similar for these two species.

Distinguishing between Cultivars

In India, two groups of cultivars are distinguished: those with small coloured seeds are known as 'Desi'; those with larger white seeds as 'Kabuli' (Langer and Hill 1991, pp. 255–6). The following lists of characteristics are based on IBPGR (1985b) and OECD (1995b, pp. 132–3).

Main features used for distinguishing cultivars in the field:

At the beginning of flowering
  Time of flowering: early, medium, late
  Pigmentation of stems and leaves: absent, present
  Number of leaflets per leaf: very few (3–9), very many (13+)

Leaflet – glossiness or hairiness of upper side: absent, present

Leaflet size: small, large

Leaflet shape of apex: ovate, elliptical, obovate

Petiole length: short, long

At flowering

Canopy height: high, low

Plant width: narrow, wide

Stem length: short, long

Number of flowers per peduncle: few, many

Flower colour: blue, pink, white

Number of flowers per peduncle: few, many

Length of first flowering node: short, long

Number of primary and secondary branches: few, many

Other features used for distinguishing between cultivars:

Pod characteristics

Number of seeds per pod

Pod constrictions

Surface texture of pod

Pod tip shape

Length and breadth of pod

Shape of pod cross-section

Pod stringiness

Pod dehiscence

Seed characteristics

Seed colour

Black dots on seed coat

Seed shape

Smoothness of seed surface

Length and shape of beak

Colour of beak

IBPGR (1985b) describes 21 different flower colours.

Environmental Requirements for Seed Production

Chickpea is a temperate crop which originated in eastern Mediterranean countries and can be grown as a winter annual in Mediterranean climates. In India it is grown in the cool season. It may require vernalisation and is either long-day or day-neutral.

Chickpea is tolerant of drought and does not grow well in humid conditions. It is suited to heavier soils provided they are well drained, and will tolerate a rough seed bed. It responds to small dressings of phosphorus and potash and in some areas has benefited from a small starter dressing of nitrogen.

Previous Cropping and Isolation

It is advisable to leave at least two years free from chickpea or other legumes before planting a seed crop. As well as the risk of volunteer plants, there are several fungal diseases which can be soil-borne.

The species is regarded as self-fertilising although some cross-fertilisation may occur. It may, therefore, be advisable to provide more isolation during the early generations of seed production, but for later generations 5–10 m is adequate.

Seed Crop Establishment and Management

The crop is normally established in rows spaced 30–120 cm apart. Seed rate is 50–70 kg/ha for small seed or 75–120 kg/ha for large seed. It is important to use only disease-free seed; both *Ascochyta* and *Fusarium* can be seed-borne.

There are no particular management practices which are peculiar to a seed crop; management similar to that for a food crop is suitable.

Seed Crop Inspection and Roguing

Inspection and roguing are best undertaken at flowering, when colour and other characteristics can be seen.

Harvesting, Drying, Storage and Seed Cleaning

Small areas are harvested and handled subsequently by hand. Larger areas may be windrowed and picked up later or combined direct.

If a forced-air drier is used, the temperature of the air should not exceed 35°C.

An air/screen cleaner will normally clean seed satisfactorily.

Lupinus species. Common Name: Lupin

There are five species of lupin which are used in agriculture, generally as grain crops, but also to a lesser extent as green fodder crops. Their use is restricted by the toxic lupin alkaloids which must be removed from the seed by soaking or boiling before consumption, and can cause sickness in sheep grazing the green plant. Modern breeding has resulted in the production of cultivars with reduced alkaloid content. The species are:

Scientific name	Common name
L. angustifolius L.	Narrow leaf lupin, blue lupin
L. albus L.	White lupin
L. cosentinii Guss.	Western Australian common blue or sandplain lupin
L. luteus L.	Yellow lupin
L. mutabilis Sweet.	Pearl lupin

L. cosentinii is used as a fodder plant in Western Australia. The seeds can lie dormant in the soil for long periods and therefore, once established in a field, they are constantly replenished from volunteer plants. *L. mutabilis* has been cultivated less than the others, principally in South America, and has recently received attention from plant breeders.

The other three species are grown more extensively for seed. In OECD (1995a, pp. 28–9) there are 34 cultivars listed for *L. albus*, of which 16 originate in France; for *L. angustifolius* there are 21, over half from Australia and South Africa; and for *L. luteus* there are 20, 16 from Germany and Poland.

Distinguishing between Cultivars

The following is based on UPOV (1979), IBPGR (1981b) and OECD (1995b, p. 57).

Main features used for distinguishing cultivars in the field:

Length of terminal leaflet measured at flowering on a leaf just below the uppermost branch carrying flowers: short, medium, long
Flower colour: white, bluish white, blue, pink, brimstone coloured, chrome yellow
Time of beginning of flowering: early, medium, late

Other features used for distinguishing between cultivars:

Plant height three weeks after seedling emergence
Plant growth habit three weeks after seedling emergence
Leaf colour at flower bud stage
Anthocyanin colouration of stem at flower bud stage
Plant height at beginning of flowering
Plant height at green ripe stage
Terminal leaflet width, at flowering, leaf just below uppermost branch
Colour of tip of carina of flower: yellow, blue-black
Pod length at green maturity, pod in midst of main inflorescence
Seed ground colour at full maturity: white, grey
Seed ornamentation at full maturity: absent, present
1000 seed weight
Time of green ripening
Time of maturity

Laboratory tests:

Presence or absence of alkaloids

Environmental Requirements for Seed Production

Lupins have originated in temperate climates and are not particularly frost hardy. However, cultivars may require some vernalisation from cold though not frosty periods. *L. albus* and *L. angustifolius* are either long-day or day-neutral. *L. luteus* is long-day.

L. albus is better adapted to the heavier soils and is more tolerant of alkalinity than the other two species. *L. luteus* is the most tolerant of light, acid soils and *L. angustifolius* is intermediate.

Previous Cropping and Isolation

Lupin seed will remain viable in the soil for long periods. There are several fungal diseases which attack lupins and other legume species (for example, *Fusarium, Botrytis* or *Sclerotinia* spp.). It is therefore advisable to grow lupins for seed on fields which have been free of legumes for four years.

Both self- and cross-fertilisation occur. *L. angustifolius* is predominantly selfed, but the other two cross-fertilise freely. Belteky and Kovaks (1984) suggest that cross-fertilisation may reach 40%, while Langer and Hill (1991, pp. 261–4) suggest 9%. Lupins are attractive to bees and it is an advantage to site hives in or near the crop.

There is the problem that bitter cultivars will cross with sweet cultivars of the same species, and since the sweet cultivars typically have a content of the bitter principle of less than 0.05% it is important not to allow any chance of crossing.

For fields of less than 2 ha an isolation distance of 200 m is recommended for seed crops to produce seed for further multiplication, and 100 m for crops to produce seed for grain crops. For larger fields the distances are 100 m and 50 m.

Seed Crop Establishment

Early sowing in spring provides opportunity for vernalisation and there is some evidence that early sown crops ripen more uniformly. The crop is sown in rows 10–17 cm apart. Seed size varies between species and cultivars, and seed rate has to be adjusted accordingly. The target plant population should be 50–60/m². In thin crops the stems may branch excessively, which can cause a prolonged ripening period.

Inoculation with *Rhizobium* may be required on soils which have not carried lupins before.

Seed Crop Management

Management of a seed crop does not differ from that required for grain crops. Lupins are not demanding of fertiliser, and only maintenance application of phosphate and potash are required; nitrogen is not needed. Weed, disease and pest control follow the same pattern as for a grain crop.

Seed Crop Inspection and Roguing

The best time for inspection and roguing is when flower colour can be seen. Some of the sweet cultivars have been provided with genetic markers of distinct flower colour to assist in the identification of off-types.

Harvesting

The crop is usually direct combined. Little adjustment is needed to a combine used for cereals. However, care is needed to avoid damage to the seed coat which may injure germination. Desiccation is not recommended as there is some evidence that this may injure germination.

Determinate cultivars of *L. albus* are becoming available and these ripen more uniformly than indeterminate.

Modern cultivars are usually resistant to seed shedding and seed is harvest ripe when it is hard and difficult to mark with the thumb-nail. However, *L. luteus* has a tendency to drop whole pods as the crop ripens and it is advisable to examine this species carefully and to harvest at a somewhat earlier stage.

Moisture content of the seed at harvest can be 18–20%; combining below 14% may damage germination.

Drying and Storage

Seed moisture content must be reduced to 14% or less for safe storage until the next sowing season. The temperature of the drying air should not exceed 40°C.

Seed Cleaning

An air/screen cleaner will normally clean seed satisfactorily.

Lathyrus sativus L. Common Names: Grass pea, Chickling pea, Vetch or Khesari dahl

Dwarf chickling vetch (*Lathyrus cicera* L.) is also grown in some Mediterranean countries. Sometimes grown as a fodder crop, grass pea is grown as a pulse in India and the Near East. It suffers from the disadvantage that too much in the diet gives rise to 'lathyrism' (a paralysis of the lower limbs), but it is a very dependable crop in dry areas. It does less well on acid soils. Seed crops are eligible for certification in India (Tunwar and Singh 1988, pp. 81–4).

Grass pea is mainly self-fertilised and requires only enough isolation to ensure that mixture with other crops is avoided at harvest (5–10 m) when grown for seed.

Grass pea is grown as a winter annual. Crops are usually small in area and are sown broadcast at a seed rate of 35–40 kg/ha. They are usually harvested by hand, although the crop is suitable for combining.

Macrotyloma uniflorum (Lam.) Verde. (Syn. *Dolichos uniflorum* Lam.). Common Name: Horse gram

Horse gram is grown in India as a pulse crop and in Australia as a forage crop. It is eligible for seed certification in India (Tunwar and Singh 1988, pp. 79–80).

Horse gram is a short-season, summer growing plant (Langer and Hill 1991, p. 249). It is self-fertilised and requires only 5–10 m isolation.

References

Belteky, B. and Kovecs, I. (1984). *Lupin, the New Break Crop.* (Ed. English edition J.G. Edwards). Panagri, Bradford-on-Avon.

Cooke, L.A. and Marchylo, B.A. (1992). Rapid electrophoretic detection of a lentil-like vetch (*Vicia sativa* L.) in commercial samples of French dark speckled lentils (*Lens culinaris* Medik.). *Plant Varieties and Seeds*, **5**, 1–12.

FAO (1995). *FAO Production Yearbook.* Food and Agriculture Organisation of the United Nations, Rome, 41.

IBPGR (1981a). *Descriptors for Pigeonpea.* International Board for Plant Genetic Resources, Rome.

IBPGR (1981b). *Lupin Descriptors.* International Board for Plant Genetic Resources, Rome.

IBPGR (1985a). *Lentil Descriptors.* International Board for Plant Genetic Resources, Rome.

IBPGR (1985b). *Chickpea Descriptors.* International Board for Plant Genetic Resources, Rome.

Langer, R.H.M. and Hill, G.D. (1991). *Agricultural Plants.* Cambridge University Press, Cambridge.

OECD (1995a). *List of Cultivars Eligible for Certification, 1995.* Organisation for Economic Co-operation and Development, Paris.

OECD (1995b). *Control Plot and Field Inspection Manual.* Organisation for Economic Co-operation and Development, Paris.

Purseglove, J.W. (1984). *Tropical Crops: Dicotyledons.* Longman, London.

Tunwar, N.S. and Singh, S.V. (1988). *Indian Minimum Seed Certification Standards.* The Central Seed Certification Board, New Delhi.

UPOV (1979). *Guidelines for the Conduct of Tests for Distinctness, Homogeneity and Stability: Lupins.* Doc. TG/66/3, International Union for the Protection of New Varieties of Plants, Geneva.

Further Reading

IBPGR (1982). *Revised Wing Bean Descriptors.* FAO, Rome.

LEGUMINOSAE: AGRICULTURAL CROPS – OILSEEDS

Glycine max (L.) Merr. Common Names: Soya bean, Soyabean, Soybean

Soyabean is a major source in the world for oil and protein. It originated in China, which is still a major producer, but the main source for export of soyabean products is the USA, with Brazil in second place. It may also be grown as a pulse crop.

Distinguishing between Cultivars

Soyabean is very sensitive to day length. There are cultivars which are short-day (both obligatory and facultative) and others which are day-neutral (Copeland and McDonald 1995a). Cultivars have been selected which are adapted to long days and can be grown in relatively high latitudes where the growing season is short. Such cultivars are not particularly early to mature when grown in short-day conditions in the tropics, where most cultivars will mature at about the same time. In North America cultivars have been placed into 12 groups designated 00, 0 and I to X, in which the former (earlier maturing groups 00 and 0) are adapted to the areas with the longest days. However, these North American groupings may not apply in other parts of the world since soyabeans are also sensitive to temperature and moisture conditions at flowering time. IBPGR (1984) gives five groups which correspond to the 12 North American groups as follows:

IBPGR Group	USA/Canadian Group
1	00, 0
2	I, II
3	III, IV
4	V, VI, VII
5	VIII, IX, X

Nine groups, which have not been related to the groups given above, are given in UPOV (1983): 1, very early; 2, very early to early; 3, early; 4, early to medium; 5, medium; 6, medium to late; 7, late; 8, late to very late; 9, very late.

The following cultivar characteristics are based on OECD (1995a) and UPOV (1983).
Main features used for distinguishing cultivars in the field:

Colour of hairs on plant: grey, tawny
Shape of lateral leaflet: lanceolate, rhomboidal, ovoid, elliptic
Leaflet size: small, large
Leaf colour: light green, dark green
Flower colour: white, violet
Beginning of flowering (one flower open on 10% plants): early, late
Plant growth type: determinate, indeterminate
Plant growth habit: erect, prostrate

Other features used for distinguishing between cultivars:

Hypocotyl anthocyanin colouration: absent, strong
Time of maturity: early, late
Plant height at maturity: short, long
Pod colour: light brown, dark brown
Seed size: small, large
Seed shape: spherical, spherical/flattened, elongated, elongated/flattened
Colour of seed testa: yellow, green, brown, black
Colour of seed hilum: grey, yellow, brown, dark brown, black

Laboratory tests:

Cooke (1995) mentions high performance liquid chromatography (HPLC) and polymerase chain reaction (PCR) as useful techniques for distinguishing between soyabean cultivars.

Environmental Requirements for Seed Production

In general, seed crops can be grown anywhere that crops to produce oil and protein are grown.

The crop is grown in the subtropics and extends between latitudes 52°. However, seed grown in tropical conditions is difficult to store and may lose germination, particularly in humid areas. Seed crops grown in higher latitudes are very sensitive to rainy, cold periods during flowering which will cause flowers to abort (Shuster and Bohm 1981). The crop grows best at temperatures of 27–32°C and yield can be reduced by cool nights or periods above 38°C (Langer and Hill 1991). During the growing season the crop does best with 500–750 mm water, either from precipitation or irrigation. Fertile soils with pH 6.0–6.5 give the highest yields.

Previous Cropping and Isolation

Soyabean seed is not long-lived in the soil and an interval of one year free from any soyabean crop is usually considered satisfactory.

The species is self-fertilised and cross-fertilisation is generally accepted as being less than 1%. Isolation by a barrier or a gap of 3 m is usually required.

Seed Crop Establishment

Seed crop establishment is not different from establishment of a crop for oil and protein production. The seed is sown in rows 45–100 cm apart at a seed rate to give about one plant per 2.5 cm. There is some evidence that denser stands ripen more uniformly, but this is dependent on the conditions under which the crop is grown. Soil temperature at sowing of 3°C causes reduced emergence and yield compared with sowing at 10°C (Jones and Gamble 1993). *Rhizobium japonicum* is specific for soyabean and seed should be inoculated if the area to be sown has not carried soyabean before.

Seed Crop Management

Management as for a grain crop is also suitable for a seed crop. The crop responds well to phosphate and potash and requires lime on soils with low pH. Weed control is necessary to avoid competition with the soyabean plants, either by cultivation or by chemical means.

There are numerous pests and diseases which may attack soyabean. For many of them there are resistant cultivars available. Insecticides and fungicides can also be used. However, because conditions vary in different parts of the world it is not possible to list all the pests and diseases and the treatment of them. The following are some of the main pests and diseases.

Scientific name	Common name
Pests	
Melangromyza sojae Zehnt.	Stem miner
Leguminivora glycinivorella Matsmura.	Soyabean pod borer
Nezara viridula L.	Green stink bug
Aphis craccivora Koch.	Groundnut aphid
Tetranychus spp.	Spider mite
Epilachna varivestis Mulsant.	Mexican bean beetle
Empoasca fabae Harr.	Leaf hopper
Anticarsia gemmatalis Hub.	Velvet bean caterpillar
Cerotoma spp.	Bean leaf beetle
Spodoptera spp.	Army worm
Meloidogyne incognita (Kofoid and Wood) Chitwood.	Root-knot nematode
Fungal diseases	
Cercospora kikuchi T.Masu & Tomoyasu. Gardner.	Purple seed stain
Cercospera sojina Hara.	Frog-eye leaf spot
Cornespora spp.	Target spot
Peronospera manschurica (Acum.) Syd ex Guam.	Downy mildew
Phialophora gregata (All. & Chambert) W.Gams.	Brown stem rot
Phytophora megasperma (Drechs) var. *sojae* A.A. Hildebrand.	*Phytophora* root rot
Rhizoctonia solani Kuehn.	*Rhizoctonia* root rot

Scientific name	Common name
Bacteria	
Pseudomonas glycinea Coerper.	Bacterial blight
Xanthomonas campestris pv. *phaseoli* (Smith) Dye.	Bacterial pustule
Virus	
	Soyabean mosaic
	Soyabean yellow mosaic

Seed Crop Inspection and Roguing

Most of the characteristics listed above can be seen at maturity, but many are also best seen at flowering. Two inspections are therefore often specified for soyabean seed crops.

Harvesting

When small areas are harvested by hand, the plants should be threshed as soon as possible and the seed taken to good storage. Seed which is left too long in the pod may lose germination, particularly in moist climates.

Large areas are usually direct combined, the soyabean being very well suited to this method.

Soyabean reaches physiological maturity about two to three weeks before it is ready for harvest, and at this stage the moisture content of the seed is about 50%. However, to enable seed to be placed in store without immediate drying, the seed moisture content must be below 15%.

At physiological maturity the seed begins to turn yellow and this is followed by yellowing of the leaves, which begin to drop. Seed moisture drops rapidly as this process proceeds and the pods dry. In moister climates it may be advantageous to spray the crop with a fungicide to reduce fungal infection on the maturing pod. Mature seed sheds easily and harvest should not be delayed after seed moisture content has fallen to 14%.

The seed is easily damaged and therefore the combine should be set carefully. Seed above 14% moisture content will be soft and will bruise easily; seed below this moisture content will be hard and will crack easily.

Desiccation should be avoided for a seed crop, as yield and seed quality may be reduced. However, it may be necessary to use a desiccant in those areas where early frost or inclement weather may be expected if harvest is delayed.

Drying and Storage

The seed is easily damaged if handled roughly, and deterioration is rapid if conditions are not suitable. It is therefore essential to ensure that the seed receives attention immediately it is brought in from the field. The seed does not store well and even good seed will start to deteriorate after a few months when temperature or humidity is not controlled.

Seed moisture content must be reduced to 12% in order to maintain viability. Drying must be gradual and in forced-air driers the air temperature should not exceed 40°C, or less if seed moisture content is above 20%.

Treatment with a fungicide immediately after drying and cleaning will prevent the spread of storage fungi, but should not be used if it is intended to inoculate the seed later with *Rhizobium*.

For storage up to six months the seed moisture content should not be allowed to exceed 12% and if possible should be maintained at 10%. It is necessary to examine the seed at regular intervals and to redry if moisture content starts to rise. If air-conditioned storage is available, the seed should be placed in store with a moisture content of no more than 10%. The store should be maintained at 60% humidity and 20°C.

Small quantities of seed can be placed in a refrigerator at 5°C. Storage in moisture-proof containers such as sealed heavy-duty polyethylene bags or cans is also possible.

Seed Cleaning

An air/screen cleaner is usually all that is required. Careful handling is needed to avoid mechanical damage to the seed.

Treatment with a fungicide to reduce seed-borne fungi will not affect *Rhizobium* in the soil, but treated seed cannot be inoculated before sowing.

Arachis hypogaea L. Common Names: Peanut, Groundnut

Peanuts are reported as originating in South America (Gregory and Gregory 1984) but the largest areas are now in the tropical and warm temperate regions of Asia and Africa. Gregory and Gregory (1984) distinguish two subspecies comprising four botanical varieties:

subsp. *hypogaea*: no floral axes on main axis; alternating pairs of vegetative and floral axes along lateral branches
  var. *hypogaea*: type Virginia, less hairy, branches short
  var. *hirsuta*: type Peruvian runner, more hairy, branches long
subsp. *fastigiata*: floral axes on the main axis; continuous runs of one/many floral axes along lateral branches
  var. *fastigiata*: type Valencia, little branched
  var. *vulgaris*: type Spanish, more branched

In UPOV (1985), three main types are distinguished as 'commercial groupings': Valencia, Virginia and runner. Other classification characteristics given by UPOV (1985) are:

Time of maturity: early, medium, late
General pattern of flowering: alternate, sequential
Seed weight per 1000 seeds: low, medium, high (at 7% moisture)

Distinguishing between Cultivars

The following characteristics are based on UPOV (1985) and OECD (1995b).

Main features used for distinguishing cultivars in the field:

Plant growth habit at flowering: erect, semi-erect, prostrate
Pod constrictions: absent or very shallow, shallow, medium, deep, very deep
Prominence of pod beak: absent or very inconspicuous, inconspicuous, medium, prominent, very prominent
Shape of pod beak: straight, curved
Colour of uncured mature testa: monochrome, variegated

Other features used for distinguishing between cultivars:

Prostrate cultivars only: main stem growth habit
Prostrate cultivars only: side branches growth habit
Plant branching
Time of maturity (for curing)
Leaflet size and colour
Flowering pattern of main stem
Pod texture of surface
Pod number of kernels
Cultivars with variegated testa only: colour of mature uncured testa
Kernel shape and size
Fresh matured kernel dormancy period
Fresh matured kernel percentage of shell
Resistance to fungal diseases

Environmental Requirements for Seed Production

Because the seed of peanut is matured below ground, the crop is not suited to heavy soils nor to wet conditions at harvest. A deep, friable soil with good status for calcium is most suitable. However, the crop usually does not respond well to fertiliser applications and is often grown on the residual fertiliser from previous crops. Peanuts will not withstand frost and can only be grown in frost-free areas. They are day-neutral, not photosensitive, and can therefore be grown in tropical and the warm temperate regions. Purseglove (1984) gives a range from 40°N to 40°S.

Previous Cropping and Isolation

Peanuts generally have a rather short dormancy period and seed in pods left in the soil after harvest soon start to germinate, so that the seedlings can be destroyed by cultivation. An interval of one year free from peanuts is therefore usually sufficient between seed crops. However, in India (the world's largest producer) an interval of two years is specified in the seed certification standards (Tunwar and Singh 1988).

Cultivars in the groups Virginia and Peruvian runner are self-fertilised almost without exception. Those of Valencia and Spanish are also self-fertilised but are slightly more prone to crossing than the other two. For seed production, isolation sufficient to prevent mixture during harvest is considered sufficient. The Indian regulations specify 3 m (Tunwar and Singh 1988).

Seed Crop Establishment

Seed crops are established in the same way as crops for oil or protein production. Seed is sown in rows and closer spacing helps to produce a more compact plant for harvest. Seed may be sown with a precision drill; for plate planters a special peanut plate is available.

Inoculation with *Rhizobium* is not usually necessary except in areas which have not grown the species before. When seed is to be inoculated it should not be treated chemically.

Seed is sometimes sown on ridges. While this makes lifting the crop at harvest easier, it may also reduce the amount of pegs which the plant is able to bury, so reducing seed yield.

Seed Crop Management

Management of a seed crop is not different from that of a commercial crop. Weeds are not a special problem for seed production, but the peanut plant is a poor competitor until well established and it is necessary to control weeds in the early stages.

Fertiliser is usually applied sparingly. The seed is sensitive to fertiliser placed in close proximity to it. The species is susceptible to some fungal diseases for which genetic resistance or fungicides are available (e.g. *Cercospora* spp). The rosette virus may also cause poor growth; the aphid vector should be controlled by aphicide.

Seed Crop Inspection and Roguing

The most important period for crop inspection and roguing is between flowering and harvest.

Harvesting

After pollination, the fruit is formed at the tip of a stalk or peg which grows downwards; the peg elongates and the developing pods are forced under the soil where they remain as they ripen. Some flowers may be fertilised underground. Since the foliage may remain green when the pods are mature, it is necessary to dig samples from below ground to follow the progress of ripening. Mature pods are dark-coloured inside and the seed assumes its final colour as determined by the cultivar (see the distinguishing features of cultivars). When mature, the seed will have a moisture content of 30–40% and at this stage the crop can be lifted. Cultivars which are erect and closely bunched may be prone to sprouting and must be harvested before this can happen.

Small areas are dug by hand. Special equipment is available for lifting the crop mechanically, which cuts the tap root and removes the plants from the soil. After lifting by either method the plants are left to dry in the field with the pods turned uppermost in windrows. Alternatively the crop can be placed on tripods to dry, but this requires much labour.

It is important to handle the crop with care as seed will be lost through shedding. The pods are removed from the plants first when moisture content falls to about 20%. This can be done by hand, but special equipment is available, either small stationary or portable machines or those that are capable of picking up the crop from the windrow. Combines can also be converted for this purpose.

It is usual to store the seed in the pods until near sowing time as this provides some protection from storage pests and fungi. De-hulling is therefore usually a separate operation.

Drying and Storage

Unhulled seed will have a moisture content of about 20% when delivered to store and this must be reduced to below 10% for safe keeping. Drying must, however, be done slowly as pods dried too quickly become very brittle and may easily be damaged; for this reason also, pods should not be dried much below 10%.

When forced-air driers are used, the temperature of the drying air should be 35°C and must not exceed 38°C. Cultivars differ in the ease with which they may be damaged during drying and it is necessary to check the seed carefully during drying. In many tropical and subtropical areas the ambient air will effect drying, but the seed should not be exposed to direct sunlight and must be protected from rain. According to Copeland and McDonald (1995b), peanut seed has one of the lowest seed moisture contents when in equilibrium with the ambient air at relative humidity of 75% or above and temperature of approximately 25°C.

Seed Cleaning

Pods delivered to store may contain trash and this should be removed, preferably before drying and storing. A pre-cleaner with aspiration will do this.

After de-hulling the main task will be to remove broken pieces of pod, a task which can be achieved in an aspirator. If air/screen cleaners are used it is essential to avoid damage to the seed coat during handling, as this will reduce viability.

Size grading is only necessary when the seed is to be sown by precision planters. Inoculation with *Rhizobium* is not usually necessary, but it may be needed if the seed is to be used in an area where peanuts have not been grown before. When seed is to be inoculated it must not be dressed with a fungicide.

References

Cooke, R.J. (1995). Variety Identification: modern techniques and applications. In *Seed Quality: Basic Mechanisms and Agricultural Implications* (Ed. A.S. Basra). Food Products Press, New York, Chapter 9.

Copeland, L.O. and McDonald, M.B. (1995a). *Principles of Seed Science and Technology*. Chapman and Hall, New York, 3.

Copeland, L.O. and McDonald, M.B. (1995b) *Principles of Seed Science and Technology*. Chapman and Hall, New York, 187.

Gregory, W.C. and Gregory, M.P. (1984). Groundnut. In *Evolution of Crop Plants* (Ed. N.W. Simmonds). Longman, Harlow, 151–153.

IBPGR (1984). *Descriptors for Soyabean.* International Board for Plant Genetic Resources, Rome.

Jones, D.J. and Gamble, E.E. (1993). Emergence and yield of soyabean as influenced by seedlot vigour, seed moisture and soil temperature. *Plant Varieties and Seeds*, **6**, 39–46.

Langer, R.H.M. and Hill, G.D. (1991). *Agricultural Plants*. Cambridge University Press, Cambridge, 256–259.

OECD (1995a). *Control Plot and Field Inspection Manual*. Organisation for Economic Co-operation and Development, Paris, 81.

OECD (1995b). *Control Plot and Field Inspection Manual*. Organisation for Economic Co-operation and Development, Paris, 99.

Purseglove, J.W. (1984) *Tropical Crops, Dicotyledons*. Longman, Harlow, 225–235.

Schuster, W. and Bohm, J. (1981). Experience in Soyabean Breeding in Middle Europe. In *Production and Utilization of Protein in Oilseed Crops* (Ed. E.S. Bunting). Commission of the European Communities, Luxemburg.

Tunwar, N.S. and Singh, S.V. (1988). *Indian Minimum Seed Certification Standards*. The Central Seed Certification Board, New Delhi, 101–102.

UPOV (1983). *Guidelines for the Conduct of Tests for Distinctness, Homogeneity and Stability: Soya bean.* Doc. TG/80/3, International Union for the Protection of New Varieties of Plants, Geneva.

UPOV (1985). *Guidelines for the Conduct of Tests for Distinctness, Homogeneity and Stability: Groundnut.* Doc. TG/93/3, International Union for the Protection of New Varieties of Plants, Geneva.

LINACEAE: AGRICULTURAL CROP

Linum usitatissimum L. Common Names: Flax, Linseed

This species has two distinct groups of cultivars which are grown for different purposes. Flax is grown to provide fibre from the stems of the plants which is made into linen. Linseed is grown for seed used for oil extraction and protein feed for livestock; the oil is mainly industrial, used in paints and other products which require a drying oil, but recently plant breeders have succeeded in selecting cultivars which produce edible oils. Another recent development has been the selection of cultivars which are reasonably frost hardy and can be sown in autumn in higher latitudes.

L. usitatissimum is grown throughout the temperate regions but the main centres of production are in Argentina, Canada, India and the former Soviet Union. The latter is the major producer of fibre flax.

Flax cultivars are tall and with few branches; linseed cultivars are shorter and branch more, particularly at the upper part of the stem, and so provide more flowers and hence seed.

Distinguishing between Cultivars

The following lists are based on UPOV (1980) and OECD (1995).

Main features used for distinguishing cultivars in the field:

  At time of flowering
    Time of beginning of flowering: early, medium, late
    Natural height of plant including branches: very short, medium, very tall
    Petal colour: white, light blue, blue, pink, red violet, violet
    Longitudinal folding of petal: absent, present
    Size of corolla: small, medium, large
    Sepal dotting: weak, medium, strong

  Colour of stamen filament top: white, blue, violet
  Anther colour: yellowish, bluish
  Style colour at base: white, blue

Other features used for distinguishing between cultivars:

  At green maturity
    Stem length: short, medium, long
    Boll size: small, medium, large
    Ciliation of false septa of boll: absent, present
    1000 seed weight: low, medium, high
    Seed colour: yellow, olive green, brown, variegated
    Intensity of brown colouration of seed: weak, medium, strong

Environmental Requirements for Seed Production

Both flax and linseed are grown in temperate regions. Although flax is grown in the cooler areas for the production of fibres, so that longer stems are encouraged, for seed production both types do better in warmer areas where there is the prospect of dry weather at harvest.

The species is tolerant of most soil types, but does best on those with pH over 6.5.

Previous Cropping and Isolation

When cultivars of the same type are grown, a gap of one year free from crops of the same kind is required. However, when changing from flax to linseed (or *vice versa*) or from one type of linseed to another (winter/spring, industrial oil/food oil), a longer gap should be allowed.

The species is self-fertilised and therefore separation from other crops sufficient to prevent mixture at harvest (i.e. a fence, hedge or ditch, or a clear gap of 3–5 m) is all that is required, However, when crops of different types of cultivar are adjacent, it is advisable to allow a greater distance between them.

Seed Crop Establishment

The establishment of a seed crop does not differ from that of a commercial crop. Most cultivars are sown in the spring as early as weather will allow; however, the crop is generally slow to establish and sowing should be delayed if temperatures are very low. Winter linseed should also be sown as early as possible (after mid-September in UK).

Good seed bed conditions are necessary, i.e. a firm, fine tilth, although for winter sowing some surface roughness may help to protect from frost. It is best to delay sowing to achieve a satisfactory seed bed to allow the crop to make a quick start.

It is advisable to aim for a plant population of about 500–700 plants/m^2; depending on seed size this will require a seed rate of between 30 and 90 kg/ha. Flax for fibre is normally sown at a higher seed rate.

Seed Crop Management

Management of a seed crop is not different from that for an oil or fibre crop. Maintenance levels of phosphate and potash should be applied. The crop generally responds to nitrogen applied both in the seed bed and as a later top-dressing. However, it is important to avoid lodging, which can make harvest very difficult. Growth regulators are not advised for seed crops as they may be damaging.

Seed Crop Inspection and Roguing

Inspection and roguing are best carried out during flowering. Flowers open in the early morning and fade by mid-day; effective roguing or inspection can therefore only take place in the early morning.

Harvesting

Seed is ripe when the capsules have turned yellow brown and the leaves withered, although the stems may remain green. The capsules will be dry and the seed inside rattles when the plant is shaken. Shedding is not normally a problem.

The stems of the species are very tough and can be difficult to cut, particularly in lodged crops. It is usual to combine direct but the knives must be sharp and the combine sealed to avoid loss of seed through the casing (particularly around elevators).

An alternative to combining is to use a stripper header to remove the capsules from the standing crop, leaving the straw to be removed later.

Desiccants can be used to hasten ripening in oil or fibre crops, but should be avoided if possible for seed crops as damage to germination may occur.

Flax is normally harvested for maximum fibre production and this does not favour seed production. Combining a flax crop for seed will leave fibre which is not suitable for linen production, but it can be used for other industrial purposes.

Drying, Storage and Seed Cleaning

Seed at harvest will usually contain 12–16% moisture and must be dried to 9% or less for safe medium-term storage. For storage for longer than six months, 7% or less is recommended. The seed is difficult to handle in store as it flows very freely; equipment should be sealed to prevent seed loss.

For on-floor drying the air temperature should not exceed 5°C above ambient, and for ventilated bins 7°C. When using continuous-flow driers the seed depth must be adjusted so that air-flow is not restricted but is sufficient to prevent bubbling. At moisture content of 29% seed temperature should not be allowed to exceed 32°C; below 19% moisture seed temperature may be allowed to reach 60°C.

An air/screen cleaner is normally all that is required to clean flax or linseed seed.

There are several fungal diseases which are seed-borne and treatment is possible. For an account of these see Mercer et al. (1994) and Mercer and Ruddock (1996).

References

Mercer, P.C. and Ruddock, A. (1996). The effect of variety and fungicide on yield and the incidence of pathogens in linseed in Northern Ireland. *Plant Varieties and Seeds*, **9**, 101–110.

Mercer, P.C., Hardwick, N.V., Fitt, B.D.L. and Sweet, J.B. (1994). Diseases of linseed in the United Kingdom. *Plant Varieties and Seeds*, **7**, 135–150.

OECD (1995). *Control Plot and Field Inspection Manual*. Organisation for Economic Co-operation and Development, Paris, 55.

UPOV (1980). *Guidelines for the Conduct of Tests for Distinctness, Homogeneity and Stability – Flax, linseed*. Doc. TG/57/3, International Union for the Protection of New Varieties of Plants, Geneva.

MALVACEAE: AGRICULTURAL CROPS

Gossipium spp. Common Name: Cotton

Purseglove (1984a) gives the following key to the cultivated cotton species.

A. Bracteoles entire or dentate; teeth usually less than three times as long as broad: Section *Herbacea*

 B. Bracteoles flaring widely from flower, usually broader than long; upper margin usually with six to eight teeth; capsule rounded or with prominent shoulders: *G. herbaceum* L.

 BB. Bracteoles closely investing flower, longer than broad, entire or with three to five teeth near apex; capsule tapering: *G. arboreum* L.

AA. Bracteoles deeply lancinate; teeth usually more than three times as long as broad: Section *Hirsuta*.

 B. Leaves deeply lancinate for two-thirds length into three to five lobes; petals usually bright yellow with basal reddish spot; anthers compactly arranged; pollen deep yellow; capsule coarsely pitted with black oil glands: *G. barbadense* L.

 BB. Leaves less deeply lacinate for half length or less into three or rarely five lobes; petals usually pale yellow or cream without basal reddish spot; anthers loosely arranged; pollen pale yellow or cream; capsule surface smooth: *G. hirsutum* L.

The majority of the world's cotton is grown from cultivars of *G. hirsutum*, with a medium/long staple. Cultivars with long staple are of *G. barbadense* and these provide the best quality lint. Those with short staple are of *G. herbaceum* and *G. arboreum*. Hybrids within and between species are produced by hand emasculation and pollination.

Trade in cotton seed is largely controlled by the ginneries where the seed capsules are delinted to provide cotton fibre. These companies or co-operatives are able to control the cultivars which may be grown because they control this essential process.

Distinguishing between Cultivars

The following list of characteristics is based on OECD (1995) and UPOV (1985).

Main features used for distinguishing between cultivars in the field:

 Colour of petals at flower opening
 Plant shape
 Plant height
 Leaf shape
 Leaf size
 Boll size

Other features used for distinguishing between cultivars:

 Density of foliage
 Number of nodes to first fruiting branch
 Length of longest vegetative branch
 Length of first fruiting branch
 Pubescence of lower side of mid-rib
 Gossipol glands
 Nectaries
 Boll shape
 Pitting of boll surface
 Length of peduncle
 Prominence of boll tip
 Density of fuzz on seed
 Boll: content of lint
 Fibre length, strength and fineness
 Seed size
 Boll opening

Laboratory tests:

Electrophoresis of seed proteins (see, for example, Agrawal et al. 1988; Rao et al. 1990).

Environmental Requirements for Seed Production

Cotton is grown throughout the tropics and sub-tropics. The plant is day-neutral. The main producing countries are in Asia (China, India, Pakistan, the former Soviet Union and Turkey), Africa (Egypt and Sudan), Europe (Greece), North America (Mexico and the USA), South America (Argentina and Brazil) and Australia.

Low temperature increases the production of vegetative growth. For seed production there should be a frost-free growing period of 200 days with temperature above 22°C; below 15°C the plant makes little or no growth.

Cotton requires sun and will not tolerate shade. At least 500 mm rainfall is required during the growing season; the crop does well under irrigation. A dry period during flowering is desirable; excessive moisture during flowering causes higher boll shedding.

Soil type is not critical but it is essential to produce a good seed bed which should be fine, firm and free from weeds. Cotton is said to be sensitive to boron deficiency.

Previous Cropping and Isolation

Fields chosen for seed production must be free from volunteer plants; normally this will require at least one year free from cotton.

While mostly self-pollinating, cotton flowers are visited by insects and so a varying amount of cross-pollination does take place. The extent of cross-pollination ranges from 6 to 25%. There are varying requirements for isolation in different seed certification schemes, but the consensus seems to be as follows.

(i)   Between cultivars of the same type: early generation, 50 m; Certified Seed, 20 m.
(ii) Between cultivars of different types: early generation, 200 m; Certified Seed, 100 m.

(iii) Between cultivars of different species with different ploidy levels: all generations, 5 m.

Cytoplasmic male sterility was reported by Percival and Kohel (1990) as having been isolated by Meyer in 1973, but it has been difficult to use. Hybrid cultivars are produced by hand emasculation and pollination in countries where cheap labour is available. Isolation as for other cultivars is required, except that isolation from other crops of the male (pollen donor) parent can be 5 m.

Seed Crop Establishment

After harvest, seed of cotton is first de-linted (the longer hairs are removed to produce cotton fibre). This leaves ' fuzzy' seed (seed covered by short hairs). Fuzzy seed can be sown by hand but is not suitable for mechanical drills; fuzz must be removed either mechanically or by treatment with sulphuric acid, the latter method being preferred.

Rows are usually approximately 90 cm apart with 30 cm between plants, but this spacing can be varied depending on water supply and soil type. Because of the branching habit of the plant, spacing is less critical than with other species. Seed rate varies from 20 to 30 kg/ha with the type of seed (fuzzy, de-fuzzed mechanically or by acid); the lowest rates are used for seed de-fuzzed by acid.

Early sowing is essential to obtain a long growing season; cotton usually requires about 200 days, with 50–70 days between fertilisation and the opening of the boll.

Bacterial blight (*Xanthomonas malvacearum* (E.F. Sim) Dowson) is seed-borne and it is essential to use disease-free seed or seed which has been treated with suitable anti-bacterial dressing.

Seed Crop Management

Management of a seed crop does not differ from that required for a crop to produce fibre. The crop responds to nitrogen and requires adequate phosphate and potash (see, for example, Sawan et al. 1989). Boron deficiency symptoms have been noted in some areas.

There are numerous fungal diseases and insect pests in different parts of the world where cotton is grown. Cultivars which are resistant to many of them are available, and fields with a history of disease should be avoided for seed growing.

Weeds may compete with the seedlings during the establishment phase and should be controlled. As the plants cover the ground, weeds are not usually a problem. In the USA weeds listed as 'objectionable' are sandbur (*Cenchrus patuciflorus*), cocklebur (*Xanthium* spp.) and Johnson grass (*Sorghum halepense*).

Seed Crop Inspection and Roguing

Roguing is possible once the crops are established, but will be most effective when the plants begin to flower.

Most seed certification schemes require more than one inspection to be made between early flowering and harvest.

Harvesting

The crop is mature when the bolls open and the lint is exposed. Harvest begins when the lint is dry. Many crops are picked by hand and it is then possible to visit the crop on several occasions to harvest the bolls as they ripen.

Equipment for mechanical harvesting has been developed, particularly in the USA. Two types of mechanical harvester are available: the spindle picker and the stripper. When mechanical harvesting is intended, cultivars should be chosen which are as uniform in ripening as possible. With some cultivars it is possible to apply defoliants before harvest.

Seed harvested by machine contains much trash (broken leaves and pieces of capsule) which has to be removed before ginning.

Drying, Storage and Seed Cleaning

Cotton seed will lose viability rapidly if stored at moisture content above 10%.

Seed after harvest is de-linted in a gin, normally a saw gin, but for high quality fibres a roller gin may be used. The latter is less likely to cause mechanical damage to the seed and hence less damage to germination.

After de-linting the seed retains the short, fuzz hairs and fuzzy seed may be used for sowing. However, for mechanical handling the fuzz must be removed, but it is often advisable to delay this treatment until the seed is required for sowing. Removal of fuzz may be mechanical or by sulphuric acid. The latter process is generally preferred, but there is some evidence that some cultivars may be sensitive to acid treatment (Khah and Passam 1994).

Hibiscus cannabinus L. Common Name: Kenaf

Kenaf is used to produce fibre which is similar to that obtained from jute (*Corchorus* spp.). Production has increased over the last 50 years, starting during World War II.

Distinguishing between Cultivars

The species is very variable and cultivars are reported to differ in the following characteristics (Purseglove 1984b; Cobley 1977):

Habit of growth
Colour of stem: red, purple, green
Leaf type: cordate, divided (five to seven lobes)
Colour of petals: with or without a crimson spot

Environmental Requirements for Seed Production

Kenaf is grown in the tropics and subtropics between latitudes 45°N and 30°S. It is photosensitive and flowers at day length of 12.5 or less; consequently, for seed production the range within which it can be grown is more limited than that for fibre production.

Kenaf requires humid conditions with temperatures of 20–30°C. However, it is less demanding of rainfall or temperature than jute. Kenaf will not tolerate waterlogging and gives the best yields on free-draining loams which are rich in humus.

Previous Cropping and Isolation

The seed of kenaf is not normally long-lived in the soil and may be expected to lose germination after one or two years.

The species is self-pollinating and is usually treated as such for isolation. However, up to 4% out-pollination has been reported and therefore it is advisable to provide more isolation in the early generations of production of improved cultivars.

Seed Crop Establishment and Management

For seed production, crops are sown on a wider spacing than those intended for fibre. The requirement is to encourage flowering branches rather than long straight stems. Spacing of 30 x 60 cm is generally recommended.

After establishment, the crop requires little attention until harvest.

Seed Crop Inspection and Roguing

Crops may be inspected and rogued when vegetative characters can be observed. A later inspection during flowering allows for an additional check on trueness to cultivar and general condition of the crop.

Harvesting, Drying, Storage and Seed Cleaning

Seed harvest will be later than harvest for fibre. The capsules change colour to dark brown. The calyx and epicalyx persist and become hard. The brown or grey seeds are hairy and these hairs can cause irritation if handled.

Hand harvesting is usual. The hairs should be removed from the seed by rubbing to allow the seed to flow freely. Subsequent cleaning may be by hand winnowing or on an air/screen cleaner.

Urena lobata L. Common Names: Aramina, Congo jute

This fibre crop is grown mainly in Brazil and the Congo. The species is said to be very variable, but there is no published information on cultivars.

The crop is grown in a similar manner to kenaf (*Hibiscus cannabinus*). Purseglove (1984c) states that the seeds are normally left in the carpels before sowing but that the hooks on the carpels should be removed by rubbing or treatment with sulphuric acid, otherwise the seeds will tend to cling together.

Crane and Walker (1984) were unable to find any information on mode of pollination, but state that honey bees collect nectar from the flowers and may contribute to pollination.

References

Agrawal, P.K., Sing, D. and Dadlani, H. (1988). Identification of cotton hybrid seeds using PAGE. *Seed Science and Technology*, **16**, 563–569.
Cobley, L.S. (1977). *An Introduction to the Botany of Tropical Crops* (Revised by W.M. Steele). Longman, London, 273–274.
Crane, E. and Walker, P. (1984). *Pollination Directory for World Crops*. The International Bee Research Association, London, 78.
Khah, E.M. and Passam, H.C. (1994). Sensitivity of seed of cotton cv. Zeta-2 to damage during acid delinting. *Plant Varieties and Seeds*, **7**, 51–58.
OECD (1995). *Control Plot and Field Inspection Manual 1995*. Organisation for Economic Co-operation and Development, Paris, 51–52.
Percival, E.A. and Kohel, R.J. (1990). Distribution, collection and evaluation of *Gossyfium*. *Advances in Agronomy*, **44**, 225–256.
Purseglove, J.W. (1984a). *Tropical Crops, Dicotyledons*. Longman, Harlow, 345.
Purseglove, J.W. (1984b). *Tropical Crops, Dicotyledons*. Longman, Harlow, 365–368.
Purseglove, J.W. (1984c). *Tropical Crops, Dicotyledons*. Longman, Harlow, 374–376.
Rao, T.N., Nerkar, Y.S. and Patil, V.D. (1990). Identification of cultivars of cotton by sodium dodecylsulphate polyacrylamide gel electrophoresis (SDS-PAGE) of soluble seed proteins. *Plant Varieties and Seeds*, **3**, 7–14.
Sawan, Z.M., Maddah El Din and Gregg, B. (1989). Influence of nitrogen, phosphorus and growth regulators on seed yield and viability and seedling vigour of Egyptian cotton. *Seed Science and Technology*, **17**, 507–519.
UPOV (1985). *Technical Guidelines for the Conduct of Tests for Distinctness, Homogeneity and Stability: Cotton*. Doc. TG/88/3, International Union for the Protection of New Varieties of Plants, Geneva.

MALVACEAE: VEGETABLE CROP

Hibiscus esculentus **L. Syn.** *Abelmoschus esculentus* **(L.) Moench. Common Names: Okra, Lady's finger**

Okra is an annual crop cultivated for the production of its immature fruit (commonly referred to as 'lady's fingers') which are cooked as a vegetable. The immature pods are also dried and stored for cooking in the hot, dry seasons. The crop is widely cultivated in Africa, Asia and America.

Classification of Cultivars

General use: fresh market, early production, drying and storage
Plant character: response to day length, height, degree of branching, duration of cropping, tolerance of rainy season, stem pigmentation
Leaf characters: specified according to position on plant, lobing
Flower: intensity of yellow, relative size and basal pigmentation of petals
Pod: relative length, shape of 'beak', colour, externally glabrous or spiny, shape of transverse section, degree of internal mucilage and fibre
Resistance to mosaic virus

Environmental Requirements for Seed Production

This is a tropical crop; it requires a minimum day temperature of 25°C and will succeed up to 40°C. It is not frost hardy.

Okra will succeed under slightly acid conditions, and will tolerate a soil pH from 6.0 to 6.8. The crop responds to bulky organic manures incorporated during site preparation and will also succeed when following a green manure crop. The optimum ratio of base fertilisers incorporated during preparation is NPK 1:2:1.

Previous Cropping and Isolation

A minimum of two years should be observed between crops, and if there has been an incidence of wilt disease or nematodes, this period should be extended.

The crop is partly self-pollinated and partly cross-pollinated by insects, therefore the minimum isolation distance is 500 m, although this may be increased by some authorities.

Seed Crop Establishment

Seed is sown at the rate of approximately 8 kg/ha in rows 50 cm apart; the inter-row spacing is increased to 60 cm for the more vigorous cultivars. The plants are thinned to approximately 15–30 cm within the rows, depending on the cultivar's vigour.

Seed Crop Management

Cultivations to control weeds should commence by the time the seedlings are singled out. The crop should be irrigated on a regular basis when rainfall is sparse. A nitrogenous fertiliser should be applied as a top-dressing at the rate of approximately 50 kg/ha about one month from seedling emergence, unless the leaching rate is thought to be negligible.

Seed Crop Roguing

The first roguing is done when the plants are well developed but before the start of anthesis. At this stage the general plant habit and vigour are checked, and in addition petiole and stem pigmentation are checked. The second roguing is done after the first flowers open to confirm their size and colour. The final check is made when the first fruits are developed to confirm fruit characters. Plants showing virus symptoms are removed from the crop at all stages.

Harvesting

Individual fruits are ready for harvest as they become dry; this is indicated when they turn a light brown colour. The maturity of pods is in sequence on individual plants, therefore hand harvesting will secure the maximum yield, especially as the ripe fruit have a tendency to shatter. However, in large-scale production the crop is combined when the majority of fruit have ripened.

Threshing and Cleaning

The small-scale crops are hand threshed, otherwise a stationary thresher is used. The extracted seed can be further upgraded by passing it through an air/screen cleaner.

Drying

Okra seed is dried to approximately 12% moisture content for short-term storage or 9.0% for storage in vapour-proof containers.

ONAGRACEAE: AGRICULTURAL CROP

Oenothera biennis L. Common Names: Evening primrose

Evening primrose is a temperate crop and has recently been developed as a source of oil with a high gamma linolenic content for use in the pharmaceutical industry. Seed crops should be grown on contract. Several improved cultivars have been bred and UPOV (1993) gives detailed descriptors. Cultivars are classified by the following characteristics:

Stem – anthocyanin pigment at base of hairs: absent, present
Flower bud – anthocyanin pigment: absent, present
Pod – diffuse anthocyanin pigment: absent, present

For further characteristics see UPOV (1993).

The crop is difficult to establish. The plant is biennial and is planted in late summer/early autumn for harvest in the following year. The seed is small and can be sown in rows 40–60 cm apart at a seed rate of 1 to 1.5 kg/ha. Alternatively the plants may be raised in soil blocks or other suitable medium for transplanting; 30–50 transplants are required per square metre. It is essential that the seed bed is weed free and that soil moisture is adequate to ensure good establishment; irrigation may be necessary.

The species is insect pollinated and suitable isolation should be provided.

The plant is indeterminate in flowering pattern so that harvest timing is difficult to judge correctly. The ripe seed sheds easily and harvest should begin as soon as possible after the first seed has dropped. Desiccants may be used. The crop can be swathed and picked up after a few days or it can be combined direct. The seed is very small and equipment should be sealed against loss. Minimum air draught is needed.

At harvest the seed will usually have a moisture content of about 20% and this must be reduced to 10% for storage.

Reference

UPOV (1993). *Guidelines for the Conduct of Tests for Distinctness, Homogeneity and Stability: Evening Primrose*. Doc. TG/144/3, International Union for the Protection of New Varieties of Plants, Geneva.

PAPAVERACEAE: AGRICULTURAL CROP

Papaver somniferum L. Common Names: Poppy, Opium poppy

Because of the social problems caused by the addictive drug, the cultivation of opium poppy is subject to strict control in many countries. However, there is legitimate use of the plant for the production of the painkiller morphine and the seeds can be used for cooking and for oil extraction.

The crop can be grown over a wide area in temperate and warm, dry climates. It will not grow well in the humid tropics. It is not frost hardy.

Improved cultivars have been produced, but because of the restricted use permitted for the crop they are usually controlled by the pharmaceutical companies.

The seed is very small and is mixed with some inert matter (often heat-killed poppy seed) for sowing. The recommended plant population is 65 plants/m^2.

Poppies are self-pollinating and require only isolation sufficient to prevent mixture with other crops.

For seed, the crop is harvested by combine. When ripe the capsules change colour and the seed can be heard to rattle within (Langer and Hill 1991).

Reference

Langer, R.H.M. and Hill, G.D. (1991). *Agricultural Plants.* Cambridge University Press, Cambridge, 304–307.

PEDALIACAE: AGRICULTURAL CROP

Sesamum indicum L. Common Name: Sesame

Sesame is said to be one of the oldest species grown for extraction of cooking oil of high quality. Production is now mainly centred in Myanmar (Burma), China and India, which together produce about half of the world supply; there is also considerable production in Sudan. Until recently the majority of production has been from local cultivars, of which there are a very large number. Plant breeding is now providing cultivars with improved yield, disease resistance and adaptation to mechanical harvesting through elimination of the dehiscent character.

Distinguishing between Cultivars

No descriptors have yet been published by the international organisations. The main characteristics given by Purseglove (1984) are as follows:

> Season of planting and time of maturity
> Degree of branching of the stem
> Number of flowers per leaf axil: one or three (but note that Cobley (1977) also mentions the possibility of two)
> Number of loculi per capsule: four, six or eight
> Colour of corolla outside: white, pink, purplish
> Colour of capsule: brown, purple
> Colour of seed: white, yellow, grey, red, brown, black (it is thought that seed with white testa yields the best quality oil)

Environmental Requirements for Seed Production

Sesame is essentially a crop for hot and dry areas in the tropics. It is grown in areas with rainfall of about 400 mm distributed throughout the growing season, or in areas with about 1000 mm where it is restricted to a limited period, usually at the start of the growing season.

Sesame will not withstand waterlogging and is best adapted to well drained, deep soils with a sandy texture which allows penetration by the plants' long tap roots. It does not require high applications of fertiliser and will tolerate moderate fertility conditions.

Both long- and short-day cultivars are available.

Previous Cropping and Isolation

Volunteer plants from previous crops can usually be controlled within one year. There are, however, some soil-borne and seed-borne diseases (for example, leaf spot caused by *Cercospora sesami* Zimm.); in areas where these occur it is advisable to allow a longer interval between sesame crops. Chung and Choi (1990) report the possibility of controlling damping-off in seedlings biologically by coating the seed with *Trichoderme viride*.

Although sesame is mostly self-fertilised, a variable amount of cross-fertilisation does occur. Insect cross-pollination has been variously reported as between 1 and 10%. The Indian seed certification standards require 100 m for Foundation and 50 m for Certified Seed production (Tunwar and Singh 1988).

Seed Crop Establishment and Management

Establishment and management of a seed crop does not differ from that employed for a crop to produce oil. The usual seed rate is 4 kg/ha for sowing in rows 40–80 cm apart, depending on available moisture. Avila et al. (1992) found that 4 kg/ha in rows 50–80 cm apart gave the highest seed yields and provided the highest number of plants per square metre at harvest.

A weed-free seed bed and good control of weeds during early growth are essential. Grass weeds such as *Sorghum* and *Digitaria* are particularly troublesome because they have seeds similar in shape and size to those of sesame; they cannot be controlled after the crop is established.

Once established, the crop requires little attention until harvest. In some very dry areas irrigation may be necessary.

Seed Crop Inspection and Roguing

Roguing normally takes place when the crop is flowering or when the capsules have formed. These are also suitable times for inspection. In India three inspections are required: before flowering, during flowering, and at maturity before harvest (Tunwar and Singh 1988).

Harvesting

The crop is ripe when the capsules change colour and the leaves begin to drop.

Local cultivars often have a prolonged ripening period and the dehiscent capsules ripen from the bottom of the plant upwards. Such crops are usually harvested by hand. Plants are cut when the lower capsules are ripe and before they start to dehisce. The cut plants may be placed on racks to dry with mats below to catch the seed as it falls.

Indehiscent cultivars are now available and these can be harvested mechanically. Cutting by binder and stooking in the field for subsequent threshing is possible, but it is now more usual to combine, either direct or after windrowing. Some cultivars have a prolonged ripening period and these are best windrowed. Mechanically harvested seed usually contains some immature seed which must be removed during cleaning.

The seed is very delicate and must be carefully handled to avoid damage to the seed coat which might reduce germination. It also flows very freely and is easily lost during handling. All equipment should be sealed to prevent loss through gaps.

Drying, Storage and Seed Cleaning

Seed is normally grown in areas which are dry and warm during harvest and therefore usually contains about 8% moisture and can be stored safely without further drying. However, if seed is to be stored in moisture-proof containers, moisture content should be reduced to 5%.

Seed is also normally relatively free from green matter after threshing and can be cleaned satisfactorily on an air/screen cleaner.

References

Avila, A., Hernandez, J. and Acevedo, T. (1992). Effect of sowing distance between rows on performance of four sesame (*Sesamum indicum* L.) cultivars. *Agronomia Tropical* (*Maracay*), **42**, 307–320.

Chung Hoo-Sup and Choi Woo-Bong (1990). Biological control of sesame damping off in the field by coating seed with *Trichoderme viride. Seed Science and Technology*, **18**, 451–459.

Cobley, L.S. (1977). *An Introduction to the Botany of Tropical Crops*. (Revised by W.M. Steele). Longman, London, 299–301.

Purseglove, J.W. (1984). *Tropical Crops, Dicotyledons*. Longman, Harlow, 430–435.

Tunwar, N.S. and Singh, S.V. (1988). *Indian Seed Certification Standards*. The Central Indian Seed Certification Board, New Delhi, 115–116.

POLYGONACEAE: AGRICULTURAL CROP

***Fagopyrum esculentum* Moench. Common Names: Buckwheat, Sweet buckwheat**

Buckwheat is grown throughout the temperate regions as a grain crop, but is of only minor importance. Another species, *F. tataricum* Gaertn., is favoured in cool tropical regions, at high altitudes, and is known as bitter buckwheat.

The crop is sensitive to frost. It will grow on poor, infertile and acidic soils, but prefers well drained sandy soils and responds to nitrogen and phosphate in moderate applications.

According to Campbell (1984), *F. esculentum* is heteromorphic and self-incompatible, but *F. tartaricum* is homomorphic and self-fertilised. Seed crops of the former should therefore be grown in good isolation (200 m).

The crop may be sown in rows 20 cm apart or broadcast. Seed rate is 35–40 kg/ha in rows or 40–50 kg/ha when broadcast.

Joshi and Rana (1995) state that ripening is not uniform and suggest that crops should be swathed when 75% of the grains are mature, for later picking up with the combine. Shattering is a problem and up to 22% loss has been recorded.

For storage, the moisture content should be below 16%. When using forced-air drying the air temperature should not exceed 43°C.

References

Campbell, C.G. (1984). Buckwheat. In *Evolution of Crop Plants* (Ed. N.W. Simmonds). Longman, Harlow, 235–237.

Joshi, B.D. and Rana, R.S. (1995). Buckwheat, *Fagopyron esculentum*. In *Cereals and Pseudocereals* (Ed. J.T. Williams). Chapman and Hall, New York.

SOLANACEAE: AGRICULTURAL CROP

Nicotiana spp. Common Name: Tobacco

There are two species which are grown for leaf production used for smoking, snuff or chewing. These are (based on characters in Purseglove 1984):

> *N. rustica* L.: petiole not winged; corolla greenish-yellow
>
> *N. tabacum* L.: leaf sessile or with decurrent petiole which is usually not winged; corolla white, pink or red

N. rustica is less widely grown than *N. tabacum*, being largely confined to the former Soviet Union and northern India. It has a higher nicotine content than *N. tabacum*.

Tobacco seed is very small and therefore very small quantities of seed are required. Each plant may produce one million seeds. Much of this is grown locally in small quantities for on-farm use and is controlled by the tobacco manufacturing companies.

Distinguishing between Cultivars

There is a wide range of cultivars, particularly in *N. tabacum*, which is very morphologically variable. Many of these cultivars are quite local. Purseglove (1984) gives the following classes of cultivars:

(A) flue-cured: basic type Orinoco
(B) fire-cured: basic type Pryor
(C) air-cured: basic type Burley
(D) cigar
    (1) cigar-filler: type includes Pennsylvania broad-leaf, Spanish, Gebhart and Dutch
    (2) cigar-binder: principal type Havana
    (3) cigar-wrapper: principal type Cuban
(E) Turkish: very variable, but all are small-leaved with a distinct aroma when cured.

Although originally a subtropical plant, tobacco has spread to many areas through the selection of adapted types. It is day-neutral but is not frost hardy and requires a frost-free growing period of about 100 days; the optimum temperature for the crop is about 25°C. The crop grows best in areas with sunny days.

Soils should be well drained and preferably of light sandy texture, slightly acid but with good organic content for flue-cured, or rather heavier for other types.

Previous Cropping and Isolation

Tobacco seed is very long-lived and can survive for up to 20 years. However, crops grown for the leaves are usually topped (i.e. the inflorescences are removed to promote leaf growth) and this prevents seed being deposited on the ground. Against this, the crop is very susceptible to eelworm and therefore a good rotation is required with alternate species which are not susceptible (e.g. Gramineae).

Tobacco is self-fertilised, although the flowers are visited by nectar-collecting insects and up to about 4% out-pollination may occur. Generally, an isolation distance sufficient to prevent mixture with other crops is satisfactory.

Seed Crop Establishment and Management

Tobacco seed is very small and is therefore sown in a seed bed for later transplanting. A level teaspoon of seed is sufficient to sow a seed bed for 0.4 ha of transplanted crop.

Seed beds should be carefully prepared and should be sterilised before use. The seed should be cleaned before sowing and treated with an effective sterilising agent. Seed should be sown thinly by first mixing it with an inert filler such as sand or ash.

The seed beds must be shaded from direct sunlight and once the seedlings are established the covers can be gradually removed to harden them off. The beds should be kept damp, but not overwatered.

When the seedlings are growing strongly and are about 18 cm high they may be transplanted to the field. The soil must be well cultivated and of good nutrient status. Spacing varies depending on the district and cultivar to be grown, but 90 × 90 cm is about average. The crop must be kept moist and responds to irrigation but requires a dry period as it ripens.

Seed Crop Inspection and Roguing

Seed crops may be quite small and can be kept under observation throughout the growing season so that off-type plants can be removed as they become obvious.

Inspection is most useful when the crop is in flower.

Harvesting, Drying, Storage and Seed Cleaning

The capsules are two-valved and dehisce longitudinally. Seed is usually harvested by hand, removing the inflorescences for drying and threshing. The moisture content is not normally a problem as the seed matures during dry weather. An air/screen cleaner is sufficient for cleaning. Because the seed is so small it must be handled with care otherwise much may be lost through spillage.

Reference

Purseglove, J.W. (1984). *Tropical Crops, Dicotyledons.* Longman, Harlow, 539–555.

SOLANACEAE: VEGETABLE CROPS

Lycopersicon lycopersicon (L.) Karsten ex Farw.
Syn. *Lycopersicon esculentum* Miller. Common
Name: Tomato

The tomato is cultivated in most regions of the
world. It is consumed as a salad and is also widely
used in the preparation of cooked dishes. In
addition to the use of the fresh crop from either
field or protected cropping, it is an important
processed commodity, either canned, dehydrated
or made into paste.

Classification of Cultivars

Hybrid, non-hybrid
Production season, suitability for specific types
of cropping systems
Market outlet, fresh salad, cooking, processing
Plant type: determinate or indeterminate
Leaf: type, relative width and length, colour
and pigmentation
Stem: hairiness, colour
Inflorescence: type, flower colour, relative size
and type of style
Fruit: relative size, shape, external features,
transverse section and number of locules,
colour (pre-ripe and ripe), greenback or
greenback-free
Resistance to specific pathogens, e.g. *Verticillium dahliae* Kleb. (*Verticillium* wilt), *Fusarium oxysporum* Schlecht ex Fr. (*Fusarium*
wilt), *Cladosporium fulvum* Cooke (leaf
mould)
Resistance to specific nematode species, e.g.
Meloidogyne sp.

Environmental Requirements for Seed Production

The tomato is not frost hardy and will not survive
prolonged periods below 10°C. Although it will
usually survive high temperatures, the plant's
development is best when temperatures do not
exceed 25°C. Bacterial canker (*Corynebacterium
michiganense* (E.F. Smith) Jenson) is a major
pathological problem when the crop is exposed to
prolonged high relative humidity and high
temperatures.

The optimum soil pH is 6.5 and the soil should
have adequate liming materials added if levels are
lower than this. The NPK ratio for base fertilisers
is 1:2:2 applied during final site preparation.

Previous Cropping and Isolation

Most seed legislation authorities stipulate that a
minimum period of two years must elapse
between a tomato seed crop and any previous
tomato or related crop; this requirement varies
from one country to another. The minimum
requirement is usually not required if the soil or
substrate has been partially sterilised since the
previous crop. There are several important soil-
borne pathogens and pests, such as the wilt
diseases and nematodes, for which there should be
longer rotations to ensure an economic crop.

The tomato is generally accepted as being self-
pollinated, although cross-pollination has been
reported in some instances (Rick 1978). The
minimum isolation distance between crops is
therefore only 50 m, although some authorities
require greater distances.

Seed Crop Establishment

Tomato plants are raised in nursery beds and
transferred to the field when they have approximately three leaves. The young plants can be
produced in protected structures in areas with a
relatively short growing season.

Sufficient plants for 1 ha can be produced by
drilling approximately 60 g of seed in 250 m² of
seed bed.

The indeterminate cultivars require supports,
either as canes, stakes or strings. Plants of this
type are planted approximately 50 cm apart, with
1 m between the rows. The determinate ('bush')
types are planted 45–120 cm within the rows and
90–150 cm between the rows; the decision on
spacing within these parameters depends on the
cultivar's vigour and the method of irrigation.

Seed Crop Management

The plants should be irrigated to assist plant establishment, especially in arid regions, although only light applications should be made in the early stages. Once the plants are established the irrigation frequency will depend on local conditions, but as a guide this would be every five to six days in arid conditions, while in the humid tropics it would be weekly.

The fields should be cultivated to control weeds; in many areas this is achieved by hoeing. An interval should pass between cultivating and irrigating to minimise weed regeneration.

The indeterminate cultivars require weekly attention, either to tie them to their stakes or to twist the supporting strings around each plant, depending on which plant support system has been adopted.

The prevailing pathogens and pests vary from one region to another and care must be taken to control them with appropriate methods. Those of global importance, in addition to those listed in 'classification of cultivar' include, *Alternaria solani* Soraner (early blight), *Phytophthora infestans* (Mont.) de Bary (late blight), *Phytophthora nicotiana* B. de Haanvar. (buck-eye rot), *Corynebacterium michiganense* (E.F. Smith) Jenson (bacterial canker) and tobacco mosaic virus.

Aphis spp. and whitefly are vectors of viruses in addition to causing direct plant damage. The fruit borer is a common pest in the tropics and subtropics.

Seed Crop Roguing

The tomato seed crop is rogued at three stages. The first is before the start of anthesis; at this stage the general plant habit, vigour and foliage details are confirmed. The second roguing stage is after the start of anthesis when the first immature fruit are checked for the greenback or greenback-free character, according to the cultivar's description. The foliage characters and general form of the plants are also confirmed. The final roguing is done when the first fruit are ripe. At this time the mature fruit characters are checked and any plants showing off-type characters are removed

from the crop. Plants showing symptoms of tobacco mosaic virus are removed as soon as identified.

Harvesting

The ripe fruit is either gathered by hand over a period as it becomes ripe or by a single once-over operation using a specially designed mechanised harvester. The machines are mainly used for large-scale operations to harvest the determinate cultivars, especially those which have been developed for processing; in these seed crops it is usual practice for seed production to be combined with utilisation of the fruit pulp and juice. Hand harvesting is used for the trained indeterminate types and also in areas where there is adequate hand labour, or for cultivars with sequential ripening.

The seed is extracted from the ripe fruit using a fleshy fruit seed extractor which separates the gelatinous coated seed from the fruit debris. Crops which have been mechanically harvested are transferred to the processing plant where the seed is extracted as a separate stage in the overall processing and conservation of the fruit material.

There are three alternative methods for removing the gelatinous layer from the extracted seeds: fermentation, sodium carbonate or hydrochloric acid.

Fermentation method   The seed and pulp mixture is left to ferment for up to four or five days in non-metallic containers (the seeds become stained if metal containers are used). The duration of fermentation depends on ambient temperatures: in relatively high temperatures (25–30ºC) the operation is complete in approximately 24 h. The material is stirred every 2–3 h to ensure uniformity in the container and to avoid discolouration. Each time it is stirred a check is made to determine how far the fermentation has progressed: the clean seed tends to sink in the container. It is usually necessary to cover the containers in order to exclude fruit flies. At the end of fermentation the seed is washed, the dross material is floated off and the cleaned seed is collected.

Sodium carbonate method In this method the pulp and seed mixture are mixed with an equal volume of 10% sodium carbonate. The mixture is left at ambient temperature for up to two days. After this period the mixture is washed in a sieve and the cleaned seed is retained. This method is especially useful for dealing with small quantities, for example as produced in breeding programmes; it is also useful where ambient temperatures are too low for the fermentation process.

Hydrochloric acid method This method is frequently used by large seed companies employing very experienced staff. Details of the method used by each are usually regarded as company secrets. The use of hydrochloric acid can impart a very good appearance to the finished seed lot and commercial companies therefore frequently combine it with the final stages of the fermentation process.

Great care must be taken in the handling and use of the hydrochloric acid: protective clothing should be worn and in any procedure the acid is added to water – water must *not* be added to the acid.

Workers extracting relatively small quantities of seed find that the rate of 567 ml concentrated hydrochloric acid stirred into 10 litres of seed/pulp mixture and left for 30 min is successful. The seed mixture is washed at the end of the process.

Drying

The drying operation should commence immediately the seed extraction has been completed. On a commercial scale the wet seed is initially spin dried to remove the surface water. The seed lot is then transferred to a rotary paddle hot air drier with the initial air temperature of 37–40°C; as the seed starts to lose excess moisture the temperature is reduced to 32–35°C. The duration of this drying operation is approximately 10 h.

In some areas the seed is dried on mats in the sun; the seed is brushed every 2–3 h to ensure that the batch dries evenly.

The seed is dried down to approximately 7%, or for vapour-proof storage to 5.5%.

Cleaning

Normally, further processing is not necessary when the extraction process has been done efficiently. Further cleaning of the dried seed lot can be done with an air/screen cleaner. If there are groups of seed adhering together the seed lot is brushed or rubbed to separate the individual seeds.

Production of F1 Hybrid Seed

The male and female parents are grown in separate rows or blocks. A ratio of 1:5 male to female is the usual proportion. Information from the maintenance breeder will assist in determining the sowing date of each parent so that the start of anthesis in the male coincides with the stigma being receptive in the female. It is sometimes necessary to make a second sowing of one of the parents in order to extend the period that hybridisation can continue.

The female parent flowers are emasculated as they approach the late bud stage. This operation is done in advance of the pollination transfer, either early in the day or one or two days before, depending on location. The closed bud is carefully opened with a pair of fine pointed forceps and the cone of fused anthers removed. This operation ensures that the buds which have been emasculated cannot self-pollinate. At the same time, part of the calyx of the emasculated bud is removed or the bud is tagged. Any developing fruit or flowers not previously hybridised are removed and other buds which are at the same stage are removed if not emasculated.

The pollination operation commences with the collection of pollen from the open flowers of the male parent. This is done by either removing whole flowers or vibrating the flowers and collecting the pollen on a small plate or disc, or in a tube. A little of the pollen is then transferred by a small, fine-haired brush to each of the stigmatic surfaces of the prepared flowers on the female parent plants.

The fruit is harvested when ripe and seed extracted; care must be taken that only the tagged or marked fruits are harvested.

Solanum tuberosum L. Common Name: Potato

The potato crop is mainly produced from tubers, often referred to as 'seed potatoes', although this planting material is not botanically true seed.

The International Potato Center in Peru indicated that there was an increasing interest in producing crops from seed taken from the berries of potatoes (CIP 1979). The term adopted for the seed is 'true potato seed' (TPS).

Agronomic techniques for TPS production and the husbandry for growing food crops from it are in relatively early stages compared to many other crops. The main advantage of the use of true seed is that a valuable food source, i.e. potato tubers, can be used as food instead of planting material. This is extremely important for areas where there are food shortages. The other main advantages include the fact that using TPS can result in significant savings on the importation of potato tubers for planting. It has been indicated by Jones (1982) that some of the economically important potato viruses are not seed transmitted.

The science and technology for mother plant maintenance, seed production and agronomy were reviewed by Umaerus (1989).

Seed Extraction

The fruits are harvested approximately 40 days from pollination. The mature berries are picked and stored at room temperature until they have become soft. The seed can be extracted by hand. The following method for extraction of larger quantities was described by Sadik (1982). The seed is extracted in a funnel which has a cylinder attached to the top and a rubber bung with coiled copper tubing is fitted into the funnel's neck (the apparatus is very similar to the funnels used for the extraction of nematode cysts from soil suspensions). The potato berries are macerated and the material placed in the funnel. Water is gently passed through the funnel and as it passes through the coiled tube there is a cyclonic action. The scum and fruit debris float off with the overflowing water at the top while the separated seeds collect at the bottom of the apparatus.

When the separation is complete the seeds are collected by removing the bung.

The extracted seeds are dried at room temperature.

Solanum melongena L. Common Names: Aubergine, Eggplant

This is a widely grown crop in the tropics, subtropics and warmer temperate areas of the world. It is also cultivated as an early protected crop in some areas, such as around the Mediterranean, for export to northern Europe. It is usually referred to as brinjal in the Indian subcontinent.

Classification of Cultivars

Hybrid; non-hybrid
Production season: early, maincrop, suitability for protected cropping, number of days to harvesting
Plant habit: relative height, vigour, degree of branching
Leaf characters: shape, size and pose, pigmentation
Stem characters: pigmentation, degree of downiness, incidence of spines in relation to plant age
Inflorescence: number of flowers on inflorescence
Calyx characters: shape, pigmentation, size relative to fruit
Fruit characters: colour before and after maturity, bicolour if cultivar specific, shape, relative length and breadth, exposure and incidence of sun-scald
Resistance to specific pathogens, e.g. *Phomopsis vexans* (Sacc. and Syd.) Harter (fruit rot)
Resistance to specific pests, e.g. *Meloidogyne* spp. (eelworm)

Environmental Requirements for Seed Production

This species is not frost tolerant and requires a long, warm growing season with day temperatures of 25–35°C and night temperatures of 20–27°C. There is no specific day length requirement.

The crop requires a soil pH of 6.5; soils with a lower pH should receive adequate levels of liming materials in the early stages of preparation. The crop responds to base fertiliser applications with an NPK ratio of 1:1:1.

Previous Cropping and Isolation

Eggplant has many soil-borne pests and pathogens in common with other members of Solanaceae and rotations should therefore exclude the related crops. There should be at least four years between eggplant crops.

Although eggplant is generally considered to be self-pollinated, a significant amount of cross-pollination by insects occurs, therefore the recommended isolation distance is 500 m.

Seed Crop Establishment

The young plants are usually raised in seed beds or modules. A seed rate of 60 gm sown in 250 m^2 will provide sufficient plants for 0.5 ha. The young plants are transferred to their final quarters when approximately six weeks old. They are planted 50–75 cm apart in rows 60–120 cm apart; the plant density depends on both the cultivar's vigour and the irrigation system.

Seed Crop Management

The crop is maintained weed free, especially in the early stages of crop establishment, and irrigated frequently to assist plant establishment. Subsequent water applications should be weekly, or less when there is rainfall.

Seed Crop Roguing

The seed crop is rogued at three stages. The first is when the plants are established, but before the start of anthesis. At this stage the crop is checked for general plant habit and foliage characters, including leaf type being in accordance with the cultivar's description.

The second roguing is at the start of anthesis when the same characters are checked as for the first stage, and in addition the spine characters of the plants are checked.

The final roguing is done when the first fruits are fully developed. At this time plants are checked for yield potential and fruit characters.

Harvesting

The fruit is not picked until it has passed the normal maturity of the market crop. Fruit of a seed crop is mature when its colour has started to fade. For example, the dark colour of a 'black' fruited cultivar would have started to become a bronze colour.

Seed Extraction

There are two alternative extraction methods, wet or dry.

Dry extraction   After picking, the fruit is left to dry further; during this period the fruit shrivels and the colour continues to fade. The seed is then extracted by hand, although machine crushing is possible. The dry extraction method is mainly confined to small-scale operations where there is plentiful hand labour.

Wet extraction   In this method the seed is extracted from the fleshy fruit; the process is the same as for *Lycopersicon lycopersicon*, except that extra water is added to the mixture to assist the separation of the seed from the relatively dry fruit material.

There is no fermentation or other chemical cleaning process for eggplant.

Drying

The drying process is either done in driers or in the sun, as described for tomato. The seed is dried to approximately 11% moisture content. For vapour-proof storage the seed moisture content is reduced to 6%.

Processing

This is as described for *Lycopersicon lycopersicon* seed.

F1 Hybrid Seed Production

The procedure is as described for hybrid seed production of *Lycopersicon lycopersicon*. As the eggplant flower buds and flowers are larger than those of tomatoes, hand emasculation is easier. It is usually necessary to emasculate the flower buds on the female parent two to three days before pollination; this is done while the petals are still white.

Capsicum annuum L. Common Name: Sweet pepper

Capsicum frutescens L. Common Name: Chilli pepper

The sweet and chilli peppers are widely grown in the tropics, subtropics and warmer temperate areas of the world. The sweet peppers are also cultivated as an early season protected crop in temperate areas. The fruits of both species are used as fresh and processed vegetables.

The seed production techniques are the same for both species.

Classification of Cultivars

Hybrid; non-hybrid
Season and use: cropping period, fresh market, processing, e.g. pickle, sauce or dehydration
Type: sweet (mild), pungent (hot), intermediate
Plant habit: erect, prostrate or intermediate
Plant pigmentation
Leaf characters: relative size and width: length ratio, leaf edge and surface
Flower characters: duration of anthesis, distinctive flower characters, e.g. colour
Fruit characters: colour before and at maturity, pose, shape, transverse shape, apex shape, number of locules, sweet or degree of pungency
Resistance to specific pathogens, e.g. tobacco mosaic virus

Environmental Requirements for Seed Production

The *Capsicum* spp. are not frost tolerant, and are most productive in a temperature range of 20–28°C. Although grown as annuals, they are relatively long season crops. The peppers require a soil pH of 6.0–6.5. The NPK ratio applied as a base-dressing is 1:1:1.5; excessive nitrogen will delay the start of anthesis.

Previous Cropping and Isolation

Pepper crops should not be in the same rotation as other members of Solanaceae. There should be a break of four years in order to minimise the risk of soil-borne pathogens.

The *Capsicum* spp. are largely self-pollinating, but cross-pollination can occur between cultivars of different types as well as within the same type of cultivar, therefore an isolation distance of 500 m is recommended, although some authorities may stipulate greater distances than this.

Seed Crop Establishment

Plants are either produced in seed beds as described for *Lycopersicon lycopersicon* or sown *in situ*. The planting distance is 30–60 cm apart in the rows with 50–100 cm between rows; the exact plant distances depend on the vigour of the cultivar and the method of irrigation.

Seed Crop Management

The irrigation and cultivation for weed control are the same as *Lycopersicon lycopersicon*. Thrips and aphids should be controlled, especially as they can be vectors of seed-transmitted viruses.

In areas where there is a high leaching rate, up to approximately 60% of the nitrogenous fertiliser allocated for the base-dressing is applied as a top-dressing after plant establishment, but not after the start of anthesis.

Seed Crop Roguing

There are three roguing stages. The first is after plant establishment but before the start of anthesis. At this stage the general morphological characters are checked, and also stem pigmentation and characters.

The second stage is after the start of anthesis but while the first fruits are still immature. During this roguing the plants' general morphology and leaf characters are again confirmed; it is also checked that the number of flowers is in accordance with the cultivar's potential.

The third stage is when the first fruits are mature. In addition to the general morphology, the fruit characters are confirmed. Plants with virus symptoms should be rogued out as soon as identified at any of the three stages.

The following pathogens are seed-borne and infected plants should be removed from the crop during roguing if they have not been controlled by routine crop protection methods: *Colletotrichum piperatum* (Ell. and Ev.) Ell. and Halst (ripe rot, anthracnose), *Phytophthora capsici* Leonian (*Phytophthora* blight, fruit rot), *Pseudomonas solanacearum* (E.F. Smith) E.F. Smith (bacterial wilt, brown rot).

Harvesting, Seed Extraction, Cleaning and Storage

The fruit of the sweet and chilli types are harvested when they become fully ripe. The sweet peppers are wet extracted by the same process as *Lycopersicon lycopersicon*; the material is not fermented. The extracted seed is dried to approximately 10% moisture content.

The chilli pepper fruit are first dried, either in the sun or in a batch drier. The dried fruit is then threshed. When there is sufficient hand labour the dried material is flailed. Further cleaning is done with an air/screen cleaner.

Care must be taken when handling fruit of the chilli types because their pungency causes severe irritation to mucus membranes.

The safe moisture content for seed storage in vapour proof conditions is 4.5%.

F1 Hybrid Seed Production

The same technique is used as for *Lycopersicon lycopersicon*. In order to ensure an adequate supply of pollen, the ratio of male to female parents is 1:6.

References

CIP (1979). *Report of the Planning Conference on the Production of Potatoes from True Seed*. International Potato Center, Lima, Peru.

Jones, R.A.C. (1982). Tests for transmission of four potato viruses through potato true seed. *Annals of Applied Biology*, **100**, 315–320.

Rick, C.M. (1978). The tomato. *Scientific American*, **239**(2), 67–76.

Sadik, S. (1982). A method for seed extraction. *True Potato Seed (TPS) Letter*, **2**(3).

Umaerus, M. (1989). *True Potato Seed*. International Potato Center, Lima, Peru.

Further Reading

Accatino, P. and Malagamba, P. (1982). *Potato Production from True Seed Bulletin*. International Potato Center, Lima, Peru.

UPOV (1974). *Guidelines for the Conduct of Tests for Distinctness, Homogeneity and Stability, Potato*. Doc. TG/23/2, International Union for the Protection of New Varieties of Plants, Geneva.

UPOV (1976). *Guidelines for the Conduct of Tests for Distinctness, Homogeneity and Stability, Tomato*. Doc. TG/44/3, International Union for the Protection of New Varieties of Plants, Geneva.

UPOV (1980). *Guidelines for the Conduct of Tests for Distinctness, Homogeneity and Stability, Sweet Pepper*. Doc. TG/76/3, International Union for the Protection of New Varieties of Plants, Geneva.

TILIACEAE: AGRICULTURAL CROP

Corchorus species. Common Name: Jute

There are two species which are grown to produce fibre (Purseglove 1984).

(i)   *C. capsularis* L. Common name: White jute. Characterised by globose capsules and brown seeds.
(ii) *C. olitorius* L. Common name: Tossa jute, Jew's mallow. Characterised by long cylindrical capsules and dark greyish-blue seeds.

The majority of the crop is grown in Bangladesh, India and China. Of the two species *C. capsularis* is the most important, but *C. olitorius* has a somewhat better quality fibre. Production fields are mostly small and much of the work is done by hand.

Distinguishing between Cultivars

There are numerous local cultivars of both species. Some improved cultivars of each species have been released in India. Cultivars are distinguished by degree of hairiness, stem colour, plant height and leaf shape (Cobley 1977).

Environmental Requirements for Growing the Crop

Jute is grown in the deltas of large rivers such as the Ganges or Brahmaputra. Climate is tropical with high rainfall and humidity. Jute seedlings will not tolerate waterlogging, but once established the plants of *C. capsularis* will withstand flooding and are often standing in 2 m of water when mature. *C. olitorius* is less tolerant of flooding.

The delta soils are fertile alluvials which normally receive additional deposits each year.

Previous Cropping and Isolation

Jute is frequently grown in rotation with rice and the fields are subject to flooding, making seed survival in the soil unlikely.

Although mainly self-pollinated, some cross-pollination does occur. For seed production the Indian seed certification standards specify 50 m for Foundation and 30 m for Certified Seed from other cultivars of the same species. As species do not cross-pollinate to any extent, the isolation distance between species is reduced to 5 m (Tunwar and Singh 1988).

Seed Crop Establishment and Management

Jute is normally sown broadcast onto a well prepared moist, but not wet, seed bed. The seed is small and must not be sown too deep.

When the seedlings reach about 20 cm they may be thinned to about 12 cm apart. Weeds should be controlled (usually by hand weeding).

Seed Crop Inspection and Roguing

Some stem and leaf characteristics can be seen prior to flowering. During flowering and subsequently, the capsule characteristics are used to detect off-types.

Harvesting, Drying, Storage and Seed Cleaning

Harvesting for seed is some four to six weeks after the time for harvesting for fibre. In small-scale production it is usual to leave a part of the fibre crop to produce seed. The crop may be cut by hand and the plants allowed to dry before threshing.

For cleaning an air/screen cleaner is generally satisfactory or the seed may be winnowed by hand.

References

Cobley, L.S. (1977). *An Introduction to the Botany of Tropical Crops* (Revised by W.M. Steele). Longman, London, 269–272.

Purseglove, J.W. (1984). *Tropical Crops, Dicotyledons*. Longman, Harlow, 613–619.

Tunwar, N.S. and Singh, S.V. (1988). *Indian Minimum Seed Certification Standards*. The Central Seed Certification Board, New Delhi, 133–134.

UMBELLIFERAE: VEGETABLE CROPS

Daucus carota L. subsp. *sativus* (Hoffn.) Thell.
Common Name: Carrot

This species is grown for its roots which are mainly used as a cooked vegetable, or to some extent as a salad. The roots are also processed as frozen, canned, pickled and dehydrated commodities. Purée and juice products are also produced. One of the main reasons for the crop's popularity is the root's high carotene content. The root and top may be used as stock feed but this is relatively limited compared with production as a vegetable.

Carrots are essentially a crop of temperate regions, although there is limited production in other climatic areas.

Classification of Cultivars

Hybrid; non-hybrid
Production season and use: suitability for early crops, maincrop and storage, fresh market, pre-packaging, storage, processing
Root type: colour, length and shape in longitudinal section, shape of shoulder and root tip, external surface, transverse section and core colour
Foliage: leaf form, amount of leaf division, petiole, basal pigmentation

Environmental Requirements for Seed Production

The carrot is biennial, although there is sometimes a tendency for a few plants in a population to have an annual habit and bolt prematurely. It is important to exclude any of these early bolters from the seed production programme. The seed crop requires vernalisation.

Carrots do not tolerate acid soils; the optimum pH is 6.5 and an NPK ratio of 1:2:2 is incorporated as base fertiliser during site preparations. The root crop requires a relatively deep, stone-free soil with satisfactory drainage.

Previous Cropping and Isolation

There should be a break of at least three years between carrot crops. This interval should be increased when soil-borne pathogens such as *Helicobasidium purpureum* Pat. (violet root rot) and *Sclerotinia sclerotiorum* (Lib.) de Bary (watery soft rot) have been observed in previous crops.

The crop is largely cross-pollinated and the recommended minimum isolation distance between seed crops is 1000 m, although some authorities may stipulate different distances from this.

Seed Crop Establishment

There are two methods used for carrot seed production, namely 'seed-to-seed' and 'root-to-seed'.

Seed-to-seed This method is normally used only for the production of the final seed category; maintenance of the cultivar's genetic quality relies on the quality of the stock seed used. It is not normally practised in areas where winter temperatures fall below freezing, although in some areas, such as parts of the USA, the crop is protected by a snow cover.

Stock seed is sown in late summer and the plants remain *in situ*. Seed is sown at the rate of 2–3 kg/ha in rows 90 cm apart.

Root-to-seed In this system, young plants, usually referred to as 'stecklings', are produced from a late summer sowing. The sowing rate is approximately 5 kg/ha; the ratio of seed bed to replanted steckling area ranges from 1:5 to 1:10, the actual ratio depending on the proportion of stecklings remaining after roguing. The stecklings are either overwintered *in situ* or lifted and stored during the winter. If stored, the stecklings' tops are cut off approximately 2–3 cm above the crown after removal from the soil. The trimmed roots are then stored in cool rooms or cellars at a temperature of approximately 3–5°C with the humidity controlled at 90–95%. The stored roots are replanted at the end of the winter 20 cm apart in rows 30 cm apart. The stecklings which over-winter in the field are also planted at this density after they have been lifted and rogued.

There is an alternative root-to-seed method which is usually only adopted for Basic Seed production. The crop is sown early in the summer (the timing is the same as for a commercial root crop). The developed roots are lifted and rogued in the autumn of the first growing season and stored in cellars or clamps. The roots are replanted at the end of the winter at the same distances as stecklings.

F1 Hybrid Seed Production

Hybrid carrot seed is produced by the use of cytoplasmic male sterility.

One of the advantages is the increased root uniformity. The male to female parent rows are usually in the ratio of 2:4 or 2:6. Reduced pollinating insect activity is often observed in the crop from the start of anthesis; this is attributed to the smaller petals on the male sterile flowers of the seed-producing parent. Therefore supplementary bee hives are usually placed in the crop.

Seed Crop Management

The crop is cultivated to maintain a weed-free environment unless herbicides are used. Irrigation may be necessary, especially following summer sowings or to assist establishment after planting, depending on local conditions.

Carrot fly should be controlled, especially in crops being grown for root selection. Other pests include capsids and aphids, which should also be controlled. The willow-carrot aphid is responsible for the transmission of carrot motley dwarf virus (a complex of two viruses, carrot red leaf and carrot mottle) which is also seed transmitted. Capsids are common pests of carrot in some areas.

There are several pathogens causing carrot diseases; important ones which are seed-borne include *Alternaria radicina* Meier, Drechsl. and Eddy (black root rot, seedling blight) and

Xanthomonas carotae (Kendrick) Dowson (bacterial blight, root scab). These should be controlled when they are first identified.

Seed Crop Roguing

Seed-to-seed crops The only roguing is to remove early bolters in the first season and also any plants which have atypical foliage characters seen before the first flower shoots emerge in the second season.

Root-to-seed crops In the first year, premature flowering plants are removed and the crop is checked to confirm foliage characters; plants with off-type leaf characters are removed. The second roguing stage is when the roots are lifted. At this time the cultivar's root characters, including shape, colour and size, are confirmed.

Harvesting and Threshing

The primary, secondary and tertiary umbels ripen in succession. The seeds in an umbel are ready for harvesting when they have turned brown; this is usually associated with the umbel curling and becoming brittle. When the plants in the seed crop are at a relatively wide spacing the successive ripening of umbels on individual plants presents a harvesting problem if the crop is to be mechanically harvested. Traditionally, the seed crop was hand harvested, enabling workers to take only those umbels with ripe seed. However, research work has led to two important changes to the way that carrot seed production is considered. Firstly, it has been demonstrated that seed position on the mother plant influences seed quality and seedling performance (Gray 1979). Further work has indicated that if carrot seed crops are grown at higher plant densities there is a shortening of the overall anthesis period and an increase in seed yield (Gray 1981).

When the crop is mechanically harvested it is cut when the earliest maturing seed on the primary umbel is mature. The crop is left in windrows to dry further. The seed is separated from the remainder of the plant material with a combine harvester. Seed can be separated from umbels which have been hand harvested by passing them through a thresher.

Cleaning

The initial process following threshing is removal of the seeds' spines. This operation is usually done with a barley de-bearder or similar machine which rubs the spines off the seed. The seed lot is further cleaned with an air/screen cleaner. If the seed lot is still not clean it can be improved with a gravity separator.

Drying

The seed is dried to 10% moisture content, or 7% when stored in vapour-proof containers.

Pastinaca sativa L. Common Name: Parsnip

The parsnip is cultivated for its edible root which is used as a cooked vegetable. It is a temperate region crop, mainly cultivated in northern Europe and to a lesser extent in North America. The market crop can be produced at higher altitudes in the tropics.

Classification of Cultivars

Parsnip cultivars are usually only classified according to their root characters.

    Season and use: fresh market, pre-packaging,
        processing
    Root: overall shape, relative length, depth of
        shoulder, shape of shoulder, surface, colour
        and texture
    Resistance to specific pathogens, e.g. *Itersonilia
        pastinaceae* Channon (canker).

Environmental Requirements for Seed Production

The optimum soil pH is 6.5 and an NPK ratio of 1:2:2 is incorporated as base fertiliser during site preparations. The plants require a relatively deep, stone-free soil with satisfactory drainage in order to produce their characteristic roots.

Parsnip is a biennial; a cold period is required at the end of the first growing season prior to flower production in the second year.

Previous Cropping and Isolation

The minimum rotation between parsnip and related crops is three years. This species is mainly cross-pollinated and the minimum isolation distance is 500 m. The seed crop should also be isolated from any ware crops of parsnip by the same distance.

Seed Crop Establishment

Parsnip seed is produced by either the seed-to-seed or root-to-seed method. Only the latter is used for production of Basic Seed.

Seed-to-seed   The crop is sown in the Northern Hemisphere during May or the first half of June at the sowing rate of 4 kg/ha in rows 1 m apart. The young plants are subsequently thinned to approximately 50 cm apart.

Root-to-seed   The stock seed is sown early in the season, at the same time as for a market crop of the cultivar. This ensures that the mature roots are showing their cultivar characters by the end of the growing season. The sowing rate and row spacing are the same as for a seed-to-seed crop.

Seed Crop Management

Parsnip seed is relatively slow to germinate. It is therefore important that weed control is efficient in the early stages of establishment. This can be achieved by early cultivations or pre-emergence herbicides. The seed-to-seed crops remain *in situ* until the crop matures the following year. The root-to-seed crop is lifted in the late autumn for root selection before replanting the selected roots. The retained roots are replanted 30–60 cm apart in rows 1 m apart. Work by Gray et al. (1985) and Gray and Steckel (1985) has demonstrated that there is a higher proportion of primary umbels at higher plant densities.

Pests of seed crop include carrot fly, celery fly and aphids, which should be controlled.

Harvesting

The parsnip seed crop is ready to harvest when the primary umbels turn a light brown. Close examination of the umbels will indicate that the schizocarps have split, each displaying two seeds. The seed is harvested as described for *Daucus carota*.

Threshing and Cleaning

The seed is threshed with either a combine or a stationary thresher. The dry umbel material is very susceptible to fracturing during threshing, therefore this operation is often done while the cut crop is damp with dew. The seed lot is upgraded with an air/screen cleaner.

Drying and Storage

The seed is dried to 10% moisture content or 6% for storage in vapour-proof containers.

Petroselinum crispum (Miller) Nyman ex A.W. Hill. Common Name: Parsley

This crop is cultivated for its leaves which are widely used as a fresh or dried herb; the fresh foliage is also used for garnishing. Although mainly grown as a commercial crop in temperate areas, it is also cultivated in the tropics and subtropics. There is a type which has a swollen tap root, usually referred to as 'Hamburg parsley'; it is mainly cultivated in private gardens for the edible root in addition to its leaves.

Classification of Cultivars

There are two main types of parsley cultivated for leaf production: plain leaved and curled leaved. The curled leaf cultivars are described by the degree of leaf crenulation. Other characters of both types include length of petiole and intensity of leaf colour.

Environmental Requirements for Seed Production

Parsley is a biennial with the same soil and nutrient requirements as *Daucus carota*.

Previous Cropping and Isolation

The site for a seed crop should not have been used for parsley production for at least three years.

Parsley is cross-pollinated by insects; the minimum isolation distance for seed crops is 500 m for cultivars with the same leaf type or 1000 m between cultivars with different leaf types.

Seed Crop Establishment

The majority of seed is produced by the seed-to-seed method, although stock seed is produced by the root-to-seed method.

Seed-to-seed The crop is sown after mid-summer, in July or early August in the Northern Hemisphere, at the rate of approximately 3 kg/ha in rows 50 cm apart. The crop is not normally thinned.

Root-to-seed For this method the seed is sown in the Northern Hemisphere in May at the rate of 6 kg/ha in rows 50 cm apart. The plants are lifted and rogued in the late autumn and the selected plants are replanted 25 cm apart in rows 1 m apart. This is a higher plant density than was used traditionally; the closer spacing results in a more condensed umbel maturity period in the seed crop.

Seed Crop Management

Parsley can take up to four weeks to germinate, although in warm soils the seedlings emerge in three weeks from sowing. It is important to maintain a weed-free site either by cultivations or the use of appropriate herbicides.

Aphids should be controlled, and the cultivars with very curly foliage should be carefully checked for insect pests.

Seed Crop Roguing

The young plants in both seed-to-seed and root-to-seed systems are checked for foliage characters when their leaves have formed a rosette. Early bolting plants are also removed in the first growing season.

When the root-to-seed crop is lifted in the autumn, any plants with off-type leaves are discarded.

Hamburg parsley is assessed for root shape and quality before replanting; roots which are relatively straight and not misshapen or split are retained.

Harvesting, Threshing, Drying and Storage

These are the same as for *Daucus carota*. Parsley seed does not have spines.

Apium graveolens L. var. *dulce* (Mill.) DC. Syn. *Apium dulce* Mill. Common Name: Celery

Celery is cultivated in temperate areas for use as a salad or cooked vegetable. The long, relatively tender petioles are the main part eaten, but the leaves are also consumed, especially as a flavouring in soups. The crop is also produced in parts of Asia and Africa, but mainly for use as a flavouring.

Classification of Cultivars

    Suitability for specific crop production systems, self-blanching or requires blanching
    Resistance to bolting in specific environments
    Overall plant height
    Petiole: colour, length, transverse section
    Leaf: colour, texture, character of margin

Environmental Requirements for Seed Production

Celery is a biennial seed crop with a low temperature requirement for flower initiation. The seed crop is not frost hardy.

It is a relatively shallow rooted crop and is more successful on organic soils, or mineral soils which have had bulky organic materials incorpo-

rated during the early stages of preparation. It has a high water requirement, especially in the vegetative stage, and if not grown in areas with sufficient rainfall, irrigation is essential. The optimum soil pH is 6.5–7.5 and the base fertiliser NPK ratio is 1:2:4.

Previous Cropping and Isolation

There should be a minimum period of three years between celery crops. Although celery is self-pollinated, some cross-pollination occurs and the recommended minimum isolation distance is 500 m. This should be extended to 1000 m between cultivars which have different petiole colours; some authorities may specify alternative isolation requirements.

Seed Crop Establishment

The celery seed crop has traditionally been produced in areas where winter frost does not present a problem and also in areas, such as northern Europe, where the plants require protection through the winter. Winter protection has been provided either by covering the plants *in situ* or transferring them to stores such as cellars or pits where they remain until replanting in the field as soon as the winter has ended. The advent of plastic tunnels coupled with the demand for high quality seed has largely led to the replacement of winter storage by overwintering the plants in protected structures.

Seed crop establishment in areas where the plants overwinter in the field   The seed is either sown at the same time as a commercial crop (usually spring or early summer, depending on the cultivar) or immediately after mid-summer. In both cases the plants are raised under protection and subsequently planted out. One gram of seed will produce approximately 15 000 plants. The process of plant raising has been improved by sowing seed direct into modules. The young plants are planted in the field at approximately the five-leaf stage, 30 cm apart with 60 cm between the rows.

Seed crop establishment in areas where the plants overwinter under protection This method was developed for the production of Basic Seed and has been described by Faulkner (1983). In the Northern Hemisphere seed is sown in greenhouses during August; the seedlings are grown in pots and planted at a spacing of 90 cm by 90 cm in the plastic tunnel in January. Further protection is provided as necessary to prevent frost damage. Anthesis commences the following June.

Seed Crop Management

The crops should be maintained weed free, especially during the post-planting period. Blow-flies are usually put into the structures to ensure satisfactory pollination, unless the isolation is sufficient for the doors to be open throughout anthesis (Smith and Jackson 1976). The main pathogens which may occur include *Cercospora apii* Fresen. (early blight, leaf spot), *Phoma apiicola* Kleb. (celery root rot, black neck, scab, seedling canker) and *Septoria apiicola* Speg. (late blight, leaf spot).

Pests which should be controlled before populations build up include *Acidia heraclei* L. (celery fly) and aphids, but care must be taken not to harm the populations of pollinating insects.

Seed Crop Roguing

The first roguing stage is when the young plants are planted and they are checked for leaf and petiole characters. They are next rogued while still vegetative, before the start of anthesis. Early bolting plants are discarded and the general plant vigour, petiole and leaf characters of the remaining plant population are confirmed; plants with excessive basal shoot development are rogued out.

Seed crops which are being produced under the traditional system of lifting and storing mother plants are also rogued to discard plants which have developed basal shoots.

Harvesting, Cleaning, Drying and Storage

The seed crop is mature and ready for harvest when the plants have become yellow and the seed is hard and a grey-brown colour. Harvesting and processing are as described for *Daucus carota*. Celery seed does not have appendages. Crops which have matured in protected structures are usually cut by hand, placed on tarpaulins and extracted with small threshers. The threshed seed lot can be upgraded with an air/screen cleaner.

Seed is stored at 10% moisture content or 7% for vapour/proof containers.

Apium graveolens L. var. *rapaceum* (Mill.) DC. Common Name: Celeriac

This vegetable crop is botanically very similar to celery. The leaves are smaller and petioles shorter than those of celery. It has a swollen, edible root for which it is mainly cultivated. The root can be used fresh or stored, and is used as a salad or in cooked dishes. The leaves and petioles are used but not as widely as those of celery.

Classification of Cultivars

Suitability for specific crop production systems and seasons
Main use: fresh or storage
Resistance to bolting in specific environments
Plant height
Petiole: colour, length, pose
Leaf: colour, size, character of margin
Root (tuber): relative size, height or proportion above soil level, pigmentation

Seed Crop Requirements, Production, Processing and Seed Drying

These are the same as for *Apium graveolens* var. *dulce*.

References

Faulkner, G.F. (1983). *Maintenance testing and seed production of vegetable stocks at the National Vegetable Research Station*. Vegetable Research Trust, Wellesbourne, Warwick.

Gray, D. (1979). The germination response to temperature of carrot seeds from different umbels and times of harvest of the seed crop. *Seed Science and Technology*, **7**, 169–178.

Gray, D. (1981). Are the plant densities currently used for carrot seed production too low? *Acta Horticulturae*, **111**, 159–165.

Gray, D. and Steckel, J.R.A. (1985). Parsnip (*Pastinaca sativa*) seed production: effects of seed crop plant density, seed position on mother plants, harvest date and method, and seed grading on embryo and seed size and seedling performance. *Annals of Applied Biology*, **107**, 559–570.

Gray, D., Steckel, J.R.A. and Ward, J.A. (1985). The effect of plant density, harvest date and method on the yield of seed and components of yield of parsnip (*Pastinaca sativa*). *Annals of Applied Biology*, **107**, 547–558.

Smith, B.M. and Jackson, J.C. (1976). The controlled pollination of seeding vegetable crops by means of blowflies. *Horticultural Research*, **16**, 53–55.

Further Reading

Gray, D., Steckel, J.R.A., Jones, S.R. and Senior, D. (1986). Correlations between variability in carrot (*Daucus carota* L.) plant weight and variability in embryo length. *Journal of Horticultural Science*, **61**(1), 71–80.

Gray, D., Steckel, J.R.A., Dearman, J. and Brocklehurst, P.A. (1988). Some effects of temperature during seed development on carrot (*Daucus carota*) seed growth and quality. *Annals of Applied Biology*, **113**, 391–402.

Thomas, T.H. and O'Toole, D.F. (1981). Environment and chemical effects on celery (*Apium graveolens*) seed production, *Acta Horticulturae*, **111**, 131–138.

Tucker, W.G. and Gray, D. (1986). The effects of threshing and conditioning carrot seeds harvested at different times on subsequent seed performance. *Journal of Horticultural Science*, **61**(1), 57–70.

UPOV (1976). *Guidelines for the Conduct of Tests for Distinctness, Homogeneity and Stability, Carrot*. Doc TG/49/3, International Union for the Protection of New Varieties of Plants, Geneva.

UPOV (1976). *Guidelines for the Conduct of Tests for Distinctness, Homogeneity and Stability, Celery*. Doc. TG/82/3, International Union for the Protection of New Varieties of Plants, Geneva.

UPOV (1980). *Guidelines for the Conduct of Tests for Distinctness, Homogeneity and Stability, Celeriac*. Doc. TG/74/3, International Union for the Protection of New Varieties of Plants, Geneva.

SECTION III

Appendices

APPENDIX 1. GLOSSARY OF TERMS USED

admixture material added to a seed lot which does not correspond to the specification for purity; it usually refers to genetic material inadvertently added during harvesting or processing.

AFLP Amplified fragment length polymorphism: a method of DNA profiling involving the selective amplification of restriction fragments via PCR.

allele alternative forms of a gene at a locus.

anther terminal part of the stamen which contains the pollen sacs.

anthesis the duration of pollen release from anthers.

anthocyanin a group of pigments responsible for additional colouration on stems, leaves or flowers.

apomixis development of seed without pollination having taken place.

awn a stiff or bristle-like appendage on the seed of some species.

combine harvester a machine which cuts and threshes the seed crop in a single operation; it can also be used as a stationary thresher or to pick up and thresh material left in windrows.

defoliation removal of excess vegetative growth from a potential grass or fodder seed crop, usually by grazing or cutting.

desiccant a chemical drying agent, applied to a seed crop to kill the green leaves prior to harvesting.

de-tasselling removal by hand, or machine, of the male inflorescences ('tassels') before the start of anthesis, usually referring to maize (*Zea mays*).

DNA deoxyribonucleic acid: the genetic material of all organisms.

electrophoresis a biochemical separation method, especially useful for proteins and nucleic acids.

F1 hybrid the first filial generation resulting from a cross between two selected parents.

farm gate value the value of a crop or commodity as received from the producer before any further processes which may increase its value.

field inspection the inspection of a potential seed crop by a crop inspector.

gel the inert porous matrix used in the electrophoretic separation of proteins and nucleic acids.

GLC gas–liquid chromatography: a biochemical separation method.

ground-keeper a plant derived from seed which has been shed or dropped on the field and is not part of the seed crop.

HPLC high-performance (or pressure) liquid chromatography: a biochemical separation method.

IA image analysis: the extraction of numerical data from an image acquired via a video camera or similar means.

IEF isoelectric focusing: a type of electrophoresis.

isozyme an enzyme that exists in two or more forms, separable by electrophoresis.

land race local plant material which contains a range of genotypes, derived as a result of selection pressures under local conditions.

locus a site on a chromosome that has been defined genetically.

lodging the collapse of a seed crop before cutting or harvesting.

maintenance breeder the organisation or person responsible for the continuing maintenance of a cultivar, and for production of seed up to and including Basic Seed; or, in the case of hybrid cultivars, responsible for maintaining parental material.

microsatellite regions of DNA composed of a tandem repetition of a short sequence; also known as simple sequence repeats (SSR).

moisture content the proportion of water in a given seed lot as found at the time of determination, expressed as a percentage of the original weight of the sample.

objectionable weed an undesirable or unwanted plant which can cause particular problems in the seed crop, or the seed of which can be especially difficult to separate from the harvested seed crop. The seed of the objectionable weed may cause special problems if left in the seed lot finally used by farmers.

off-type a plant occurring in a seed crop which is not in accordance with the description of the cultivar being multiplied.

organic farming production of crops without the use of inorganic fertilisers or crop protection chemicals.

PAGE polyacrylamide gel electrophoresis.

PCR polymerase chain reaction: a method for amplifying specific DNA sequences *in vitro*.

pellet a single seed enclosed in material to form a convenient size and shape to facilitate sowing by hand, seed drill or vacuum sower.

poach compaction of soil surface by trampling, especially when done under wet conditions.

polymorphism existence in many forms.

primer a short strand of DNA of defined sequence, used in PCR.

probe a specific small piece of DNA, which when labelled can be used in RFLP analysis to detect complementary sequences.

RAPD random amplified polymorphic DNA: a DNA profiling technique that utilises PCR and arbitrary primers.

restriction enzyme an enzyme that cuts DNA molecules at a specific sequence.

RFLP restriction fragment length polymorphism: a method of DNA profiling.

roguing the act of identifying and removing off-types (rogues) in a seed crop.

SDS sodium dodecyl sulphate: commonly used in certain types of electrophoresis.

seed lot a quantity of seed which is homogeneous, of one cultivar, derived from known material and treated uniformly throughout its production, harvesting and processing.

shedding loss of seed from the fruit or seed-bearing organ before threshing or seed extraction.

sod-bound forming a dense mat.

stigma surface extending from the style which receives pollen when pollination occurs.

STMS sequence tagged microsatellite site: a method of DNA profiling that uses specific primers in PCR to amplify a defined microsatellite.

stook a group of bundles (sheaves) of cut material stood upright to dry further in the field.

testa the outer covering of a seed, the seed coat.

threshing separation of seed from the remaining mother plant material.

tramlines parallel and regular spaces in a broadcast or close-spaced row crop left at sowing time for later access.

umbel an inflorescence in which the flower stalks all originate from the same point on the stem.

volunteer a plant derived from dormant seed left from a previous crop.

windrow a row of cut or pulled plant material left loosely on the ground for further drying *in situ*.

winnowing separation of seed from lighter loose plant material by the action of air movement.

APPENDIX 2. USEFUL ADDRESSES

ALADI
Asociacion Lattinoamericana de Integracion, Cebollati 1416, Montevideo, Uruguay

AOSA
Association of Official Seed Analysts, 201 N 8th Street, Suite 400, PO Box 81152, Lincoln, Nebraska, 68501-1152, USA

AOSCA
Association of Official Seed Certifying Agencies in North America and New Zealand, c/o Idaho Crop Improvement, 600 Watertower Lane, Suite D, Meridian, Idaho, 83642-6286, USA

APSA
Asia and Pacific Seed Association, P.O. Box 1030 Kasetsart Post Office, Bangkok 10903, Thailand

ASSINSEL
Association Internationale des Selectionneurs Professionels pour la Protection des Obtenions Vegetal, Chemin du Reposoir 7, 1260 Nyon, Switzerland

ASTEC
Australian Seed Technology and Education Centre, Queensland Agricultural College, PO Box 720, Lawes (Gatton), Australia

AVRDC
Asian Vegetable Research and Development Center, PO Box 42, Shanua, Tainan 741, Taiwan

BBP
Biotechnology for Biodiversity Platform, c/o Dr J.C. Reeves, NIAB, Huntingdon Road, Cambridge, CB3 0LE, UK

CAASS
Chinese Association of Agricultural Science Societies, No. 11, Nongzhauguan Namli, Beijing 100026, Peoples Republic of China

CFIA
Canadian Food Inspection Agency, Plant Products Division, 59 Camelot Drive, Nepean, Ontario, K1Y 0Y9, Canada

CGIAR
Consultative Group on International Agricultural Research, 1818 H Street NW, Washington, DC 20433, USA

CGIAR-supported centres

CIAT
Centro Internacional de Agricultura Tropical, Apartado Aereo 6713, Cali, Colombia

CIP Centro Internacional de la Papa, Apartado 5969, Lima, Peru

CIMMYT
Centro Internacional de Mejoramiento de Maiz y Trigo, Londres 40, Mexico 6, DF, Mexico

ICARDA
International Center for Agricultural Research in the Dry Areas, Box 5466, Aleppo, Syria

ICRISAT
International Crops Research Institute for the Semi-arid Tropics, Patancheru PO, Andhra Pradesh 502 324, India

IITA
International Institute of Tropical Agriculture, PO Box 5320, Ibadan, Nigeria

IPGRI
International Plant Genetic Resources Institute, Via della Sette Chiese 142, 00145 Rome, Italy

IRRI
International Rice Research Institute, PO Box 933, Manila, Philippines

ISNAR
International Service for National Agricultural Research, Laan van Nieuw Oost Indie 133, PO Box 93375, 2593 BM, The Hague, The Netherlands

WARDA
West African Rice Development Association, E.J. Roye Memorial Building, PO Box 1019, Monrovia, Liberia

CSGA
Canadian Seed Growers Association, Box 8455, Ottawa, KIG 3TI, Canada

CTA
Technical Centre for Agricultural and Rural Co-operation, Postbus 380, 6700 AJ, Wageningen, The Netherlands

DISP
Danish Government Institute for Seed Pathology in Developing Countries, Ryvangs Alle 78, DK-2900, Hellerup-Copenhagen, Denmark

EU
European Union, rue de la Loi 200, DG VI B II I, Loi 130 4/174, Brussels, Belgium

FAO
Food and Agriculture Organisation of the United Nations, Seed and Plant Genetic Resources Service, Plant Production and Protection Division, Via delle Terme di Caracella, 00100 Rome, Italy

FIS
Federation Internationale du Commerce des Semences, Chemin du Reposoir 5-7, 1260 Nyon, Switzerland

HRI
Horticultural Research International, Wellsbourne, Warwick, CV35 9EF, UK

IAC	International Agricultural Centre, PO Box 88, 6700 AB, Wageningen, The Netherlands	OECD	Organisation for Economic Co-operation and Development, Directorate for Food Agriculture and Fisheries, 2 rue Andre Pascal, 75775 Paris Cedex 16, France
ICC	International Association of Cereal Science and Technology, PO Box 77, Schwechat, A-2320, Austria	PGRO	Processors and Growers Research Organisation, Great North Road, Thornhaugh, Peterborough, PF8 6HJ, UK
IOH	Institute of Horticulture, 14/15 Belgrave Square, London SW1X 8PS, UK		
IIRB	Institut International de Reserches Betteravieres, 47 rue Montoyar, B-1040, Brussels, Belgium	RHS	Royal Horticultural Society, The RHS Garden, Wisley, Woking, Surrey, GU23 6QB, UK
ISTA	International Seed Testing Association, Rechenholz, PO Box 412, CH-8046, Zurich, Switzerland	STC	Seed Technology Centre, Massey University, Palmerston North, New Zealand
JIC	John Innes Centre, BBSRC Institute for Plant Science Research, Colney Lane, Norwich, NR4 7UH, UK	UPLB	University of the Philippines in Los Banos, Seed Technology Programme, PO Box 430, Los Banos College, Laguna, Philippines
NIAB	National Institute of Agricultural Botany, Huntingdon Road, Cambridge, CB3 0LE, UK		
NRI	Natural Resources Institute, Central Avenue, Chatham Marine, Kent, ME4 4TB, UK	UPOV	International Union for the Protection of New Varieties of Plants, 34 Chemin des Colombettes, 1211 Geneva 20, Switzerland
ODI	Overseas Development Institute, Portland House, Stag Place, London SW1E 5DP, UK	USDA	US Department of Agriculture, Agricultural Research Service, Building 005, Room 113, Beltsville, MD 20705, USA

APPENDIX 3. IMPORTANT VEGETATIVELY PROPAGATED SPECIES NOT INCLUDED IN SECTION II

	Common name	Form of propagation
Monocotyledons		
Agavaceae		
Agave sisalana Perrine.	Sisal	Bulbils
Agave fourcroydes Lemaire	Henequen	Rhizomes
Agave cantala Roxb.	Cantala	Bulbils
Furcraea gigantia Vent.	Mauritius hemp	Bulbils
Phormium tenax Forst.	New Zealand flax or hemp	Rhizomes
Alliaceae		
Allium chinense G. Don.	Rakko	Bulbs
Allium fistulosum L.	Japanese bunching onion, Welsh onion	Division
Allium sativum L.	Garlic	Cloves
Araceae		
Colocasia esculenta (L). Schott	Cocoyam, dasheen, eddoe, taro	Tuber cuttings
Xanthosoma sagittifolium (L). Schott	Cocoyam, tannia, yautia	Tuber cuttings
Dioscoreaceae		
Dioscorea species	Yams	Tuber cuttings
Gramineae		
Axonopus compressus (Swartz) Beauv.	Savanna grass	Root division
Bracharia brizantha Stapf.	Palislade grass	Root division
Cymbopogon citratus (D.C.) Stapf.	Lemon grass	Root division
Cymbopogon nardeus Rendle.	Citronella	Root division
Cynodon dactylon (L.) Pers.	Bermuda grass	Runners, rhizomes
Ischaemum indicum (Houtt.) Merr.	Batiki blue grass	Root division
Saccharum barberi Jeswiet.	Sugar cane	Stem cuttings
Saccharum edula Hassk.	Sugar cane	Stem cuttings
Saccharum officinarum L.	Sugar cane	Stem cuttings
Saccharum sinense Roxb.	Sugar cane	Stem cuttings
Marantaceae		
Maranta arundinaceae L.	Arrowroot	Rhizomes
Musaceae		
Musa textilis Nee.	Abaca, Manila hemp	Corms
Zingiberaceae		
Zingiber officinale Rosc.	Ginger	Rhizomes
Dicotyledons		
Compositae		
Cynara cardunculus L.	Cardoon	Division
Cynara scolymus L.	Globe artichoke	Suckers
Helianthus tuberosus L.	Jerusalem artichoke	Tubers
Convolvulaceae		
Ipomoea aquatica Forsk.	Kang kong, water spinach	Stem cuttings
Ipomoea batatus (L.) Lam.	Sweet potato	Stem cuttings, tubers
Cruciferae		
Crambe maritima L.	Seakale	Root cuttings
Euphorbiaceae		
Manihot esculenta Crantz	Cassava, manioc, tapioca	Stem cuttings
Labiatae		
Stachys tuberifera Naudin.	Chinese artichoke	Tubers
Leguminosae		
Derris elliptica Benth.	Derris	Stem cuttings

	Common name	Form of propagation
Polygonaceae		
Rheum rhaponticum L.	Rhubarb	Division
Solonaceae		
Solanum tuberosum L.	Irish potato, potato	Tubers
Umbelliferae		
Sium sisarum L.	Skirret	Tubers
Urticaceae		
Boehmira nivea (L.) Gaud.	Ramie	Rhizome

Index

Notes: Page numbers in **bold type** refer to tables, those in *italic type* refer to illustrations. Entries for crop species refer to their occurrence in Section I; there is a separate species index to Section II. Accepted scientific names are in *italic type*; synonyms are in plain type.

Index compiled by Colin D Will

Index of species in Section II